Larcum Kendall London
1

Hans von Bertele

MARINE & POCKET CHRONOMETERS

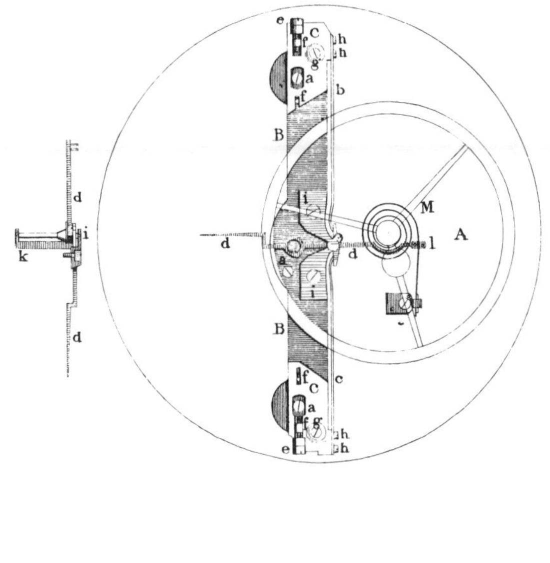

History & Development

Schiffer Publishing Ltd

1469 Morstein Road, West Chester, Pennsylvania 19380

In memory of the two pioneers of horological history H. Alan Lloyd-Hockwood, who helped me to understand the contributions of the various countries at the various times, and R.W. Symonds-Peasenhall, who stressed to me the accentuation of the essential with good, purposeful illustrations.

Idea and conception: Christian Pfeiffer-Belli, Munich
German edition Edited by: H.F. Tölke, Munich
Proofreading for the English edition: D. Bodley-Scott

Translated from the German by
Dr. Edward Force, Central Connecticut State University.

Library of Congress Catalog Number: 90-64231.

Printed in the United States of America.
ISBN: 0-88740-303-4

This title was originally published under the title,
Marine-und Taschenchronometer,
by Verlag D.W. Callwey, München.

Published by Schiffer Publishing, Ltd.
1469 Morstein Road
West Chester, Pennsylvania 19380
Please write for a free catalog.
This book may be purchased from the publisher.
Please include $2.00 postage.
Try your bookstore first.

Foreword

Until a short time ago, an active interest in chronometers was limited to a small, exclusive group of clock collectors. That was true until the most recent times, and the interest in dial clocks has already grown to a very large extent. The reason for this may be that the problems of chronometers are not easy to deal with and appeal to very different natures from the historically and artistically very comprehensible large clocks and pocket watches; in addition, its requirements in terms of training are not simple.

Until 15 or 20 years ago, the boxes of ships' chronometers, made of mahogany, ebony, or other hardwoods and all looking alike externally, could be seen in the show windows and on the shelves of every maritime antique shop in the larger and smaller seaports of England, France, or Italy; and in the drawers of those shops one could perhaps find pocket chronometers (deck watches) as well. All of these present-day rarities could be had then for the price of one or more sandwiches, and if one was especially lucky, among those chronometers that had been retired from service might be specimens from the hands of the great masters of the early days.

The beginning of the introduction seems to be the place to note the great difference between clocks in general and chronometers in particular: For a long time it was taken for granted that a clock would be called a chronometer only when it had a genuine chronometer escapement.

But whoever has thought about the subject of clocks knows that the oscillating system with the escapement is the vital part of any mechanical dial clock, because this system divides the constant flow of time into as precise a rhythm as possible, so that measuring time consists only of counting the number of successive divisions of time with hands on a dial.

The precision of a clock becomes greater as the influence of external conditions (movement, shock, temperature and altitude changes, dampness, different climates, magnetic fields, etc.) on the rhythm of the oscillating system becomes smaller.

Today, though, the viewpoint that originated in Switzerland is often heard: only the quality of the running precision qualifies for designation as a chronometer, without consideration for the type of escapement used. Years ago, the F.H. (Fédération Suisse des Associations des Fabricants de l'Horlogerie) had already established the criterion that a manufacturer or seller could use the designation "chronometer" only when presenting an official certificate from the observatories of Geneva and Neuchâtel, as well as from various specific state test centers. This requirement had been made because at that time even very inexpensive and imprecise watches were being called chronometers.

The history of chronometer development is the result of a need, which was felt more and more urgently in the course of history, to determine geographical longitude, one's position in regard to a definite point of departure, in desolate regions and, above all, on the high seas, a task that in science is called the longitude problem. Thus dealing with the longitude problem necessarily marks the beginning of the chronometer's history.

The use of various branches of technology and physics, such as the kinetic theory, the mechanism of the pendulum, the theory of elasticity, etc., played a major role in the improvement of clockmaking, essential to the development of usable chronometers. For more than three centuries the awareness that insufficient knowledge was at hand in those times led to efforts that resulted in numerous important scientific-technical advances.

The basis of all chronometer history has been a step-by-step recognition of the individual problems and a struggle to overcome them. The precision technology and mechanical-physical details of this struggle are a very vital and interesting facet of chronometer problematics. It is exciting to see how increasingly refined concepts of the operation of a chronometer have led to many changes, and to suggestions for improving the decisive mechanism, the oscillating system with its escapement, and how only very few steps have brought real progress — a phenomenon that can also be seen in most other technical developments of mankind.

In order to be able to evaluate the many designs, usually formed with fine mechanics and great devotion, their oscillating systems must be explored in great detail. In this book there is an elementary description of the oscillating mechanics of the balance in reference to the effect of the balance spring, while there is also an analysis and synthesis of the dynamics of the oscillating system and the various escapements. Only with a thorough understanding of the problematics of the balance system can one understand the driving force behind more than a century's efforts in the fine-mechanical epoch of chronometer building and the significance of the breakthrough to new superiority around 1900 through the use of modern metallurgy, as opposed to a century of mechanical striving for improvement.

The adventurous course of chronometer development begins with primitive efforts by individual forerunners, including the appearance of numerous brilliant and extraordinarily capable personalities determined to overcome errors and reverses in time, who scored the first real successes often in the course of bitter battles with each other. The next stage was the development of new inventions into broad economic advances in the mechanical production of chronometers that extended over more than a century. The end finally comes in the first half of this century with the use of unexpected new improvements, determined by metallurgical technology and followed by a completely new electronics-and computer-oriented era of technology and organization.

The practical effects of the use of chronometers involve sea travel and the success of geographical-scientific expeditions to determine the extent and division of the continents and islands and their coastlines on the globe.

It can be assumed that the readers of this book have already developed an interest in the field of chronometers for particular reasons and would like to know more about the course of its technical and economic development, and about the motivation of the individual phases or the personalities of those involved.

Despite the economic and technical consequences of the chronometer on public life, there are very few publications devoted to this wide field. Individual advances and practical achievements can often be found in the numerous articles and books that deal with clocks but not specifically with chronometers, and their scattered nature gives the impression that chronometer literature is very limited — but chronometer development is the basis for the formation of this book. The author has tried to present the great amount of material on the chronometer in the literature in its most important and significant relationships. It is understandable that many details could not be included. And yet the author feels obligated to be of help to the reader through sufficient and thorough references to the literature in the text in order to answer particular questions.

So as not to be lost in a multi-volume encyclopedia, but rather to limit oneself to one readable book, it was necessary to make many condensations and compressed portrayals, which naturally are very subjective.

In general scientific history, the chronometer and its technology have received little attention; thus it is understandable that little can be found on it in general literature. Apart from this lack, there have been remarkably many, often very penetrating, publications devoted to the importance of the use and production of the chronometer up to the latter half of the 18th Century.

The bibliography offers the chronometer publications found and used in the course of preparing this book. Each title and listing that may be helpful is divided according to the various realms of knowledge and work and listed alphabetically by their authors' names. The two most important and inclusive works on chronometers deserve to be mentioned at the start: The very instructive book by R.T. Gould,[1] which for the first time in modern times offers an instructive portrayal of chronometer problematics and history, such as is of interest to the collector and aficionado today (Gould's broad spectrum and remarkable wealth of detail have also influenced the compilation and organization of this book); and the other is the book written by E. Guyot.[2] The many publications, mainly in the specialist literature, that have dealt with chronometer problems, generally refer only to one or another of the indicated areas and thus usually contain an overabundance of details that say little of the general course of chronometer development. Thus we shall try here, though subjectively, to bring the most important points into cohesion, in order to allow the interested reader his own critique and his own insight into the individual phases of events. For this purpose, at the important points the most important sources are cited in footnotes, usually with brief notations of the applicable chapters or pages. Thus justice can be done to the researchers and personalities to whom the greatest advances and discoveries are attributable.

Seen in a wide perspective, the high level of chronometer development attained at the end, and the further scientific and cultural consequences thus made possible, can be seen as the sum of achievements by the involvement of many European nations whose unity — at least in this area — has been manifested.

Going beyond the usual short, factual biographical accounts found in the scientific literature, there are also occasional personal evaluations or emphasis on individual books or articles, which are meant to encourage the reader to search farther in the original literature.

The course of historical development has brought it about that, just as in the practical development of the chronometer and the use of the chronometer in seafaring, the German contributions in the realm of publicity have also appeared later than elsewhere in Western Europe; thus no particular apology is due for the fact that the greatest number of the writings noted in this book are in the aforementioned languages.

As for the significance of referring back to the literature, one will become fully aware of it only over a period of time of ongoing involvement with chronometers; but it is a worthwhile activity, because only thus is it possible to recognize the level of technology attained at a given time. Also, only by comparing individual publications can the justification, or lack thereof, of the evaluation of individual chronometer inventions be fully understood or criticized, and only by following the details can the human factors related to them be recognized.

When one traces many questions back historically, reference to older books and articles, some of them very old indeed, cannot be avoided. Unfortunately, locating such works is often very difficult, since they are available only in a few special libraries. Fortunately, though, several publishers have published reprints of important old works, because referring to such early literature proves to be especially rewarding for the true aficionado of chronometry.

As the interest in chronometers has increased in the

most recent years, a number of biographies of the great masters and some of the founders of the great manufacturing firms have appeared. In the same way, publications in specialist journals have increased, dealing with some of the technical problems, the construction of individual chronometers and groups of them, as well as biographies — all works that are of importance to the person involved in chronometer problematics. In this respect too, the publications of the various chronometric and horological societies, and especially those dealing with their conventions, are worth noting.

Although much valuable information on chronometers, including unusual chronometer designs, can be found in the catalogs of the great auction houses in recent years, these materials had to be excluded from the bibliography; still it must be said that this catalog material has sometimes proved to be extraordinarily valuable in writing this book.

The availability of chronometers has put seafaring and all the political and economic consequences of it on a solid foundation. It was not one sudden discovery, but laborious, self-sacrificing and conscientious mechanical, scientific, technical and imaginative developmental work that led to the emergence of chronometer making from the earlier days of clockmaking in the 18th Century. Two brilliant personalities, completely separate from each other but both wrapped up in the task they had set themselves — John Harrison in Britain and Pierre Le Roy in France — solved the technical problems of producing clocks impervious to the sea and reliable enough to solve the longitude problem.

It seems to have been intended by fate that the public reaction in Britain was much more intense than that in France, so that it was Britain that was able to vastly expand its sea power through the development of the marine chronometer and form the basis of the Commonwealth that influenced world history for a hundred and fifty years. Although the French achievements in scientific knowledge came a little earlier chronologically and appear to have had a better theoretical foundation, a far smaller extent of practical application was derived than in Britain. One reason for this may be the lesser maritime involvement of the French people, another perhaps the very strong centralization of all action around the elite of Paris rather than the few enterprising souls in the seaport cities.

The course of chronometer development and the history of the marine chronometer are significant examples of how great inventions depend on favorable conditions for their practical development, even conditions outside the actual technology of their invention.

Above and beyond the joy and satisfaction of actually observing the individual objects, involvement with chronometers leads one very quickly into the broad spectrum of world politics and economy, because their introduction into maritime activity simultaneously influenced the whole world situation. Thus the development and introduction of the methods and technology of the chronometer is an example of the influence of a new way of thinking and behaving in human society, whereby the tracing of the individual phases of chronometer history is very instructive for other future investigations of new technical undertakings and plans. The variety of inspiration that the study of chronometers offers makes the fast-growing recent interest in chronometers understandable. Even though the chance of finding good specimens today is limited, the high prices make it difficult to build up on one's own collection. Many thoughts are worth following further, ideas that can be found in the individual chapters of this book and the appendices that concern important details and the theories behind them. Altogether, though, chronometer history offers all those who take an interest in the origins of our present-day civilization and culture an array of unique and interesting viewpoints that are perhaps just as important as fine mechanics and watchmaking, namely those that particularly concern the use of the chronometer in seafaring and great expeditions. There are scarcely any reports of exploration in the tropics or the polar regions in which the decisive role of the chronometer for the expedition's cartographic pursuits has not been stressed. Even more important, perhaps, is that the awareness that the utilization and acceptance of the chronometer's potential by appropriate organizations and offices has been invaluable in making them truly effective.

Any research into the practical application of the chronometer leads to astronomical considerations, for only through them is it possible in the end to gain a reliable standard for time measurement and comparison.

In locating the literature, and particularly in assembling the illustrative material, much of it never before published, help has been provided by many personal friends, specialists and collectors, to whom I or the publisher have turned with requests. I believe I can be fairest to all those who provided this help by listing the organizations and then the individuals in alphabetical order, beginning with the museums: Arts & Mêtiers, Paris, with Director M. Cazin and Curator Mme. C. Cardinal; the British Museum, London, with Curator Mr. Stratford and specialists Mr. Good and Mr. Hutchinson; the Guildhall Museum, London, with President Mr. Toni Mercer (1979) and Curator Mr. Jagger; the L'homme et le Temps Clock Museum, La Chaux-de-Fonds, with Curator M. Curtis; the Handels-og-Sófartsmuseet pa Kronborg, Helsingór, with Director Henning Henningsen; the National Maritime Museum, Greenwich, with Director Comm. Howse and former Curator Dr. Bennett; the Historische Uhrenmuseum, Wuppertal, with Mr. Jürgen Abeler; the Museum für Zeitmessung Th. Beyer, Zürich; the Time Museum, Rockford, Illinois, USA, with Curator Mr. Williams Andrewes; the National Museum of History and Technology (Smithsonian Institution), Washington, D.C., USA, with Curator Mrs. Carlene Stephens and Director Mr. Bedini.

Along with the museums, a number of antique and art dealers, auction houses and publishers have been of great help in making photographs available, namely: Holland Press Ltd., London, for the many important pictures from Commander Gould's pioneering book "The Marine Chronometer" (J. D. Potter, 1923); Mr. Vaudry Mercer and his Antiquarian Horological Society, London, for the numerous pictures from the book "Edward John Dent and His Successors" (London,

1977); Mr. Ineichen of the auction house that bears his name in Zürich; the Galerie d'Horlogerie Ancienne in Geneva; Mr. E. F. J. Stender, Saint Michelgestel; the firm of Sotheby, Parke & Bernet, London, especially Mr. Hayward and Mrs. Tina Miller; and the firm of Algeron Asprey, London.

Under every illustration there is an appropriate reference giving its origin. For their permission to use these pictures, so significant in illustrating the development of the marine chronometer, I am very grateful to all of these publishers, museums and collectors.

From the following specialists and friends I have received advice and help with numerous questions in terms of the general presentation, the entire layout and the evaluation of individual objects: Mr. Jürgen Abeler, Wuppertal; Mr. Charles Allix, Sevenoaks; Mr. Atwood, Rockford; Mr. C. P. Briet, Rotterdam; M. Chapiro, Paris; Mr. Curtis, La Chaux-de-Fonds; Kl. Ebrich, Frankfurt; Sir William Hayter, Oxford; Mr. Jagger, London; Hart. Kiefer, Düsseldorf; Dr. Mann, Wiesbaden; Dr. P. T. R. Mestrom, Masstricht; M. Panicali, Paris; Prof. Dr. Pillowitzer, Vienna; Col. Quill, London; M. Paul Rudolf, Triel-Sur-Seine; Mr. A. Randall, Birmingham; M. Sabrier, Aquiny-Louviers; Dr. Schraepler, Bremen.

For their kind photocopying of many important works I am greatly indebted to the leader of my former Institute for Industrial Electronics, Prof. Hadwig Thim, and his associate, Frau Matyas; Herr Wilhelm Auer tirelessly prepared the technical diagrams for me despite many changes; Frau Kotulski developed many portions of the text, sometimes from the dictaphone, sometimes from my badly scribbled handwritten manuscript; and — last but not least — Herr Pfeiffer-Belli and Herr Tölke of the Callwey-Verlag have provided much generous help, not only in obtaining photographs from all over the world, but also by obtaining literature and information on chronometers otherwise inaccessible to me, so as to provide as broad a foundation as possible for this book. To all of those named here, once again my hearty thanks.

1. Rupert T. Gould, "The Marine Chronometer, Its History and Development", London 1923.

2. Ed. Guyot, "Histoire de la détermination des longitudes", La Chaux-de-Fonds 1955.

I: The Problem of Longitude

The longitude problem is the basic problem of determining location on the surface of the earth. When one is in a human settlement, it is easy to find out where one is and what connections with other important places exist. In unsettled areas, in high mountains and deserts, and especially on the high seas, there is no one present to give information, so that the technical processes are necessary to determine location. While geographical latatude has been determinable since antiquity from the height of the sun and stars, it has been difficult, indeed often impossible until recent times, to determine geographical longitude in relation to a definite point of departure or harbor. This handicap, especially on the high seas, has endangered all enterprises to a high degree.

The actual motivation in the hearts of men to determine positions was the urge to travel afar and partake of adventure. This urge goes back hundreds, even thousands of years, and is worthy of consideration for its own sake. The youthful fascination of a trip into the blue or the first trip outside the narrow circle formed by the family's living space remains unforgettable, as does the awareness that the world is not nailed together of boards, but that there are always roads and passes leading over the hills to other people and places. Even if the first trip was not planned with maps and thorough considerations, the impression lasted. Only later came the wish to get thoroughly acquainted with the new found area through the organizing power of the map, but at the same time to push even farther in search of similar new experiences. This urge, if not even stronger, is what the inhabitants of coastlands must have felt toward the unknown extent of the ocean, a magic that the inland dweller does not know but can experience from older novels and chronicles, and which the author has felt when studying the first ventures in the realm of the chronometer.

Thor Heyerdahl[1] expresses vividly the concept of the dream and the urge to go into the distance; and in most recent times this is expressed perhaps even more intensely in the book by the Irishman Tim Severin,[2] who was able to cross the Atlantic from Ireland to North America a few years ago in a boat covered with cowhide. He had heard of an adventure that took place at the beginning of the Middle Ages: According to an old chronicle, the oldest version of which dates to about the year 800, the devout bishop St. Brendan, after receiving a report of a successful trip to the West by other monks, set out with a few friends around 550 and sailed a leather boat to America and back to Ireland.

These and similar ventures into the unknown distance surely cannot be attributed only to illogical fantasies; rather a sixth sense, a kind of presentiment, probably provided the impetus as well as the later drive in the actual expedition, long before the interest in rational navigation appeared or economic or political motivation appeared.

Tim Severin's voyage was of course supported by modern equipment, but it provided not only the proof that it was possible to cross the ocean in a leather boat and attested to many details of the old chronicle as to the route, but also emphasized the veracity of the old reports.

Even in antiquity, before the birth of Christ, the Phoenicians, Greeks and Romans made enterprising voyages on the Mediterranean and sailed out through the Pillars of Hercules (the Straits of Gibraltar) as far as England and maybe even farther. In the same manner, Vikings had sailed from Norway to the French coast, to Normandy, before the year 1000, and later to Sicily; in fact, the older texts are constantly interpreted to mean that they had also reached America around the year 1000. Recently it has been learned that Chinese expeditions had sailed via India to the Cape of Good Hope. In the late Middle Ages the Hanseatic cities sent their ships into the Baltic Sea, while the Venetians and Genoese made voyages to the southeast. After 1440 a strong new wave of discoveries began. Never before the middle of the 15th Century had there been such a drive to discover new lands.

The problem of longitude, of the "terminus technicus", was a problem for every captain. It was extraordinarily difficult for a long time and only approaching solubility in special cases of the geographical distance between one given fixed point — usually a seaport — and another definite point — usually at sea — that a ship has just traversed, the captain definitely wanted to know how far east or west he was of his point of departure. To a modern man, thinking in astronomical terms, this task will seem almost trivial at first glance: one would say that he determines the sun time of his location by observation at noon, he knows the local time of his point of departure by radio signals, and this will give him the difference between the two times and thus the difference in longitude, since one hour of difference in time at a location represents a difference in longitude of fifteen degrees, or one minute of time represents fifteen minutes of arc. It is necessary to remember, though, that until not very long ago there were no wireless signals to give the time at the point of departure, that until not long before that there were no precision timepieces on which to keep the time at the point of departure, and that determining the time at a given location is not at all easy, because it requires the measurement of angles of the sun or certain stars and calculations based on these measurements, for which suitable instruments are required.

Thus the problem of longitude proves to be a very complicated matter, requiring not only geographical and astronomical knowledge but also various calculations that are not at all easy to make, including the exact determination of times and angles in order to determine the longitude either at sea or on land. For almost 1800 years the best mathematicians and astronomers have battled with this problem.

Since the middle of the 13th Century, a book called *The Spheres*[4] was the standard work used at European universities. In it the Englishman John Hollywood, who taught at the Sorbonne in Paris, presented clear concepts of the spherical form of the earth; to be sure, the earth was seen by Sacro Bosco as the center of the universe, just as Claudius Ptolemaus (ca. 137 B.C.) had stated in his Almagest, a concept that had generally been accepted as true until Copernicus published his works.

Only the possession of a definite system of coordinates allows a conception of the geographical division of important places on the earth. For the skies too, the Greek philosophers Aristyllos and Timocharios had worked out, as early as 290 B.C., a table of astronomical lengths and breadths of the most important stars of the ecliptic, on the basis of measurements.

Geographical latitude is measured in degrees of distance from the equator. Since the earth's axis points fairly accurately toward Polaris (the North Star), determining the elevation of the North Star leads directly to determining the geographical latitude, the accuracy of which is thus dependent on the precision of the available measuring instruments and processes — which have been developed and improved quickly since the end of the Middle Ages. Dreier[5] refers in this context to the fact that Portuguese seafarers could determine their latitude to within thirty nautical miles as early as the latter half of the 15th Century by using the angle of the North Star. He adds the interesting note that in 1455 the Portuguese seafarer Cadamosto did not dare to cross the equator along the west coast of Africa for fear of losing sight of the North Star and thus losing his means of determining his latitude.

The determination of the geographical longitude of a given point cannot be done by a simple process such as measuring the North Star's angle to the earth's axis, but requires a reference to a point of departure and its meridian; practically any seaport can be used, though the meridian of reference today is that of Greenwich.

Determining the longitude of a certain point in reference to the previously chosen point of departure is possible only by comparing the times of the two points; thus determining longitude is a problem in terms of a time differential (with appropriate accuracy), because the sun travels from east to west, touching every meridian and circling the earth once in 24 hours, so that it travels 15 degrees in one hour or one degree in four minutes. This basic principle was also known to the ancient Greeks. The aforementioned Hipparchos had already concluded, on the basis of observing that the shadow on the moon can be observed precisely at one specific time, though for observers at various places on the earth, it appears at different times, that for a one-hour differential in observation times at two points, there must be a difference of 15 degrees of longitude between them. With the scant data available for him to compare on the appearance of the shadow on the moon at different places, Hipparchos calculated the differences in longitude of several cities, also including in his calculations several other locations that would lie on the same meridians. Ptolemy, for example, cited observations of the same lunar shadow at Arbella at 5h and Carthage at 2h, but since the stated times are given only in hours and not broken down more precisely, was not sure whether the times were exact to more or less than a quarter-hour, so it could be that calculations based on such data might possibly have a minimum accuracy of 3 to 4 degrees. One cannot stress too strongly, though, that from ancient days to the middle of the 18th Century there were no other possible ways of determining longitude on the earth than these undependable and usually very imprecise astronomical processes, which will be explained below for the sake of thoroughness.

These historical retrospects show not only the level of knowledge in antiquity but also, and very clearly, what practical application astronomy had in those times. At the same time they let one understand how vital the accuracy of astronomical observations is and how, in particular, the exact determination of time is essential to the location of any particular point on the earth by a system of coordinates.

On the ocean there was absolutely no way to determine longitude until the beginning of modern times, so one can understand that the first suggestions in this realm were made and developed in regard to the shadows on the moon mentioned above, as well as the time-honored methods of using star and moon positions to determine longitude. They are all based on the same premise that, because of the great distances of the moon and stars, the same constellations would be seen at the same time everywhere on earth and thus the times to observe selected, readily visible constellations should be tabulated for certain points of observation, while the time at the point in question should be calculated by means of observing the sun's height or the altitude of Ursa Major to Polaris. All of this is based on complicated and laborious astronomical calculations, which were worked out only gradually in the course of modern times and depended on the possession of sufficiently accurate tables and the possibility of having good measuring instruments at hand, which formed the basis of modern astronavigation.

A British navigator, St. Pierre, was able to convince King Charles II in 1674 of the necessity of setting up accurate tables, which inspired the king to found the

Greenwich Observatory and give it the task of "correctly forming tables of heavenly movements and positions of fixed stars, in order to attain the so desired determination of longitude and to perfect the art of navigation at sea."

It is of interest that in 1713 Newton spoke in favor of these efforts, in the process of which he advocated striving for an optimal accuracy of two to three degrees.

Considerations of practical orientation on the high seas must have been taken up at Sagres shortly after 1400 by Prince Henry the Navigator (1394-1460), and must have been put to use as well, because the first great European voyages of discovery on the sea, which created the Portuguese colonial empire, were planned and sent out from there.

Insights into the scientific activity at Sagres, though, are not found in Born's book. Only scant suggestions are offered of Henry's activities with maps, about two secret rooms with maps and instruments, as well as his work with Arabic astronomers and Jewish doctors (but without any names or lists of the writings he used). So it seems appropriate to remember that since the 10th Century in Spain, and soon thereafter in southern France and subsequently in Central Europe as far as Cracow in Poland, there was great interest in astronomical, mathematical, and quantitative knowledge of celestial phenomena.[6] Noteworthy instruments and the names of scientists linked with them include: The astrolabe, known since antiquity and described in depth in 1141 by Raymond de Marseille;[7] the Arabic universal instrument Saphea Azarquiels, described in 1231 by Guillelmus Anglicus or Marseille and Jehuda bar Mosé;[8] the universal, brilliant Torquetum (Turkish device) that was made known in 1284 by Bernard de Verdun;[9] the table of planetary positions of King Pedro IV of Aragon, dating to 1320.[10]

The existence of so many and differently structured scientific efforts soon led to consequences in terms of their application to orientation on the seas, as Professor Poule (école de Chartres, Sorbonne) has been able to prove.[11] It can be assumed that some or all of the works noted above, and perhaps others as well, that we have not yet been able to relate to the Spanish-southern French area, were put to use at Sagres. Thus our present-day knowledge of the state of science at Sagres is practically nil.

In evaluating the two problems that must be solved simultaneously to determine longitude, namely determining the time at the present location (the ship) and at the point of departure (the seaport), it is clear to see that the latter task is much more difficult as long as observable astronomical phenomena serve as the basis aboard the ship, for these require stellar and planetary observations that involve much more complex laws of motion than the mere rotation of the earth. In this situation it is clear that determining longitude is simplified remarkably when one is able to carry the time of the point of departure with one by using a reliable instrument. But this is a revolutionary idea, since it replaces the mere use of astronomical methods with a new tactic.

At Sagres, apparently for the first time, the use of reliable wheel clocks was promoted by Alonzo de Santa Cruz (ca. 1510) as the most practical way of determining longitude at sea. But it was Gemma Frisius, in 1530, who first correctly expressed and documented the principle of determining longitude at sea through the use of a wheel clock. Gemma Frisius (1505-1555) was truly a remarkable Flemish scholar, Professor of Medicine in Louvain and also a very important astronomer in his day. He was the author of various articles and books on mathematical, geographical and astronomical problems, publisher of the cosmography of Apianus, the teacher of Emperor Charles V, and the inventor or developer of noteworthy devices, such as ring-shaped solar clocks, the Staff of Jacob and the Astrolabium Catholicum. In addition, he constructed his first globe in 1530 and later produced improved globes with the help of G. Mercator and G.A. Myrica.

Although the technical equipment was insufficient for its practical application at the time when this principle was proclaimed, Gemma Frisius deserves the greatest respect for his foresight. As so often in the history of discoveries and inventions, decades and sometimes centuries pass before great theories are turned to reality, but still they were the starting points for the efforts that followed.

Success at sea, the discovery of new continents and lands, and the strivings to find improved sea routes, occupied man at the beginning of modern times to a different degree from the Middle Ages. All technical advantages were mobilized to keep ahead of the developing competition. The solution to the longitude problem (determining precise geographical distances at sea) appeared to be one of the most vital factors. In order to inspire many to work intensively, the nations involved in seafaring decided to offer large prizes.

The Great Prizes

In 1598, shortly after he ascended the throne, Philip III of Spain announced a lifelong pension of many thousand ducats for whoever would find a practical solution to this problem. Soon thereafter, in 1610, the Estates General of The Netherlands followed with a prize of 100,000 gulden for the inventor of a practical, useful way of determining longitude; and in 1716 the Regent of France, in a letter to the French Academy of Sciences, set a prize of 100,000 livres.

The sharpest and clearest formulation, though, was made by the British Parliament with its description — later often called Queen Anne's Act[12] — which set very significant sums for the attainment of certain degrees of accuracy at sea, as shown in the following table.

The table of required precision of longitudinal determination at sea, and the prizes offered, were as follows, according to Queen Anne's Act of 1714:

in Miles	in Degrees	Prizes
60	1 degree	£10,000
40	40 minutes	£15,000
30	30 minutes	£20,000

The act stipulated not only the degrees of precision stated in the table, but also required:

1. That the distance in miles be measurable on a

meridian, 2. That the proof be based on a voyage to the West Indies, with a duration of approximately six weeks, 3. That a commission be established, which soon became known as the Board of Longitude.

The activity of the Board of Longitude (1713-1828), which had already begun to work before the final establishment of Queen Anne's Act, proved to be very important and progressive in the history of the chronometer. It consisted of nine persons, and in its very first sitting they included the highest Admiral, the Speaker of the House, the President of the Royal Society, the Astronomer Royal and several mathematics professors. It is not surprising that great numbers of suggestions and answers to these prize announcements were submitted in various countries, and a number of experimental devices were constructed, based on mechanical clocks, and even tested at sea. The most important of them will be described in the next chapter. Success in correct determination of longitude in the sense of the Queen Anne Act, using specially constructed mechanical clocks in unconventional form, occurred in the latter half of the 18th Century. How it was achieved is the subject of Chapter 3.

It can be noted that many suggestions and submissions were based on senseless fantasy and speculation, and do not deserve close scrutiny. Mistaken, confused ideas, such as Gould has described, were suggested in Spain between 1607 and 1626, and considerable sums of money were paid for them. For the sake of thoroughness it should also be noted that there was a whole series of efforts to construct compasses without magnetic deviation — systems that would always point correctly to the north. Even before the days of Columbus it was known, and probably also utilized by him, that on a voyage to the west the compass error — the deviation from the astronomically located north-south direction — changed. These changes occur in such a way that to about 30 degrees west of Greenwich the difference decreases to nothing (because the real North Pole and the Magnetic North Pole lie on the same meridian, the Zero Isogone = Agonic), and as one goes farther west, the deviation grows again. A table of compass errors sought to make use of this effect; this undertaking was suggested to the Board in 1674, and Charles II even nominated a special committee to study it. But determining longitude with the help of compass error was not successful; it is not hard to imagine the impossibility of attaining any noteworthy accuracy with the compass needles of the time. A determination to one degree was probably the best that could be hoped for in favorable cases.

The continuing determination of one's precise position is of the greatest importance today not only in seafaring but to an equal extent in air traffic as well. The use of electrotechnical and electrophysical phenomena for transmitting information, such as radio and computer technology, have resulted in great improvements and refinements in quick measuring of distance for long and short distances, moving and fixed points. A variety of systems, such as radar, gee, sun, loran and others, have taken navigation on the water and in the air on completely new courses since the outbreak of World War II in 1939, as Captain L. Holder[13] has shown very clearly. In spite of that, determination of longitude by comparing local time with that of the chronometer has retained its importance and is still used in the same manner since its invention when disturbances in electronic information occur, or when an amateur in his vessel does not possess or cannot use the modern means of navigation.

The precise determination of longitude with the chronometer has undergone considerable progress since the 18th Century. The modern methods of astronavigation[14] can be regarded as logical developments of the old astronomical calculations.

Among the improved procedures, the one developed in 1837 by Captain Thomas Sumner of Boston, is of major significance, but requires the possession of exact tables that give the precise location of the point on the earth from which the stars are observed. Large ships, as well as small yachts operate on this basis and in the same way today.

1. Thor Heyerdahl, "Kon-Tiki", Berlin 1949.
2. Tim Severin, "The Brendan Voyage", New York 1978.
3. Conradin Ferrari d'Occhieppo, "Kunde vom Weltall", Vienna 1952.
4. Joannes de Sacro Bosco, "Sphaera", new additions over several centuries, with various commentaries.
5. F. A. Dreier, in the section "Die Bestimmung der nautischen Breite", in "Katalog der Staatlichen Museen, Winkelmessinstrumente", Berlin 1979.
6. Em. Poulle, "Les Instruments astronomiques de l'Occident Latin aus XIe et XIIe Siècles", "Cahiers de civilisation mèdièvale", Université de Poitiers 1972.
7. Em. Poulle, "Le Traité d'Astrolabe de Raymond de Marseille, Studii medievali", Centro Italiano di Studii sull' alto Medioevo, Spoleto 1964.
8. Em. Poulle, "Un Instrument astronomique dans l'occident Latin, La Saphea", Centro Italiano di Studii sull' alto Medioevo, Spoleto 1970.
9. Em. Poulle, "Bernard de Verdun et le Turquet", Isis 1964.
10. Em. Poullo, "A Propos des Tables Astronomiques de Pierre d'Aragon", Junta de Investigaçes do Ultramar, Coimbra 1966.
11. Em. Poulle, "Les Conditions de la Navigation Astronomique au XVe Siècle", Junta de Investigaçes do Ultramar, Coimbra 1969.
12. Queen Anne's Act was later changed or expanded several times: In 1765, after the successful fulfillment of conditions 1 and 2 of the 1712 act, drawings as well as explanations of the manner of setting up and using the successful system were required to be submitted to the board, so as to enable third persons to set up similar systems. Under King George III in 1775, additional prizes for further improvements were set.
13. L. Holder, "Nine Modern Developments in Navigation", in W.E. May, "A History of Marine Navigation", Henley-on-Thames 1973.
14. Bobby Schenk, "Astronavigation", Bielefeld 1977.

II: Forerunners of the Mechanical Chronometer

The prehistory of the marine chronometer is that of its forerunners. One can set their origin at the end of the 16th Century, the period of the first Spanish and Dutch prize offerings for the solution of the longitude problem.

After winning its freedom in the 17th Century and attaining a high degree of prosperity while taking in political and religious refugees, building up a noteworthy reservoir of significant personalities, the Netherlands attempted to take on the role of a leading maritime nation in competition with Spain and Britain.

In this time of progress, Christian Huygens, born in 1629 as the son of Constantine Van Zullichem became the first great pioneer of the mechanical marine chronometer. He was a rich Dutch patrician with an interest in art and science.

In 1657, under contract and instructions from Huygens, the clockmaker Salomon Coster of The Hague produced a mechanical clock with a genuine, free-swinging pendulum. This clock shows the first correct use of a pendulum in determining time, although now and then before Huygens' time an occasional short, firmly linked pendulum-like system had been used as a running regulator. Such firmly linked apparatus could not serve as a controlling organ, independent of the power source, in an isochronically operating timepiece, as will be covered later, and thus cannot be regarded as taking away the credit for Huygens' invention.

The young Huygens, studying at the University of Leyden, surely heard much of the experimental work of Galileo Galilei, and his discovery of the law of free fall; his astronomical discoveries and his observations of oscillating pendulums must have been discussed there often. Above all, Galileo was of the opinion that a pendulum of a certain length — a heavy mass hanging on a line — always swung in the same time, regardless of the length of its swing, a law that he called the isochronism of pendulum oscillations.

The suggestion of equipping a clockwork with a pendulum as a standard of time in place of the anisochronic foliot to control the division of time was undoubtedly a truly brilliant idea, a step beyond the ordinary. Huygens' subsequent, resulting work with such pendulum clocks, probably because of the observation of noticeably greater accuracy in comparison with earlier clocks, indicates his scientific perseverance. At first Christian Huygens was interested in studying the reasons for this greater accuracy and the possibility of improving it even more. As early as 1657 he wrote that the solution of the longitude problem could be possible with a pendulum clock. Huygens' further technical steps in developing the pendulum clock are recorded in horological literature; thus it is not necessary to go into detail here. What is important is that Huygens could soon determine that Galileo's assertion of an unlimited isochronism, an unchanging time of oscillation independent of the amplitude, was mistaken. On the contrary, Huygens found that with greater amplitude the period becomes greater, so that isochronism can be presumed only for small amplitude up to +/-3 degrees.

Huygens looked for means of maintaining isochronism independent of the amplitude of the pendulum. In the process he recognized that this was possible when the body of the pendulum moves on a cycloidal course.

Huygens' first publication on his new application of the pendulum was included in the famous work "Horologium",[1] while his subsequent ideas and suggestions were the subject of numerous later publications, all of which are available in an excellent French collection.[2]

Despite the significant improvements in the accuracy of clocks by the introduction of the pendulum, Christian Huygens did not immediately attempt to test their behavior at sea. It took a special impetus from outside, from his friend Alexander Bruce, Earl of Kincardine — a Scottish refugee in Holland. It was he who, in 1662, undertook a voyage to Scotland and took along two pendulum clocks with his own improvements to reduce the rocking of the pendulum (a double-forked crutch), in order to test them at sea. It was a stormy trip, Bruce became very seasick, and both clocks stopped. When the sea had calmed, Kincardine was able to start both clocks again.

From this voyage on, testing clocks at sea was a major goal of Huygens. Through the intervention of the then secretary of the Royal Society of London at that time, Sir Robert Moray, he was able to entrust his two clocks to a Captain Holmes for a voyage from London to Lisbon and back. The clocks did not stop once in the entire trip, which gave Huygens a very positive opinion of the use of his pendulum clocks at sea, although the two clocks did differ considerably from each other in accuracy. In 1664 the same captain undertook a voyage to Guinea and Jamaica and took Huygens' two clocks along again. On this trip the two clocks proved to be suitable for use at sea, and even proved superior in determining the longitude of Barbados in comparison with the method of calculation then in use. But the results were not good enough to win real confidence from the Royal Society in London.

In 1668, one of Huygens' clocks was taken along on an expedition from France to Crete and gave excellent results in determining the geographical longitude of the island. It is particularly noteworthy that the clock did not stop during either the heavy storms at sea or the cannonade of Candia. Further testing of clocks was carried out on the Zuider Zee in 1865, then in 1867 one of the clocks was taken to the Cape of Good Hope. Reports of all these voyages show that the clocks remained in operation, but that they did not achieve the hoped-for accuracy. Finally Huygens shared his ideas for further improvements with the Marquis de l'Hospital; it is not known what came of this, since Huygens unfortunately died shortly thereafter (June 8, 1695).

The fact that his ideas did not lead to a real breakthrough is easy to understand today. A pendulum essentially is not capable of withstanding the complicated buffeting caused by the rocking of a ship at sea, especially in voyages on the high seas, without being influenced.

The original drawings shown opposite are taken from the book by Defossez. For use at sea, the #1 drawings show his first suggestions with the caption "Horologeo de Marine"; they illustrate a gimbal suspension with a heavy additional mass to stabilize the whole system, as well as the clockwork fitted with a triangular pendulum.

At the beginning of 1675 Huygens made the historically important step of linking the balance already customary for years in place of the weight pendulum, with a flat hairspring. He turned this task over to J. Thuret in Paris, who tried at first to proclaim this basic change as his own invention.

The idea of replacing the oscillations of the regulating pendulum by a hairspring was, according to Defossez, discussed by Robert Hooke of London in 1659, by the Duke de Roames and Blaise Pascal in 1660, and in 1665 by d'Esson, Seigneur d'Aimont; and soon after also by M. d'Alèsme, and probably also tested —though without any known success.

Huygens, presumably regularly informed of these tendencies, was at first so wrapped up in his own ideas about the pendulum that he did not take seriously the use of hairsprings.

From the beginning of 1675, though, he worked energetically at testing such hairsprings. In the summer of 1675 Huygens informed the secretary of the Royal Society of London, Mr. Oldenbourg, of his developments in this area. Namely the use of a ring balance with a hairspring, after adding the following as refinements for use at sea: (a) the installation of two opposed, coupled balances, as seen in #2, as in the cross-beat escapement of Jost Bürgi,[4] and (b) the installation of an intermediate wheel to attain greater amplitude of the balance (up to +/- 120 degrees) — called a pirouette.

Derham [5] reported in 1696 that the reception by the members of the Royal Society of the first clock by Huygens with a hairspring caused such a sensation that one would have thought the longitude problem had already been solved.

Huygens had already recognized the influence of the temperature on the oscillating time of clocks with balances and hairsprings, and had determined in addition that the resulting change in the elasticity of the balance spring is the cause of this behavior. He had also noted the great influence of changes in bearing friction. Now one improvement followed another: Huygens tested roller and knife edge bearings (#3). The hairspring's sensitivity to temperature bothered him, so that he returned to pendulum systems and tried to produce especially precise isochronism through geometrical means. In addition, he designed intermediate winding systems — remontoir d'ègalitè — to keep the drive momentum as even as possible over long periods of time, an idea that Jost Bürgi had already taken up and carried out.

Without detracting from the basic ideas of Galileo on the swinging properties of the pendulum, it was really Huygens who not only recognized that the pendulum, as an isochronically working time controller is inclined to offer unequally better running results and accuracies than the other anisochronic controllers such as the balance, but also shared his awareness with the world and put it to use. In 1657 Huygens received a patent on his ideas from the Estates General, and on the basis of his numerous contacts in France, Holland and Britain, he attained in just a few years the acceptance of the pendulum as a time regulator in fixed clocks.

The year 1659 is noteworthy as the first date of special application of mechanical clocks in Britain to the solution of the longitude problem. At that time Robert Hooke (1635-1703), who had already made a name for himself in various areas of science, tried to organize a syndicate to develop his new idea of determining longitude at sea with the help of mechanical clocks equipped with a spring-driven oscillating system. On this subject, Gould adds a note from Robert Hooke that in trying to accomplish this, he encountered countless technical and organizational problems in opposition to the use of a mechanical clock with spring and balance on a sea voyage. In a series of lectures (Cutlerian Lectures of 1664) he had presented his basic idea of creating an oscillating system independent of a heavy weight by combining a mass and an elastic spring, showing twenty possible structural variations, of which he admitted later —rather naively — that he had not mentioned the most practical variation so that he could utilize it later himself.

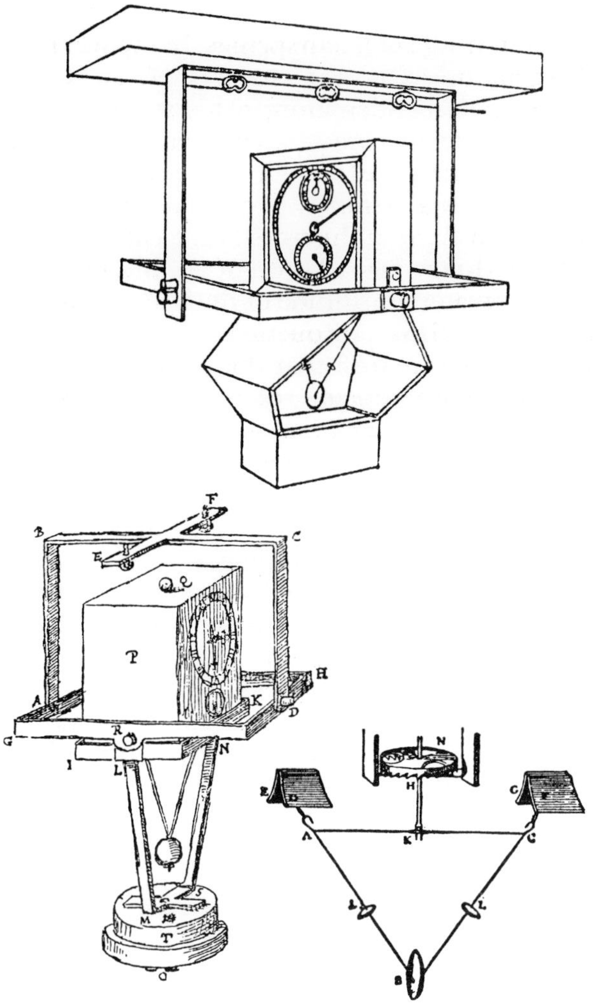

1 Huygens' first suggestions of 1661 for his *Horloges de Marine*; gimbal mounting with a counter weight to maintain a vertical position, triangular pendulum with double cycloid cheeks. (a), (b) variant designs, (c), (d) details of the variant (e), separate dials for hours, minutes and seconds.
(Defossez, Savants XVIIe)

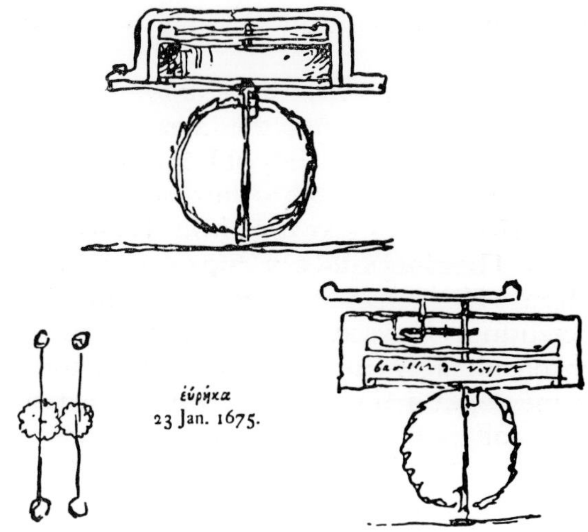

2 Huygens' suggestions for opposed double balances (cross-beat principle), with spiral spring.
(Defossez, Savans XVIIe)

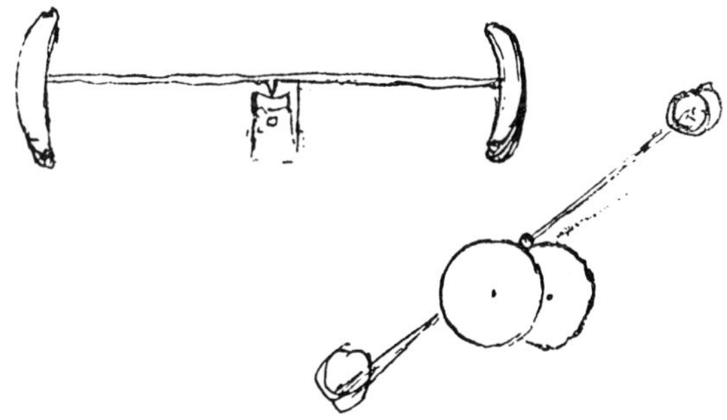

3 Huygens' suggestions for reducing friction in the balance bearings. Above: knife-edge bearings'. Below, roller bearings.
(Defossez, Savants XVIIe)

With his initial suggestion of using elastic springs with a balance, Hooke had certainly expressed a brilliant idea. But we do not know how long or how intensively he expressed his ideas or whether he ever realized them in practice. Hooke always came up with new plans and ideas, was very enthusiastic about them at first, but seldom worked steadily to make his plans become reality. It is known that in 1669 Hooke gave up his interest in the spring-controlled balance to deal with other problems. But after Christian Huygens' publication in 1675, Hooke recalled his considerations of the same subject and, stressing his intangible priority, made serious protests to the Royal Society, claiming personal discrimination. Since similar ideas have appeared from numerous thinkers at about the same time the difficulty encountered by many people in telling the difference between the expression of an idea and the achievement of practical results still occasionally leads today to discussions as to whom the priority belongs for making the spring-driven oscillating system so important to the chronometer and all other portable clocks. As a remaining fact of primary importance, Hooke's determination "as power, so tension" remains undisputed.

(The power of any spring is in the same proportion with the tension thereof). But the practitioner can still judge negatively that Robert Hooke never built a clock based on his principles for use at sea and, even later, never tested his theories at sea. Hooke thus appears as the typical example of the egocentric intellectual: Every suggestion or bright idea that he presented as an invention should, in his opinion, give him all the credit and all rights for its use in any eventual practical application. Along with this kind of attitude, Hooke was also known to be very jealous, arrogant and quarrelsome, so that it is understandable that his contemporaries viewed him with mixed feelings despite his many significant works and publications.

It is interesting to note that about the same time, in 1675, Louis XIV founded the Paris Observatory where a great number of astronomers and mathematicians worked, summoned by his minister Colbert and under the direction of the famous scientists Cassini and Perroult. From 1665 to 1681 Huygens also belonged to the academic circle of Paris.

In 1655 there appeared a suggestion from Matteo Campani for a marine clock functioning in a vacuum.[5] Campani was a scientifically inclined churchman working in Rome, who is said to have built a clock with a pendulum in 1655, thus before Huygens, at the request of Pope Alexander VII. The marine clock that he suggested in 1673 also had a pendulum, and the whole movement with the pendulum was to be housed in an airless glass container (#4). One wonders how Campani could suggest such a fantastic and, at that time, technically impossible arrangement after having built technically outstanding and now highly valued night clocks to his own designs.

Among the early and also fantastic ideas for a longitude clock is the suggestion of W. Leibniz of 1675 (#5). In the course of development of the marine chronometer, one must recognize this brilliant idea from an outsider which, although coming from a great universal thinker, was far from any practical application. The reason for this is that the running down of a spring system would not always take exactly the same length of time, so that the juxtaposition of many such intervals could not be used to achieve an even system of measuring time. Giving up a standard of time such as an oscillating system (a pendulum or combined spring and weight) affords appears to be a denial of the importance of Huygens' work, which was well known to Leibniz.

Illustration 11 shows an experimental French marine clock, probably built around 1700; unfortunately nothing is known either of its creator or of any tests that might have been made with it.

A very interesting forerunner in the course of development that led later to the very successful marine clocks, is the work of Sully. It shows how Huygens' basic ideas were taken up everywhere and developed further. Sully was a scientific adventurer on a grand scale. Born in England in 1680, he began as an apprentice to George Graham — at that time the most famous clock and instrument maker — and was able to assimilate a high degree of clockmaking skill. His restless nature, though, did not let him stay long in one place. Perhaps under the influence of Newton and Wren, he devoted his attention to the marine chronometer problem briefly in 1703. But then he became involved in founding various unsuccessful clockmaking businesses in Vienna and Paris, and not until in 1720 did he turn back to the problem of the marine clock. In 1724 he submitted such a project to the Paris Academy of Science,[6] and in the same year he sent a working model to George Graham in London, a clock still owned by the Clockmakers' Company today. Illustration 13 shows Sully's marine clock, and #13b and 13d show its movement. This clock, or a similar one, was driven all over Paris in a horse-drawn wagon in 1724 and had shown an error of 2.7 seconds in one hour, which made it obviously unusable for solving the longitude problem. An improved variation[7] was tested on the Garonne in 1726 and erred by only one second per hour, which was still not good enough to achieve a successful determination of longitude in the terms of Queen Anne's Act of 1714.

Sully's interesting escapement of 1724[6] is shown in illustration 6. Sully built numerous marine clocks of other types, one of which is shown in #7. In 1726 he undertook a sea voyage to Bordeaux with such a clock, during which the clock performed well. Thereupon Sully produced still more ideas for clocks to determine

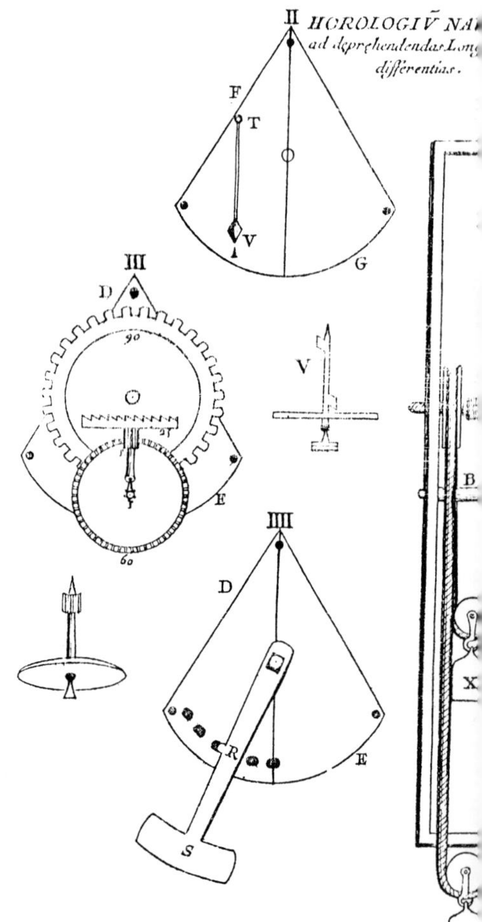

4 Matteo Campani's marine clock proposal to the King of France. Legend: G is an empty glass container that holds the actual movement, but the weight drive at left and hands at right are outside

longitude.[7] In further sea trials, all of his clocks performed considerably worse in stormy weather and finally failed completely.

Besides the aforementioned suggestions for determining longitude, there were several others from independent thinkers who deserve mention here. Some of them were intellectual speculators like J. Hutchinson (1712), Hall and Blank (1714), or William Hobbs (1714), William Palmer (1715) and Jeremy Thacker (1714), while others were practitioners with actual apparatus intended for testing at sea, such as the Abbé de Hautefeuille (1719); Jean Baptiste Dutertre (1726), a well-known clockmaker who suggested a clock with two opposed long pendulums — a misconception obviously built in ignorance of personal experience with the behavior of a ship at sea; and Pierre J. de Revaz, whose novel idea was to mount the balance on a spring band instead of a steel axle. Illustration 16 shows his presumably first test apparatus, while #17a and b show a complete clock built according to this principle (1749), which is noteworthy in that the spring band for the balance is held at one end by a bimetallic strip for the sake of temperature compensation. It is not known whether this clock was ever tested at sea.

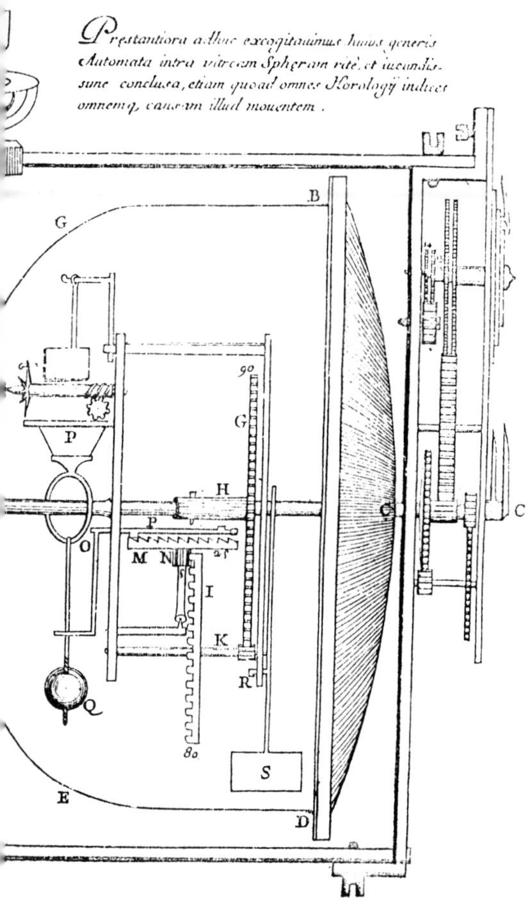

it, so that two airtight tubular arbors are required. Gimbal suspension of the whole clock and adjustment of the movement's position inside through S and R are shown in sketches III and IV.

To complete this chapter on the forerunners of chronometers meant for use on ships, it must not be forgotten that since mechanical clocks were invented, they were used to keep time on shipboard. Naturally the mechanical clocks available at the beginning of the 16th Century could not be used to determine longitude, for errors of +/- ten minutes on successive days could be expected then from mechanical clocks using a balance as a time controller; such an error of ten minutes per day, though, represents 2.5 degrees of arc, as we have seen, and thus about 190 kilometers at 45 degrees of latitude.

A relic that provides conclusive proof that clocks were taken along on ships quite early is the iron clock with weights shown in #8, which is kept at the Rijksmuseum in Amsterdam today. It was rediscovered at the winter quarters of W. Barents on Novaya Zemblya in 1871. Barents, the renowned seafarer, had discovered at the end of the 16th Century, the northern entry to the Pacific now bearing his name, and in 1596 he left this clock behind in his winter quarters.

Many outstanding clockmakers later produced significant ships' clocks for use at sea and mounting in the captain's cabin, several of which have been preserved. Generally they have a jointed gimbal apparatus for mounting in the case, but they were not usable for measuring longitude. Among them are, for example, the cabin clocks of Johannes Van Ceulen. Illustration 12 shows such a cabin clock which is presently in the Kassel Museum. These clocks made by Van Ceulen show excellent workmanship and surely gave good service, but they too could not meet the demand for high accuracy needed to solve the longitude problem, and scarcely any of their owners could have considered using them for that purpose.

The beautiful clock made by George Graham (#15) is equipped with a cylinder escapement invented by him, in which he had particular confidence. Whether this clock was ever intended for use at sea is unknown, as is the clock's history before it became part of the famous C. Ilbert collection in the British Museum.

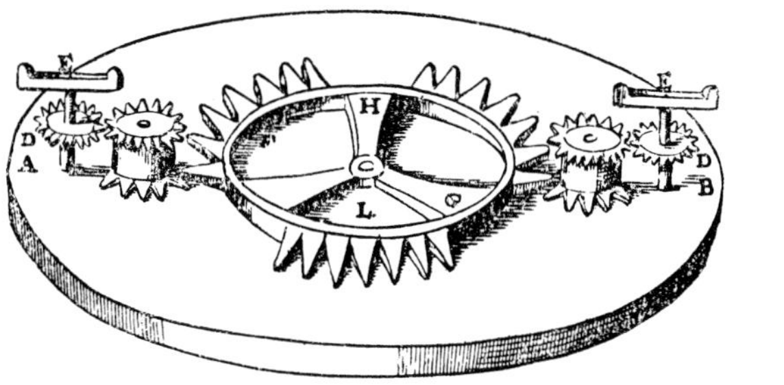

5 Marine clock proposal from W. Leibniz (1673). Only the principle of Leibniz's marine clock idea is suggested: Two wind vane systems E, E — at right and left — are driven by the small spring barrels near them. The large wheel in the center turns constantly and uses the teeth that are on its outer surface to wind first the one and then the other barrel, so that the vanes are constantly in motion.

The drive mechanism for the large wheel and any way of regulating it are not shown. Leibniz had assumed that the two vanes would run down at the same rate because of the small working area of the intermediate springs, and thus an evenly precise rate of the hands (also not shown) would result.

(Defossez, Savants XVIIe)

Other great British clockmakers also built cabin clocks, such as the bracket type with gimbal attachment shown in #10, made by Daniel Quare of London and found in Kassel today; or the very closely related clock by Thomas Tompion (#9), which was probably intended to be a traveling and marine clock, since it was equipped with a pendulum for use while in a fixed position but also with a simple apparatus for switching to balance operation when in use on a ship or land vehicle.

It must also be noted here that sand clocks (hour-glasses) were also used very early to tell time on shipboard,[8] though they were naturally suitable only for measuring short intervals of time and thus cannot be regarded as ranking among the forerunners of longitudal determination at sea.

1. Christiani Huygenii a Zutlichem, Const. F. Horologium, Hagae Comitum, Ex officine Adriani Vlacq, 1658.
2. Oeuvres complètes de Christiaan Huygens, Société hollandaise des Sciences, Den Haag 1888, new editions to 1837.
3. H. Bertele, "Precision Timekeeping in the pre-Huygens Era", London, 1953.
4. William Derham, F.R.S., The Artificial Clockmaker, A Treatise . . . London, 1696.
5. Matteo Campani, "Proposizione D"Orioli Giustissimi Utile a Naviganti", Rome, 1673.
6. Henry Sully, "Horloge inventée et executée pour une plus juste mesure du temps en mer . . .", (Mem. Ac. Royale Sciences), IV, Paris, 1724.
7. Henry Sully, "Description abrégée d'une Horloge d'une nouvelle Invention pour la juste Mesure du Temps sur Mer", Paris, 1726.
8. Com. Vivielle, Les Sabliers, horloges de mer, Paris, 1934.

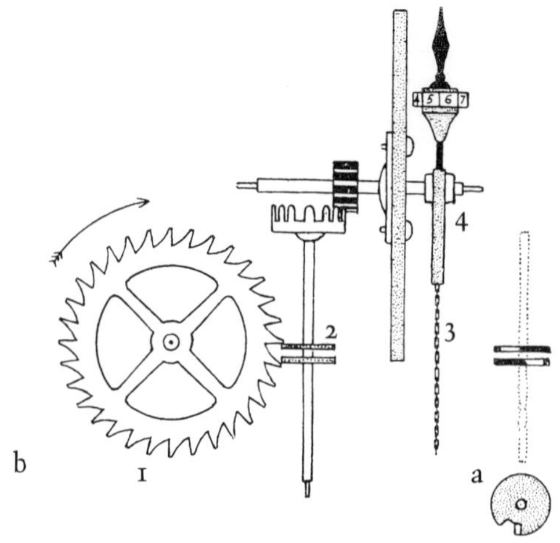

6 Details of Sully's marine clock proposal (ca. 1725): (a) Sully's escapement with two parallel notched impulse discs on the balance arbor, with which the escape wheel makes contact. Components:
1. Escape wheel
2. Impulse discs of the escapement
3. Chain to horizontal swinging lever
4. Cycloid faces

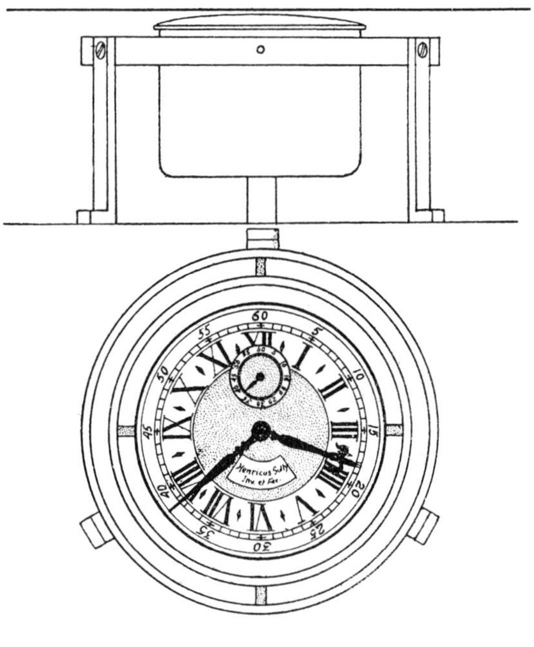

7 Sully's last marine clock (without data as to details of the movement). *(Gould)*

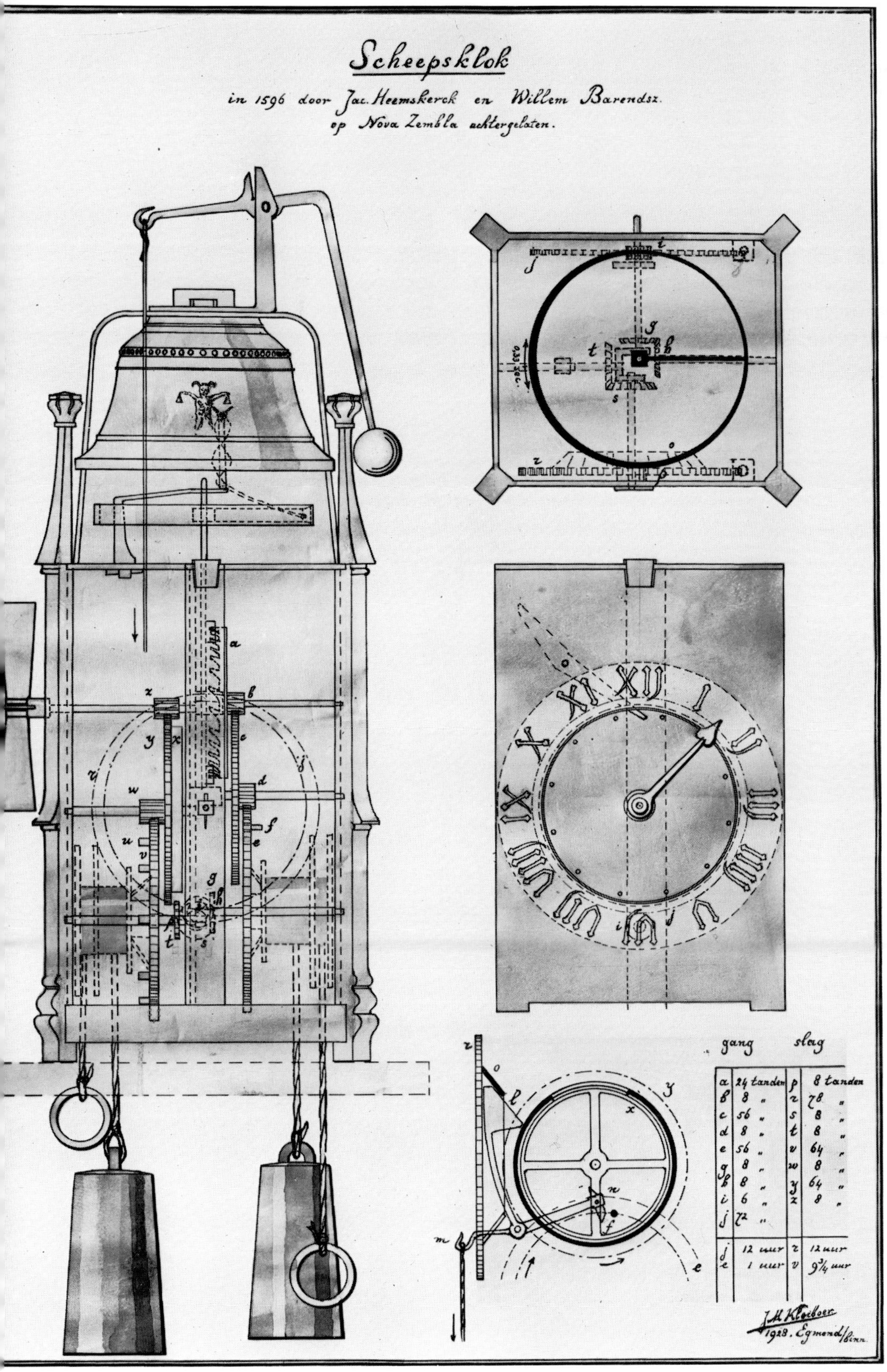

b△ c▽

8 Iron weight wall clock with verge and balance, Dutch, Gothic basic form, from around the middle of the 16th Century; taken to sea by W. Barents in 1580, found when his winter quarters on Novaya Zemlya were discovered in 1871.
a) Reconstructed drawing
b) Rear view of the movement
c) Barents' winter quarters on Novaya Zemlya, 1596

9a Bracket clock, probably intended to be a cabin clock, with striking movement and switching apparatus for optional use with pendulum (now missing) or balance. Signed: Thomas Tompion London Fecit No. 79; ca. 1690, front view.
Hessian State Museum, Kassel.

9b Movement of the clock by Thomas Tompion, back view.

9

10a English ship's clock, ca. 1700, outside view; made like contemporary English bracket clocks, but instead of a hook it has a ball-joint for hanging in the captain's cabin; case veneer of tortoiseshell; signed Daniel Quare — London.
Hessian State Museum, Kassel.

10b Back view of the movement (eight-day clock with striking). Instead of the customary short pendulum (verge escapement) found in English bracket clocks, here a double lever with one-speed hairspring and a large adjusting scale on a circular protective plate for the balance (seen at the lower right) serves as a balance; height ca. 20 cm.

10

11a Movement of a marine clock intended for use on board (case is missing), with verge escapement, no balance spring, French. Designed so the dial is horizontal, unsigned, ca. 1700.
Arts et Métiers, Paris.

11c Back view of the movement, balance with horizontal arbor that is moved over the slit plate and the vertical pin by the vertical verge.

11b Side view of the movement.

12a Dutch ship's (cabin) clock by Jean Van Ceulen Le Jeune, The Hague, ca. 1700. Drum-shaped silver case with gimbal mounting. Hour and minute hands as well as alarm dial from the center; second disc above. The movement is covered by a cut and richly engraved sound gong; similar bridge under the balance.
Hessian State Museum, Kassel.

12b Sound gong on the back.

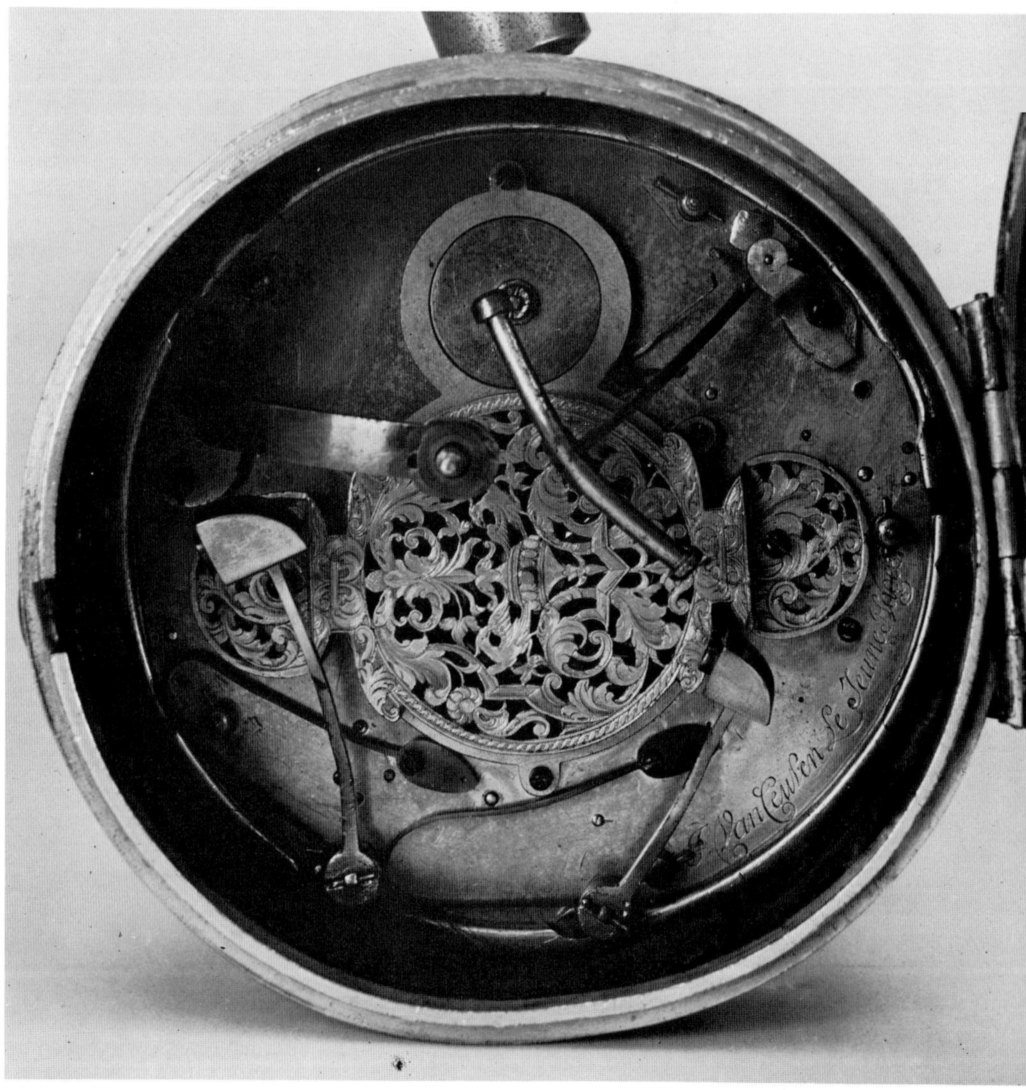

12c Rear view of the movement.

13a Experimental marine clock, signed "Henricus Sully, Invenit & Fecit;" ca. 1724, Versailles. In wooden case with carrying handle, silvered bronze dial, 127 mm diameter with separate second dial, escapement and horizontal pendulum after Sully's invention (#7a). Balance arbor with roller bearings. Flexible cord linking balance and horizontal pendulum mounted in cycloid cheeks to achieve isochronism, spring drive without chain and fusee; sent by Sully to George Graham with a description in 1724.
Guildhall, Clockmakers Company, London.

13b Back view of the movement of Sully's experimental clock with horizontal pendulum and cycloid linkage.

13d Top view of the movement showing escape wheel and Sully's unusual "double-panel" escapement.

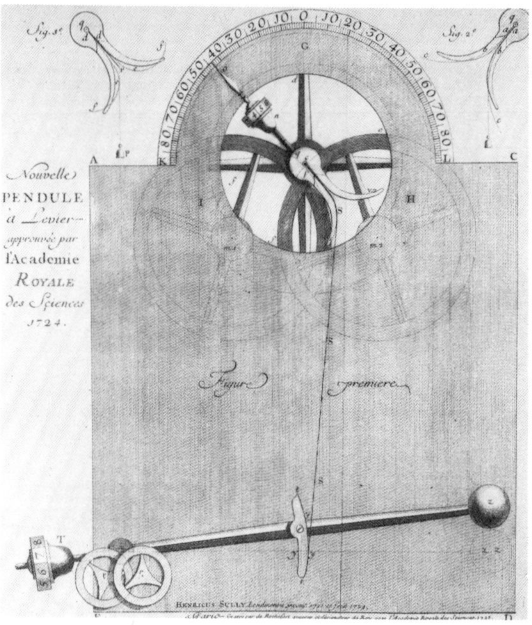

13c Working drawing of Sully's swinging system for his 1724 marine clock. The gravitational effect on the T-Z lever is used as reverse power on the mass of the balance; with the help of a line over the cycloid linkage q to the balance arbor, a proportional moment occurs. Roller bearings are used to decrease friction. The return to the center of gravity as reverse power exposes the system to all shocks and changes of location that are unavoidable at sea.

14 Experimental marine clock, signed Jean-Baptiste Dutertre, 1728 Paris; only minute and second hands; two long opposed pendulums that could scarcely have swung at sea.
Arts & Métires, Paris.

15a Drum clock, signed George Graham, ca. 1735; bronze case, 152 mm diameter, 114 mm thick, probably made to order for a prominent (Royal) family or the French Admiralty, perhaps meant to be a cabin clock; thirty-hour movement, cylinder escapement, special adjustment to suit temperature, maintaining-the-power spring; case sides covered in black leather and decorated with gilt French lilies.
British Museum, London, ex Ilbert collection.

15b Rear view of the movement of Graham's drum clock, in left foreground square block with dial for fine adjustment of the active hairspring length.

16 Experimental model of a marine clock, signed de Rivaz, before 1750; small hour and big second dials, large balance, verge on an elastic steel band, iron movement with fusee and chain, spindle escapement working on the balance via a fork and giving a certain amount of free play.
Arts & Métiers, Paris.

17a Experimental marine clock, signed Rivaz, Paris, ca. 1749-52, hour and minute hands in the usual household clock location, otherwise with a ring turning outside to indicate equitation; balance attached to a long springlike steel band, verge escapement, steel band in tension frame with bimetallic strips to change tension according to temperature; spring drive with *Remontoir d'égalité* (equalizing drive, controlled by the escapement). *Arts & Métiers, Paris.*

17b Side view of the Rivaz experimental clock.

III: The Heroic Epoch

Around 1730, the merchant and naval shipping of several West European nations had grown to considerable proportions. As those nations wanted to expand economically and geographically, interest in the form of the earth's surface and its division into mainland, islands and ocean, had become very intense, which made the solution of the longitude problem more urgent than ever.

There were two basic views of how to solve the problem: a) refining the classic astronomical methods to the necessary accuracy, b) taking the time at the point of departure along with the help of an appropriately accurate clock.

To apply the astronomical methods, more precise tables of values would have been needed for the positions of heavenly bodies. In addition, it would have been necessary to have a thorough training in the knowledge of the important constellations, along with their temporal changes, in the correct procedures of astronomical observation and the mathematical application of the tables. Also important was the readiness to apply the necessary time to this work.

As for taking along a mechanical clock set to the time of the starting point, the situation had changed significantly since Gemma Frisius' first suggestions. Thanks to the mechanical progress made by the British clockmakers, led by Tompion and Quare, there were already outstanding regulators to be had, with long, usually second, pendulums, and generally iron pendulum rods. Even with such good clocks, one had to expect irregularities in running of 20''/day when temperatures went to extremes in summer and winter, as is reported, for example, by Rees.[1]

Contemporary reports on the running rates of good standing clocks in the first two decades of the 18th Century, with average temperature changes and over periods of several weeks, have proved to be impossible for the author to find. But from personal experience with a Quare one-month clock it can be concluded that for some four weeks one could then get a rate difference in the area of 1.5 to 2 minutes per month or less. In order to receive the prize the stipulations of the 1714 act of Parliament make clear what mean daily rate deviations were acceptable in a voyage lasting some six weeks.

Allowable mean daily deviations for 40-day travel time to achieve the longitudinal accuracy specified by the Act of Queen Anne (1714):

Degrees of Great Circle	Allowable total Time Error	Daily Allowable Median Runnign Rate
1 degree	4 minutes	6 seconds per day
40 degrees	2 min., 40 sec.	4 seconds per day
30 degrees	2 minutes	3 seconds per day

Thus the running rates required by the 1714 act were no longer unthinkable, since at least equally accurate results had already been achieved with appropriately constructed clocks, but those clocks were not portable.

Now it was necessary to make a number of steps to attain the practical and mechanical means, organizational powers, and proper dissemination of information to take what had already been achieved in stationary regulators and make it possible for clocks that were subjected to disturbing influences on ships at sea, and particularly to bring the cost of such clocks to a level that made them affordable for general use.

After the factually correct but only ground-breaking steps taken by Huygens came a series of decisive but very differently inspired events between 1730 and 1785 that coalesced and thus made possible the goal's attainment.

This period can be called the "Heroic Epoch", for in it, thanks to the efforts of several extraordinary personalities, the technical and organizational steps were made that led to the development of the chronometer and its useful application in seafaring.

It was these men who, through their particular perseverance and dedication, heroically solved the problem that had often been described as insoluble. The British and French stood out in particular. On the British side it was John Harrison, Thomas Mudge, Larcum Kendall, John Arnold, and Thomas Earnshaw; on the French side, Pierre Le Roy and Ferdinand Berthoud.

John Harrison

A young carpenter left a small town in the north of England and went to London to tell the famous astronomer Halley, who was an influential member of the Longitudinal Commission that had been in operation for two decades already, that he and his brother had spent their spare time building wooden clocks and achieved unusual results that would make it possible for him to build ships' clocks with the qualities needed to solve the longitude problem. Halley listened to his story and finally recommended that Harrison present his ideas to Britain's best clockmaker, George Graham, the nephew of Thomas Tompion — the father of British clockmaking. In fact, Halley introduced Harrison to Graham personally. After overcoming some initial doubts cause by Harrison's rural appearance and unpolished manners, Graham began to develop an interest in Harrison's ideas and conclusions, which were diametrically opposed to those that prevailed in clock-making at the time. He gave Harrison advice and assistance which, in thirty years, resulted in a practical solution to the longitude problem that had not been solved until then, allowing seafaring and navigation to develop in completely new directions.

In Harrison's already mentioned work with large wooden clocks, the very unorthodox concepts and constructive details used had achieved a degree of accuracy whose historical importance has only recently been clarified by W. Laycock.[2] Harrison's independent work was, in fact a guideline in solving the problem of successfully building precisely running marine clocks with running accuracy.

Harrison's life story reflects the origin of the actual marine chronometer. Until a short time ago, not very much of it was known. It was only Colonel Quill's biography[3] that provided a fascinatingly detailed account of the historical course, the human angles, and frictions involved in Harrison's strivings that led to the successful breakthrough in solving the longitude problem.

John Harrison was born in Foulby, near Wakefield in Yorkshire, in 1693. A few years after he was born, his parents moved to Barrow-on-Humber in Lincolnshire, where his father became the village carpenter. John Harrison started early as his father's apprentice, but even as a boy he was interested and active in repairing and building clocks, and interested in music as well. Before he was twenty years old, he and his brother James had built a floor-standing clock almost completely of wood; it still exists. In their later standing clocks, the Harrisons had used a brilliant means of temperature compensation in the pendulum, with several metal staffs with different expansion coefficients working against each other in a sort of grid arrangement so as to prevent any lengthening or shortening of the pendulum length by temperature changes. Quill assumes that they had built such clocks because of the 1714 Act of Parliament and the prizes it offered. Together the two brothers worked out a concept for the construction of a seagoing precision clock, and the older brother, John, was to travel to London to meet the commissioners of the Board of Longitude, present their plans and ask for an advance.

For four years the brothers worked to build a clock according to their plans. This clock (#34a, b), today known as H1, was finished in 1735. Of particular notice are the two big vertical levers with matching arms and balls at their ends, coupled together by four springs. This was a reappearance of the cross-sect principle that Bürgi had envisaged and Huygens had later used, but was now developed further, with springs giving it its own oscillating capability. Surely in Harrison's case too the basic concept was that because of the opposing movement of the oscillating system's critical parts, the accuracy will not be disturbed by the rocking and pitching of the ship, since the effect of one balance counteracts that of the other.

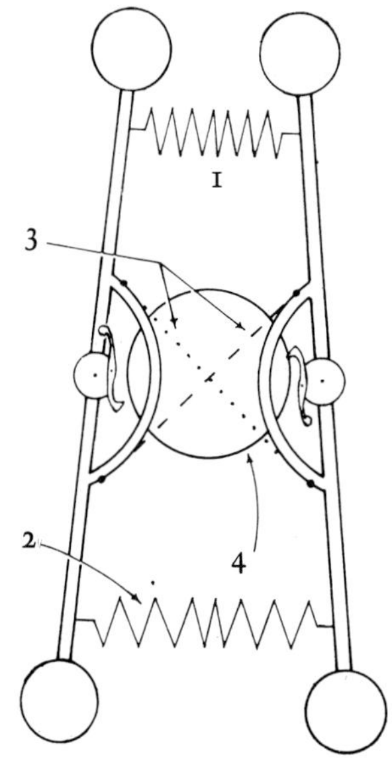

18 "Grasshopper escapement" system of Harrison's marine clocks:
a) Basic arrangement of the escape wheel, two balances with spring coupling and each balance arm with its own part of the escapement. Linkage design of the two swinging arms:
1, 2 Upper and lower coupling springs,
3 Tension lines between the two swinging arms over the shortened attachment pieces,
4 Escape wheel

b) Details of the "grasshopper escapement" for the unified functioning of two balances, used in Harrison's marine clocks H1 (1735), H2 (1737) and H3 (1757). The springs H and X, or H' and X', replace controls K or K' and the effect of gravity in the original design with a working arbor.
D-balance arbors, A-carrying panels for the pallets, B or B', H or H'-springs for pallets, J or J'-spring limiters. *(after Quill).*

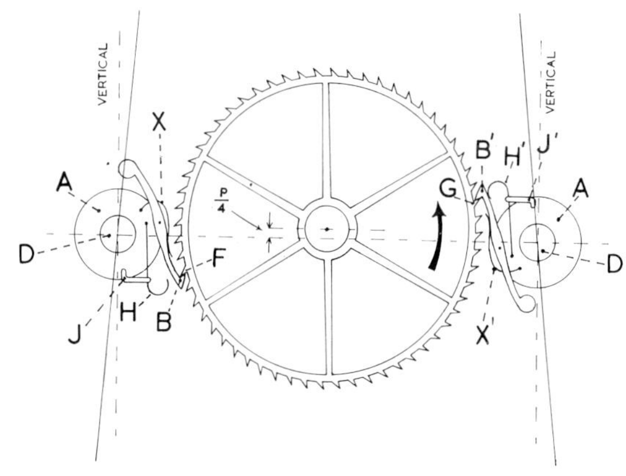

As an escapement, Harrison used in H1 a device that he invented and used very successfully in his wooden standing clocks: the grasshopper escapement. To adapt it to the double oscillating system, he developed the variation shown in #18a, b. The elimination of the drop unavoidable in other escapements was surely an advantage of it, but the permanent coupling of the oscillating system to the escapement was somewhat of a disadvantage. To equalize the effect of temperature changes on running, Harrison used the grid pendulum he invented for his standing clocks, made of iron and brass rods so that the tension of the coupling springs between the oscillating bars was influenced correspondingly.

As soon as the H1 clock was finished, the first sea test was initiated by the Board of Longitude. After laborious bureaucratic maneuvers, John Harrison set out with his H1 precision clock on board the ship "Centurion" (#33) on May 19, 1736, and reached Lisbon on May 26. H1 worked fairly well on this voyage, but did not live up to Harrison's expectations.

Around 1737 the Harrisons appear to have decided to move from Barrow to London so as to make further improvements. Here the brothers built their second precision marine clock, H2 (#35a, b), which they finished in 1739. It is very similar to H1 in design but somewhat taller and heavier. Remarkably, it does not seem to have been tested at sea.

On January 16, 1741 a reference to Harrison's third marine clock, H3, first appeared in the annals of the Board of Longitude. At that time Harrison believed that his new clock would be finished in two years; in fact it was not finished until 1760. Quill cites a variety of quite plausible reasons for the long time needed to build H3 (#37a, b). It no longer has two levers, but two round balances swinging in opposition to each other, nor does it have a grid system for temperature compensation, but rather — for the first time in the history of technology —a bimetallic strip that acts on a hairspring and thus automatically provides the necessary negative temperature compensation.

It was a return to the customary method of regulating pocket watches at that time, with a small pin (curb pin) moved along the bend in the hairspring, thus changing the effective length of the hairspring and thus influencing the rate. In the pocket watches of the time —and until very modern times — the curb pin had been pushed by hand, using a small lever, the regulator and, at times, a micrometer screw has also been provided for finer adjustment. Harrison, though, attached the curb pin to the end of a special component, made by riveting a brass and a steel strip together, for automatic temperature control. Because the two metals have different expansion coefficients and since the two strips have been riveted together, such a combination bends, to one direction or the other according to whether the temperature becomes higher or lower. Thus Harrison created a principle that has since been taken for grantedin all of technology for temperature measuring and controlling systems. Harrison gave this invention, which he made in about 1741, the name of "metallic thermometer."

The complicated coupling system between the two balances seems to have been the cause of lasting disturbances; though the components of H3 were straightforward enough, it was so hard to get to them that a simple adjustment took hours.

The decision to begin sea testing was made by the Board of Longitude on March 12, 1761. Shortly before that a letter from Harrison had arrived in which he informed them that the small clock mentioned the previous autumn had given unusually good results, and thus he asked to use it instead of H3. On March 12, 1761 this new request from Harrison was accepted. It was a decision that was later to make world history and raise this small clock, known today as H4, to a position of great significance. The break in the whole previous course of Harrison's development that this new clock (#39a, b) represents was quite incomprehensible until a short time ago. It is not only the external design that departs from the impressions made by Harrison's previous marine clocks (#34, 35, 37) in form and size, but primarily the complete difference of movement structure and the operating principles involved. H4 is basically a big pocket watch, like a carriage clock, with a two-plate movement —similar to those of pocket watches — but very different from ordinary pocket watches in several unusual details.

Naturally Harrison no longer used lignum vitae for the bearings or wheels of H4, as he had in the three large marine clocks. In the interim he had become acquainted with jewel bearings and lubricants. Jewels were used in the following important places in H4: for the balance staff bearing, the escape wheel arbor, the remontoir release, the arbors of the vane and the third, fourth and fifth wheels.

The workmanship and finish of the surfaces of all working parts equaled that of the best clocks of the time. Illustrations 19a, b show the schematic structure of the mechanism and allow one to recognize essential differences from the usual pocket-watch mechanism.

a) The remontoir (equalization drive), as already used in the bigger clocks, H1, H2 and H3, is activated by the main mechanism at short intervals of 7.5 seconds and thus gives the escapement a very steady power. In the drawing of the mechanism (#19a) it is shown that the pinion of the third wheel (4) serves to wind up the remontoir, which is done by the fusee via a wheel with internal gearing. It is released by a pivot system similar to a lantern drive, which is firmly attached to the wheel that makes direct contact with the second wheel. This always releases the remontoir after an eighth-turn of the wheel and thus tightens the intermediate spring. During the short period of remontoir release, the turning moment is made to vary in the ratio of 11:13. In addition, a safety device is included to prevent disturbances in resuming action after either the main drive or the remontoir spring has run down completely. For this purpose Harrison used a braking spring that works on the balance at the last winding of the fusee.

b) The temperature compensation (#19b) lengthens or shortens the effective length of the hairspring according to external temperature conditions. For this Harrison used, as in H3, a bimetallic strip of brass and steel, causing a curb pin at its lower end to contact the straight part of the hairspring and thus lengthen or shorten its effective length, as already was done in H3.

c) A maintaining-the-power device was built into the fusee, similar to those Harrison had already used in the big marine clocks, to allow the necessary power to be delivered to the movement during winding. The corresponding spring of H4 is big enough to function for eleven minutes, although the winding only takes a few seconds.

d) An auxiliary contact (cycloid pin) works similarly to the cycloid cups of Huygens' pendulum suspension to increase the isochronism of the oscillating system, and can be finely adjusted with an adjusting lever.

3) Illustration 19d shows this special verge escapement

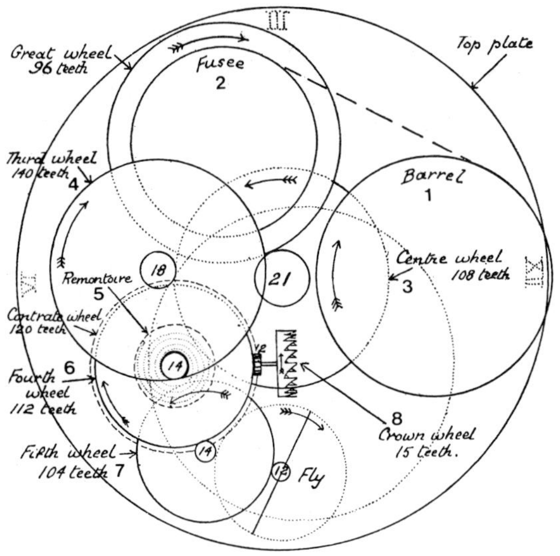

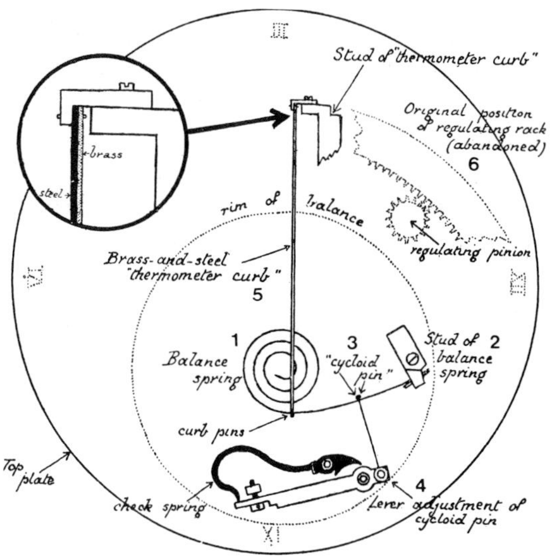

19a Arrangement of the wheels with the remontoir d'égalité in H4, cover plate removed.
1 Barrel
2 Fusee with bottom wheel, 96 Z
3 Central wheel 108 Z, with pinion 21 Z
4 Third wheel 140 Z, with pinion 18 Z
5 Remontoir (crown wheel, 112 Z), with built-in spring, winding every 7.5 seconds
6 Fourth wheel, 112 Z, with pinion 14 Z
7 Fifth wheel, 104 Z, pinion for air resistance 12 Z
8 Escape wheel, 15 Z, with pinion 12 Z
The remontoir discharge is missing from this drawing.

19b Arrangement of the temperature compensation on the upper plate of H4.
1 Balance hairspring
2 Hairspring attachment
3 Regulating pin (cycloidal pin)
4 Adjusting lever and spring for 3
5 Riveted brass-steel strip works on hairspring with curb pin, at upper left is an enlarged view of the attachment.
6 Toothed arc for adjustment of 5 (later given up)
(after Gould)

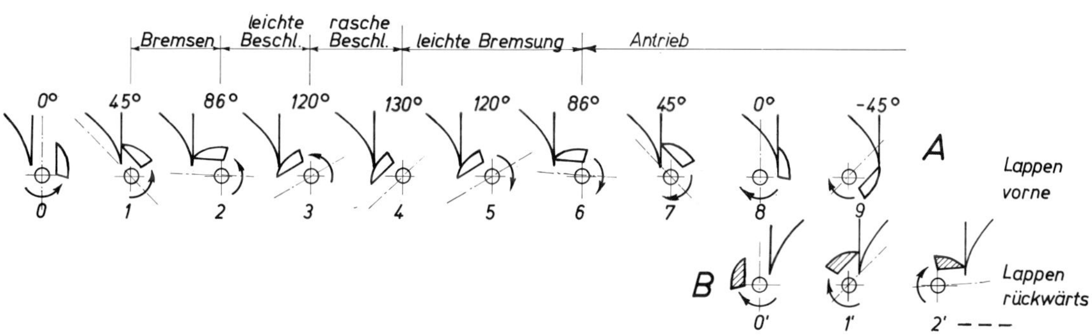

19c Sequence of pallet settings against the teeth of the escape wheel in the special verge escapement used in H4.
A: for the front pallet (cf. #19d)
B: for individual positions of the rear pallet
The escape wheel moves from left to right (0-1-2, etc.). One sees that between 1 and 2 — as in the usual verge escapement — the escape wheel falls backward, but at 2 the tip of the tooth moves out *over* the pallet angle and glides farther on the back of the pallet (from 2-3-4-5-6), thus the verge and the balance can swing out much farther than in the usual spindle escapement. This occurs along with the very slight friction of the teeth on the polished diamond pallet surface, as an important part of the good running results of H4.
From 6 to 9, power reached the straight (inner) flange as in the usual verge escapement.

in large scale. In a hitherto unknown way, it uses both sides of the pallet and thus increases the amplitude considerably. At first glance one might regard this mechanism as merely a variation of the old familiar verge escapement, but the unusual pallet location leads to an essentially different effect from the usual verge escapement. On closer observation of the pallet cross-sections it is seen that the inner flanges of the pallets are parallel to each other, instead of being at the usual 100-degree angle, but that the outsides are curved. The two pallets themselves are unusually small, their flat side measures only 0.8 mm. But the truly unusual feature of these pallets is that they are cut out of diamonds. In their small size and unusual shape, Harrison made the outer flanges of the diamond pallets take part in the operation of the escapement, in contrast to the usual verge escapement; that is, they come into contact with the hardened-steel points of the teeth of the escape wheel. This flange operation, previously not very highly regarded, was studied exhaustively by Evans in 1953. Illustration 19c shows a further development of Evans' concept, resulting from the critical pallet position and necessary to understanding the operation of Harrison's verge escapement, and the caption of the illustration offers a more detailed explanation.

As early as 1752 a watch that much resembled the renowned H4 was built for Harrison by his colleague Jefferys — a very experienced watch maker. Its main points: small dimensions, temperature compensation by means of a bimetallic strip, use of a counterlock (maintaining power); this clock (#38a, b) was very much Harrison's invention. Above all, it possessed the unusual escapement of H4 with diamond pallets used in H4, as well as a small remontoir. This watch that Jefferys made was always worn by Harrison himself. Later it passed to a grandson, who mentioned in a book he wrote[5] that the watch was made by Jeffries to special instructions from John Harrison, his grandfather. It was signed "J. Jefferys London" on the back plate and "John Jefferys, London, A. D. 1753" on the lid, which can probably be regarded as Harrison's special recognition of the maker. The temperature compensation of this watch functions in exactly the same way as in H4. Thus this may be the very first pocket watch built with automatic temperature compensation.

The fact that the idea of temperature compensation was already in the air at the beginning of the second half of the 18th Century is proved by the discussions[6] of the movement fragment no. 4759 in Paris — which lead to the conclusion that the pioneers of horology at the time, such as Julien Le Roy, were already working on temperature compensation for pocket watches.

Even so, it remains a psychological wonder that John Harrison, along with his favorite idea — the large marine clock of the H1, H2 and H3 type — also worked so intensively in the very different realm of precision watches. An explanation has recently been suggested by Laycock, who notes that Harrison had viewed the problem of using mechanical clocks to determine longitude at sea as a much broader subject than a mere mechanical system that could withstand sea travel without disturbance to its accuracy. Harrison understood the solution of the longitude problem as a complex of challenges consisting of the following individual tasks:

1. Exact knowledge of the time at the point of departure, based on having made the necessary astronomical observations there.
2. Keeping the astronomically determined time at that location on a fixed, precise regulator and having it available at all times.
3. Transporting the regulator's constant indication of that time on a ship, for which a very precise, small portable clock, is necessary. For this purpose Harrison is said to have developed H4 and the portable watch made by Jefferys; their construction was also the birth of the pocket chronometer, a type of watch that is very closely linked with the actual marine chronometer, as will be shown below, and that was steadily produced parallel to the marine chronometer by the notable chronometer makers.
4. The availability of a particularly precise, reliable clock that could be set up in a ship, the actual marine chronometer that could be placed in a captain's cabin and utilized to evaluate the astronomical observations taken on board with a pocket chronometer (deck watch).

John Harrison's confidence in the development and particular running characteristics of the Jefferys watch is deduced by Col. Quill from the fact that H4, the decisive step in winning the grand prize, was built by John Jefferys, probably along with Harrison's second works master, Larcum Kendall, under the closest supervision of John Harrison; and finally signed with the name of John Harrison because he intended to introduce this same timepiece to the public.

Testing of H4 at Sea

As soon as H4 had been presented to the Board of Longitude, an interesting period of test voyages began. The first major trip began on November 18, 1761, when H4 went to sea aboard the ship *Deptford*, accompanied by John Harrison's son William. Madiera was reached on December 9. Then the *Deptford* was at sea until January 19, 1762, anchoring on the latter day at Port Royal, Jamaica. From there it set out on an unexpectedly quick return trip after nine days, on January 28, because of the outbreak of war with Spain — without a chance to make the astronomical observations necessary for correct testing of the clock's running.

The return trip was made in rough, stormy weather. On returning to Spithead on March 26, 1762 there was —

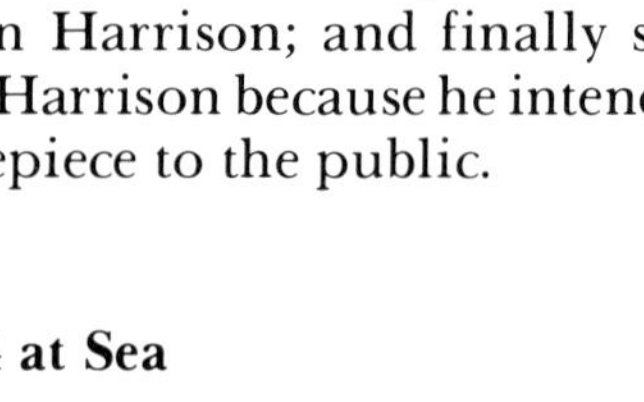

19d Pallet shape and verge arrangement as used in H4 (enlarged)
A - front pallet
B - rear pallet
S - balance arbor
(after Gould)

because of the prevailing bad weather — likewise no immediate comparison of H4 with the exact time there. Only after another week, on April 2, did this comparison with the astronomically exact local time take place. Unfortunately, there was no official certification of the difference between the time given by H4 and the astronomically correct local time. Thus despite the successful trip out and back and the uninterrupted functioning of the clock, there were no data available for an official judgment of the performance of H4 in the West Indian cruise under the terms of Queen Anne's Act of 1714, nor for a final decision. The situation was complicated further by the fact that William Harrison unofficially cited varying statistics for the running characteristics of H4 — for example, 2.36 minutes for the entire trip (to his father-in-law), and later only 5.1 seconds for the whole trip, a statistic that the Harrisons (father and son) worked out together by examining the daily running figures. The inclusion of a daily running rate is taken for granted in terms of present-day experience, but at that time when concepts of the behavior of a chronometer at sea were completely lacking, it was not at all customary. So it happened that the results of the expedition, which probably fulfilled the terms of Queen Anne's Act very well in fact and may well have been very superior to those terms, could not be considered. Despite months of work by several mathematicians at the Board of Longitude, on account of the lack of vital data and the incomprehensible procedures of the Harrisons in including running rates, no decision could be made. In July of 1762 the commissioners stated officially that the results to date would not be sufficient to determine the nautical distance in the sense of Queen Anne's Act, and that therefore a further test at sea was necessary.

To avoid this, Harrison tried various measures, such as the publication of three memoranda.[8] A great number of publications dealing with the Harrison case followed in the next few years. These can be found under the headings "Board of Longitude", "Act of Parliament", "John Harrison", "Longitude", "James Short",[9] and "Nevil Maskelyne" in Baillie's work,[10] with brief summations of their contents. For the interested public, during this period — beginning in 1763 — enough information was made public in the next few years for and against the use of a mechanical clock to determine position at sea to make the subject controversial, but also to acquaint the public with the new economic possibilities already noted.

John Harrison could not avoid a second sea voyage. The preparations for this trip, though, were made more difficult by the emergence of a new personality, namely Nevil Maskelyne, the Astronomer Royal, with his own ideas about determining longitude at sea.

It is not widely known that around the middle of the 18th Century the astronomer Johann Tobias Mayer of Göttingen, in collaboration with the renowned mathematician Leonhard Euler, had used his own very precise observations of the moon to make new tables that allowed a much more accurate determination of longitude at sea than the older conceptions had attained. J. T. Mayer had sent the new tables to London in 1755, and they were first tested by Nevil Maskelyne in 1761, at first in Greenwich and then on a sea voyage to the island of St. Helena, as a result of which Maskelyne was able to describe them as a significant advance beyond the earlier tables. On the basis of this fact, Maskelyne suggested that the Board of Longitude recognize J. T. Mayer as the first prize winner, and the Board decided to give Mayer a prize of 3000 pounds.

Only after long negotiations was the compromise accepted of testing H4 along with Maskelyne on a second sea voyage together. In August of 1763 Maskelyne first sailed to Barbados, while William Harrison set out from Spithead only nine months later, on March 28, 1764, on the ship *Tartan* and arrived in Barbados on May 13, 1764.

William Harrison made his return trip on the passenger ship *New Elizabeth.* On his return to Portsmouth in the autumn of 1764, William Harrison (unofficially) determined that H4, on the entire voyage of 156 days, showed an error of only 54 seconds. The results of this voyage were made known in London on January 19, 1765 and were a complete success for Harrison.

It was specifically and officially recognized that H4 had shown an error of only 39.2 seconds on the trip out, which limited the longitudinal calculation of the distance from Barbados to only 9.8 geographical miles, a result three times better than that which was required to win the entire prize of 20,000 pounds; unfortunately, though, Harrison could not derive the benefit he deserved from these findings for a long time.

There followed an unpleasant period in which Harrison was forced to make the important characteristics of his successful H4 clock public, inform other clockmakers of his unique experiences, and thus have his creation tested by third parties. Harrison was particularly bitter when the Board of Longitude wanted to make the mechanism of his H4 clock public. Let it suffice to say that after long hesitation, Harrison gave in on August 22, 1765 and, on the following days, dismantled the clock at his house before a six-member committee of the Board of Longitude, including the renowned clockmakers Thomas Mudge, William Mathew and Larcum Kendall, explained the mechanism, its preparation and assembly, and finally attested to the complete accuracy of his explanations.

The Board's next step was to have Larcum Kendall copy H4, so that during subsequent sea voyages H4 could be kept safely at home on land; then Harrison was finally required to turn his H4 over to the Board. As soon as this was done, H4 was taken to the Royal Observatory at Greenwich. There it was subjected to lengthy tests in various positions, during which it could be determined, although Maskelyne found a few deficiencies, that H4 was exact enough to fulfill the conditions of Queen Anne's Act in determining longitude accurately to within half a degree. In its manner of proceeding, which certainly hurt the personal feelings of John Harrison, the Board of Longitude nevertheless acted properly for its country, for through the Board's activities in introducing the H4 clock to seafaring on a broad front, Britain was able in the following decades to win a dominant position in maritime importance on the high seas. In the process, chronometer testing on Captain Cook's second and third expeditions played a decisive role, and the Board took a very decisive part in the planning and preparation for these expeditions.

In 1772 John Harrison petitioned King George III, who was interested in science and technology himself and had a special technical advisor — a Swiss named Demenbray. Between 1767 and 1771 Harrison had built H5 a marine clock of the same type as H4, and in 1772 he turned it over to the king for his personal examination. At first the results in the royal laboratory were an unexplainable failure, but this was soon explained as being caused by the influence of a large magnetic system in the room next to the chronometer test chamber. After this disturbance was eliminated, H5 ran as well as H4, and after some negotiation between the king and Parliament, John Harrison received the news on June 21, 1773 that the full amount would be paid to him for the best achievement according to Queen Anne's Act. In the process, though, certain monies were subtracted, so that he only received the sum of 18,750 pounds, still a large and significant amount.

A worthy monument to Harrison stands in Greenwich today. There, in the National Maritime Museum, stands H1, H2, H3 and H4 in chronological order, kept in splendid condition and functioning properly. A look into the depths of this display leads to a thoughtful observation of the individual pieces and awareness of John Harrison's part in this epoch. A visit to Greenwich becomes a pilgrimage for every enthusiastic horologist. And the wall regulator ultimately built by Harrison and intended as a time norm for the regulation of marine chronometers must not be forgotten. After his long struggles with the Board, Harrison had another location in mind for this regulator, but in the course of the last decades and the reconciliation they brought, it finally reached the proper place.

Pierre Le Roy and Ferdinand Berthoud

The importance and the problems that are associated with the names of these two personalities, who were working on the same problems at the same place at almost the same time, were portrayed superbly by Commander Gould in his major work.[11] The introduction states as follows:

"The true origin of the wheel chronometer of modern construction can be traced back directly to the work of Pierre Le Roy and Ferdinand Berthoud, both of Paris. A ranking of the accomplishments of these two masters is impossible, for it can be said in brief that Berthoud was an extraordinary talent who was able to turn all information on the works of others as well as the consequences of the performance of his own realized ideas into good practice very quickly, and who constantly fought his way to success through a long series of experiments, until he — considerably on the basis of improvements on

20a Pierre Le Roy's basic drawing of an escapement that largely detaches the balance (from the sealed submission to the French Academy in 1748), developed from the frictional rest cylinder escapement of George Graham 1 and 2. In the cylinder escapement, the tips of wedge-shaped parts set tangentially to the teeth of the escape wheel cause friction. During the entire swing of the balance — except the short times of impulse giving to the lips of the cylinder by the wedge flanges — they lie on the inner or outer surfaces of the cylinder and cause friction.

Pierre Le Roy's idea was to prevent the tooth tips of the escape wheel from lying on the cylinder by an additional device — an escape wheel lock. For this purpose he envisaged two elements on the balance arbor, one above the other; one to control the escape wheel lock and the other to give the impulse. The details of the manner of operation cannot be derived from the very unclear Fig. 3, but are easily understood from #20b and its caption.

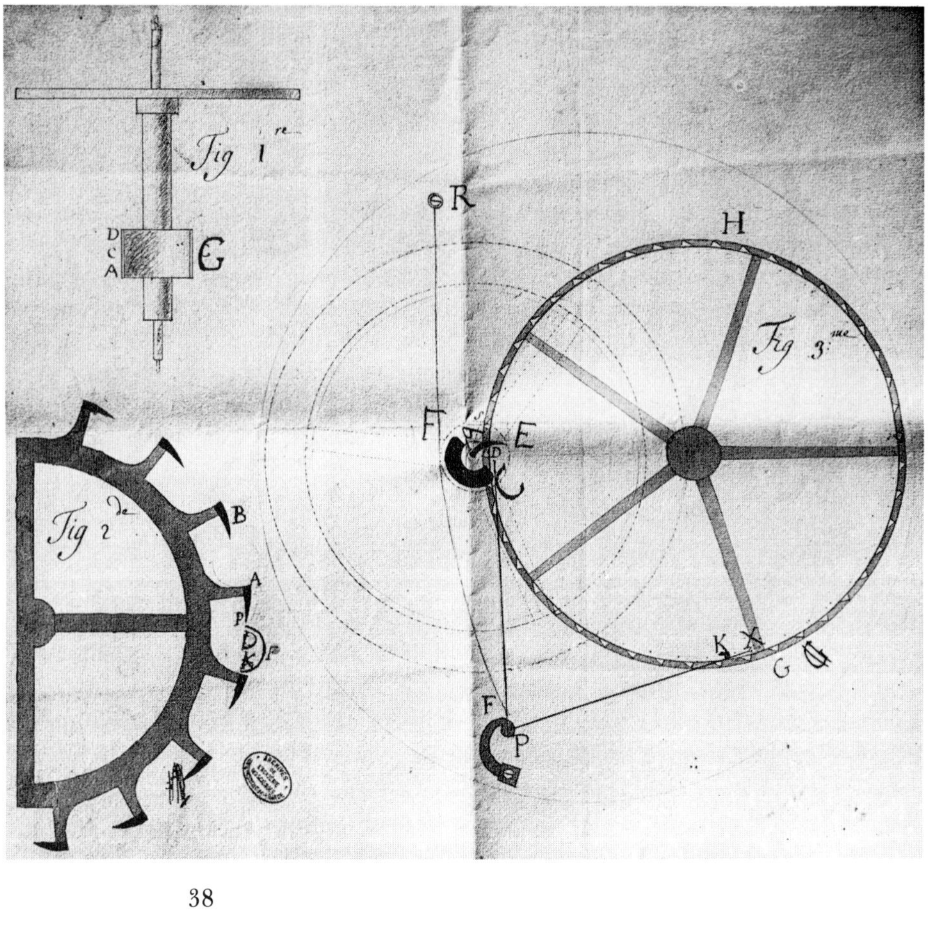

experimentally determined weaknesses — finally achieved a very successful marine chronometer. In comparison, Pierre Le Roy was a true genius to whom the solution of the marine chronometer problem was achieved in a truly scientific manner, whereby he, with much less work and expense than Berthoud, created a system that included all the essential characteristics of the modern chronometer. It must not be forgotten to mention that the two men were bitter rivals, and that throughout his life, neither was inclined to recognize the other's achievements in the least."

Pierre Le Roy (1717-1785)

Despite his significant accomplishments and progress, the achievements of Pierre Le Roy are little known to clock specialists and collectors. His own writings on the subject of chronometry have been listed in the bibliography. It is not easy to refer to them, though, since some of them exist only in manuscript form and others are found only in a few libraries. The only monograph[12] in which he was included devoted barely 30 of its 148 pages to Pierre Le Roy himself. Commander Gould, who presented a summarizing address on the subject of the marine chronometer[13] in 1923, referred briefly to the work of Pierre Le Roy, describing him as the most brilliant clockmaker who ever lived; later he devoted fifteen pages of his aforementioned major work to him, with thorough descriptions of his original work.

Pierre Le Roy's father, Julien Le Roy, was a famous clockmaker in his day and invented, among other things, a particular repeating mechanism that was used in French pocket watches. Another of Julien Le Roy's inventions was the so-called "all or nothing" striking apparatus that only repeated when the activating mechanism was turned on fully.

Julien Le Roy was the court watchmaker to King Louis XV, having his living and working quarters in the royal palace. During his apprenticeship with his father, Pierre Le Roy had the opportunity to learn watchmaking from the ground up. After his father's death in 1754, Pierre immediately succeeded him as royal court watchmaker.

Even before that, Pierre Le Roy had dealt with the basic problem of accurate timekeeping in mechanical clocks. In 1748 he had the revolutionary idea of making the oscillation of the balance essentially free from the retarding effect of the escapement and presented a diagram of the principle to the Académie Française,[14] thus assuring himself the scientific priority of his invention without making it accessible to the public. Illustration 20a shows Pierre Le Roy's original drawing in comparison with Graham's cylinder escapement. Similar escapements were probably built before that, but without the significance of the underlying principle being correctly understood or even expressed. Pierre Le Roy himself noted that the son of the famous Parisian watchmaker Dutertre once showed him a similarly functioning escapement system.

Pierre Le Roy, though, was clear as to the significance of his design. In the description he submitted in 1748, he followed a thorough discussion of the disadvantages of Graham's cylinder escapement with the following motivation for his new escapement construction:

"Therefore I believed that I must have nothing to do with any resting escapement and instead turn my attention to making the oscillating system hindered as little as possible by the various influences that we are forced to accept. It seems to me that in this context I must employ the manner of procedure used by truly experienced doctors: They respect nature and do not disturb it with radical methods of healing, because the effects thereof are almost always harmful. I believed that I had to adhere to this principle in developing the escapement that I have described here. One might permit me to name it *Echappement à détente.*"

Since then, that name has been applied to all chronometer escapements. The essence of the principle can be summed up as follows: *The less the wheel works have to do with the balance, the better.*

Whomever has worked with escapements knows well that in verge as well as cylinder and duplex escapements, the balance is never really free, and that its movements must always be either braked or absolutely stopped. Thus it is clear that a free escapement has great advantages over all other escapements. But it should not be overlooked that the older types of escapement also have certain advantages. This is because the verge escapement had continued to be used for more than a century after the invention of the first detached lever escapement by Mudge; verge watches were still being built in Clerkenwell, London, for example, in 1885. The cylinder escapement, though, was remarkably popular until the beginning of the 20th Century, and finally, as reported by Gould, for example, the duplex escapement was regarded for many years as superior to the detached lever escapement and even as an actual rival to the chronometer escapement until its fate was sealed by the misfortunes of the interesting Waterbury pocket watches around 1880. But as soon as it was a case of very precise time measurement, none of the older escapements mentioned here could compete seriously with a really detached escapement. The original drawing (#20a) scarcely lets the operation of Pierre Le Roy's idea be traced. Baillie offers a good description of how it works in the caption to a somewhat more clarified redrawing (#20b) of Baillie's original sketch. Thus Pierre Le Roy's first escapement was already a completely detached type.

The escape wheel, after giving an impulse to the escapement, is stopped from further movement by a *detent,*[15] or lock on the wheel, which holds it in a position where it is out of contact with the balance arbor. The wheel lock is formed here by a double lever mounted on a turning pivot. As in all modern chronometer escapements, and unlike the verge or lever escapements, the impulses are given only in one direction, thus only once in one full swing. We do not know how the actual constructions looked that Pierre Le Roy made according to this 1748 drawing.

Unfortunately, no watch, nor even a model of one, has been preserved. The basic principles, though, were made clear through the submission to the Academy, and they are also seen in his later works — more or less varied and simplified. Six years later (1754), Pierre Le Roy submitted to the Academy the description of a new marine timepiece

et faite par
Ferdinand Berthoud
N° 1

invented by him, but it was published only as an appendix to his "Exposé succinct . . ." of 1768. Gould and Ditisheim describe this 1754 suggestion as not being very well-developed, but in spite of that, the most important details of it can be noted:
Half-second oscillation;
Balance in ball form, weighing ca. 2.5 pounds, suspended from a straight wire spring 11 cm long;
The axis of the ball extended downward and set in a mount with four roller bearings;
Oscillating time adjustment by raising or lowering the steel wire; operating time only 6 hours;
Use of an intermediate winding (remontoir d'égalité) with a hairspring that has effect after every tenth swing.
Pierre Le Roy actually built a timepiece to these principles of 1754. A letter by a M. Petitmont, who saw this watch in December of 1756, describes it as having an unusual, star-shaped escape wheel as a special feature. Unfortunately, this marine clock of Le Roy has not survived either.

In the following years, Pierre Le Roy concentrated more and more intensively on the perfection of marine clocks. In 1763 he presented to the Academy a new marine clock that stood three feet high. In 1764 he gave the Academy another marine clock only half as big; it was tested by the Academy for almost a year.

Unfortunately, we know nothing of the performance of these clocks.
On August 5, 1766 he finally presented to King Louis XV his historically important marine clock, the design that made Pierre Le Roy one of the greatest clockmakers of all time (#43a-e).

This marine clock design of 1766 is the high point of Pierre Le Roy's creativity, the embodiment of all the technical principles that were to characterize chronometer construction for the following two centuries.

Two identical marine clocks of this design, designated A and S, were tested in three great expeditions. The first test was made by the Marquis de Courtainveaux on his yacht *Aurora;* it lasted three months. The second occurred in 1768 on the frigate *L'Enjouée,* during a 24-week voyage from Le Havre to Newfoundland and back to Brest; the third and last test took place in 1771 on the frigate *La Flore,* which ran onto the rocks of Wilmington near Antigua; Le Roy's marine clock was damaged in the process.

Most of the time during all these voyages, as well as during most land testing, the daily errors of the two marine chronometers made by Pierre Le Roy were less than a second and a half; thus the results were outstanding. To be sure, there were also individual entries that show greater errors from one day to the next, or even for several days in a row, of 5 to 10 seconds, but these cases were remarkably rare. Pierre Le Roy received several prizes from the French Academy for these outstanding achievements, but after the last voyage of 1771 he appears to have given up his activity, for there is no further evidence of it. Pierre Le Roy took a long time to publish information on his activities. In 1768, for the first time, there appeared a memorandum[16] that made his ideas and designs public, whereas his previous reports to the French Academy were only depositions and were not meant for publication.

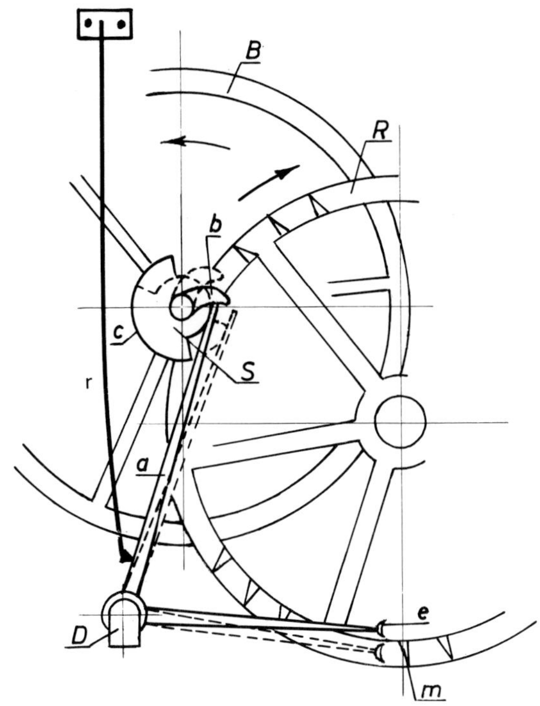

20b Diagram to explain the operation of Pierre Le Roy's detached escapement of 1748: The escape wheel R has 44 teeth vertically on its rim. The balance B has a flange b on its arbor and over it a cam S with a smaller and a larger semicircular arc c, which control the arm a of the two-armed escape wheel lock D. After the impulse is given by a tooth on the escape wheel to b, a is moved clockwise by c. The second arm D has a fork e on its end, the inside of which is bent so that when contacting the escape wheel near tooth m, it is pushed farther in a clockwise direction by it and a is lifted by c. Thus the balance can swing freely forward and back, until the flange b finally pushes the next tooth of the escape wheel in a counterclockwise direction and thus releases e by means of tooth m, so that D is turned counterclockwise by spring r, the escape wheel can give the next impulse to b, and the process is repeated anew.
(Charles Gros*, a scientifically minded horologist, built a system according to #20b and determined that the balance can indeed swing very far.
(After Charles Gros and Baillie, drawn somewhat more clearly)

* Charles Gros (Les Echappements des horloges et montres, Paris 1913).

Here we shall look only at the most important features of those two of Le Roy's clocks that were historically important, as seen in #22 and 43:
1. The large round disc-shaped balance made of steel, some 12.5 cm in diameter, weighing some 150 grams, time of oscillation one second.
2. The pivots run between two sets of four roller bearings each, while the weight is carried by a very fine (Cembalo) steel wire.
3. Two large balance springs are provided, both attached below the balance.
4. Two radially projecting screws on the balance ring are provided for adjustment.
5. For temperature compensation, the balance carries two uniquely formed thermometers, filled with alcohol above and mercury below.

The way this compensation works is that when the temperature increases, part of the mercury is moved closer to the axis and thus decreases the balance's moment of inertia. The correct design and proportions of the glass tubes allows not only all temperature influences, but also any changes in spring elasticity, to be compensated for on the moment of inertia of the balance system. Pierre Le Roy had previously used a bimetallic system (#21) for temperature compensation, but had found it not stable enough, and thus had given priority to the system with mercury. All the same, this drawing embodies the basic principle of inner balance compensation that was later in general use: an oscillating system where the temperature-induced changes of the system's inertia are compensated for by an opposite change in the moment of inertia within the oscillating system.

Details of the detached escapement system developed further by Pierre Le Roy from the détent principle of 1748 are shown in illustrations 22a, b. The originals of #22a are very unclear; the corresponding modern axionometric drawing by Auriscoste,[17] on the other hand, is very informative (#22b). The succession of impulse and lock of the escape wheel is clearly described in Auricoste's caption; the effect of power on the balance is such that the escapement almost always leaves the balance arbor free, whereby the impulse and the release take place very close to the middle position of the balance, where the influence of the impulses on the amplitude are least. An addition refinement in this design is that the impulses are given at the edge of the balance instead of near its pivot, which has a stabilizing effect on the oscillating process. In addition, it is worth noting that the lever system A is not exposed to any spring effect, which makes the recoil effect on the balance movement especially slight. Finally, it is worth noting that a case provided to house the movement which hung on gimbals, allowed +/-20 degrees of swing, and also had swing dampers.

Pierre Le Roy built not only the two big marine clocks that were tested on the expeditions of 1766 (on the *Aurora*) and 1768 (on the *L'Enjouée*), but also another marine clock in small round form for the 1771 expedition — *La petite ronde*, as it was later called — which was taken along aboard the *Flore*, under the command of Verdun de la Crenne. This clock functioned remarkably well for a long time. Very little is known about this little marine clock built by Le Roy. It or a similar one turned up in the European antique trade some years ago, but disappeared from sight again before it could be examined more closely.

Ferdinand Berthoud

His works, just like those of Pierre Le Roy, are in many ways a high point in the epoch of the chronometer's history. Berthoud's father was an architect and justiciary in Plancemont, southwest of Le Locle in the Swiss Jura, where he was born in 1727. He became a master in 1754. Even before that — in 1752 — he had given the French Academy a watch with a perpetual calendar and simultaneous indication of mean and real sun time. In 1753 the Journal Helvétique in Neuchâtel published his letter to Pierre Jaquet Droz, which includes an outline of his 1763 book, *Essai dur l'horlogerie*. The volume of work done by Berthoud is scarcely imaginable; he was, in fact, not only a successful watchmaker who created many high-quality floor clocks, regulators, wall clocks and pocket watches as well as chronometers, but simultaneously the author of numerous and many-faceted works.

Around 1764 he was named Horloger de la Marine, which assured him a yearly income of 7500 livres as well as a future pension of 3000 livres. During the course of time he also became a member of the Institute de France, the Royal Society of London, and later also a Knight of the French Legion of Honor.

The idea of developing a seaworthy mechanical clock for determining geographical longitude was taken up by him at the age of 27 and became a lifetime career. In 1754 he submitted the description of his Marine Clock No. 1, finished in 1761. He worked on it in his own independent way. First he built two heavy, relatively unpolished test systems, to which he added the numerous variations and refinements that, over a course of some fifty years, added up to an almost complete modern chronometer system. Among them one can find almost every possible design variant. At first there were chronometers with weight drives, of which he was especially fond, but soon he turned to spring drives and built them with and without fusees. Berthoud called weight-driven clocks *Horloges Marines* (HM) and spring-powered clocks *Montres Marines* (MM).

His Marine Machine (MM) No. 1 of 1761 (#41) had spring drive and an undetached lever escapement in which the impulse roller was held permanently in a slot in the lever. From this Montre Marine No. 1 and Horloge Marine No. 2 (#42), the line of development goes via an intermediate type, No. 4, with cylinder escapement, to the historically important Horloges Marines (HM) No. 6 and 8 (#45,46), which were already undergoing exhaustive testing at sea in 1768, proving themselves very well.

In Marine Clock No. 8 a cylinder escapement with a ruby cylinder was used, with an external gridiron designed for temperature compensation. Clocks No. 6 and, in particular, No. 8 show that particular care in developmental work and fine mechanical execution can lead to outstanding performance in an inherently complicated and basically unhealthy technical concept.

We are not well informed as to exactly when he began to build his spring-powered chronometers, or as to the chronological order that this development took. It seems that Berthoud began to work on them very early — parallel to his large clocks. A recent discovery in the reserves of the Paris Museum of Arts and Métiers (#36) dates Berthoud's activity in this area back to the mid-1850's. Undoubtedly #40a, b, and the later #62 and 63 belong to this era. In his *Montre Marine* No. 3 (ca. 1778), Berthoud had already foreseen a type of simple detached anchor escapement that was, to be sure, not as highly developed as that designed by Le Roy shortly before.[18] Berthoud, in fact, used the lever only to release the escape wheel.

At first Berthoud used two balances in his marine

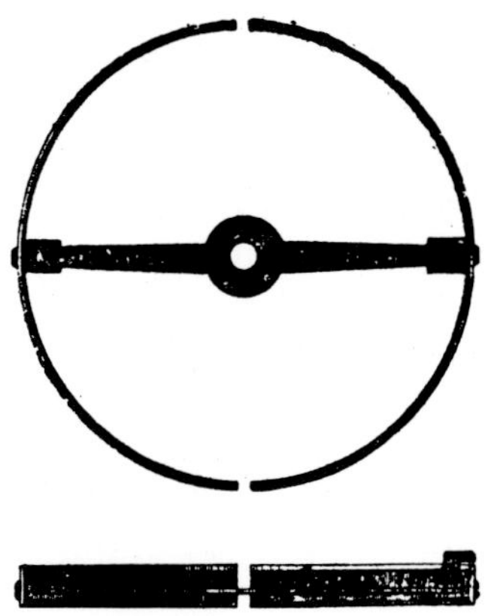

21 Pierre Le Roy's first bimetallic compensated balance. Drawing of a circular balance consisting of two bimetallic strips, from a document submitted to the French Academy by Pierre Le Roy in 1761. In this document he mentions that he used the principle, originated by John Harrison, of combining two metals of different expansion not to change the length of the spring with temperature changes but rather to influence the moment of inertia of the balance through the greater or lesser bending of the strips. Pierre Le Roy also mentions the possibility of attaching additional weights to the strips. *(ex Ditisheim)*

chronometers. For them he prescribed roller bearings for the balance arbors; sometimes both ends were mounted in this way, sometimes only one. Later Berthoud often used full bearings, but throughout his life he was opposed to jewel bearings.

For temperature compensation he originally used metal grids, later bimetallic strips that affected the balance spring from outside. Later Berthoud also built compensating mechanisms into the balance itself, using such setups sometimes alone, sometimes with additional temperature compensation on the hairspring.

The first spring detent escapement for chronometers was adopted by Berthoud between 1780 and 1782. This appears to involve Montres Marines No. 23, 24 and 25. This very important invention also runs parallel to the invention of the locking principle by Earnshaw and Arnold, who created the chronometer spring detent at about the same time.

Berthoud always had numerous apprentices, journeymen and colleagues, some of whom later won their own fame. For example, Vincent Martin became *Horloger du Roi et de la Marine* at Brest, and in 1805 his nephew Pierre-Louis Berthoud became the clockmaker of the Paris Observatory and the French Geographical Longitude Bureau; another colleague, Jean Martin, returned to Ferdinand Berthoud after a time and was assigned the completion of particularly important work.

In all, Ferdinand Berthoud created some 70 to 80 marine chronometers, of which the earlier important examples in particular are displayed at the Conservatoire des Arts et Mêtiers in Paris.

Pierre Le Roy

1717 - born
1742 - submitted notes on a frictional rest[19]
1748 - submitted notes on a marine chronometer with detached escapement (picture and description)
1754 - submitted notes on an improved marine clock (contents not available to author, but said to state the essential advances later used in Marine Clocks A and S)
1763 - Marine Clock A presented to the Academy (first described in 1770)
1764 - Marine Clock S presented to the Academy
1766 - Marine Clock S demonstrated to the king
1767 - first sea test of A and S on *Aurora* (Courtainveaux) from 5/25 to 8/27 (Le Havre-Amsterdam-Le Havre)
1768 - second sea test of A and S on *L'Enjouée* (Troujoley) during 161 days, via Le Havre-St. Pierre (Newfoundland)-Sole (Morocco)-Cadiz-Brest. Cassini Jr. acted as observer, both clocks gave good results, in particular, longitude determination with B was exact to 0.75 degrees
1770 - published *Mémoire sur la meillure manière de . . .*
1771/72 third sea test of A and S on *La Flore* (Verdun de la Crenne. Bordas, Pingré) from 10/29/71 to 10/8/72, wins Academy prize
1773 - Pierre Le Roy gives up work on longitude
1776 - receives pension by special decree of the king
1785 - died

Ferdinand Berthoud

1727 - born
1754 - submits description of design for marine clock (Horloge Marine 1)
1760 - submits Mémoire 1 on marine clock (Horloge Marine 1) design to Academy
1761 - submits Mémoire 2 to Academy, delivers Montre Marine 3 to Goupy for sea testing on *Hirondelle*
1763 - presents and submits Horloge Marine 1 to Academy; tested by Camus; presents Horloge Marine 2 and Montre Marine 4 to Academy; publishes *Essai due l'Horlogerie*
1764 - submits Mémoire 3 on Horloge Marine 2 and Montre Marine 3 to Academy; sea tests with Montre Marine 4
1766 - royal contract for two sea clocks (No. 6 and 8)
1768 - submits notes on isochronism, finishes Marine Clocks 6 & 8
1768/69 - sea tests from 11/10/1768 to 11/12/1769, via Cadiz-African coast-Caribbean Sea-Newfoundland and back, of Clocks 6 & 8 on *Isis* (Fleurieu and Pingré); both clocks show good results; sea testing of Montre Marine 3 on *Hirondelle*
1771/72 - sea expedition on *La Flore* (Verdun de la Crenne) on similar route to 1768-69 (Berthoud's clocks performed well, but were not tested again, at his request)
1773 - several important publications 1776 - sea expedition with Horloge Marine 18 and Montre Marine 4 for coastal survey via Brest-Cadiz-Canary Islands and back, with Borda as scientist
1785/88 world cruise with Horloges Marines 18 and 19 and Montres Marines 25, 27 and 28
1802 - Publication of *Historie de la Mesure du Temps*

Larcum Kendall and Thomas Mudge

Larcum Kendall's great significance is that he built the first absolutely exact copy of Harrison's H4, under

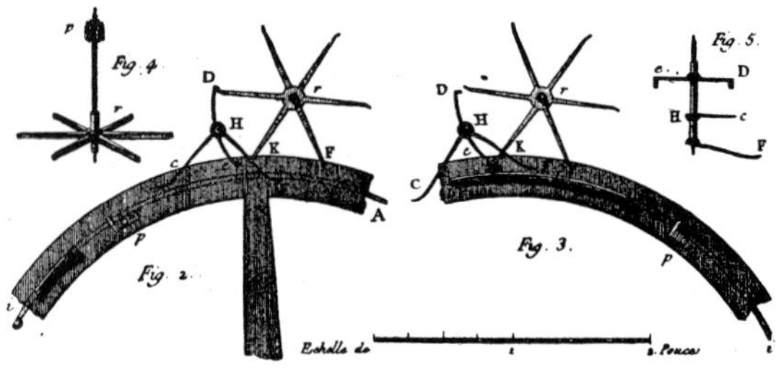

22a Drawing to explain the design and functioning of the échappement à détente of Pierre Le Roy, from Le Roy's own publication of 1770. In Fig. 2 + 3 D is the star-shaped escape wheel lock; in Fig. 2 it stops the escape wheel r. In Fig. 3 the escape wheel (star) has just been released by the arc b attached to the balance ring.
(Ditisheim/Tardy)

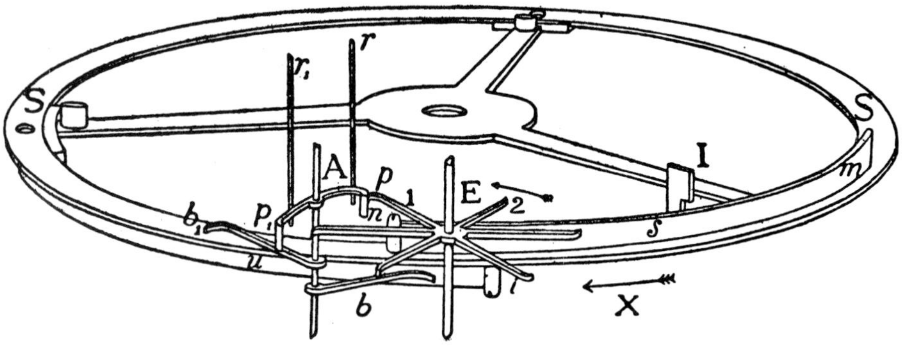

22b Drawing and legend after Auricoste to explain the detached escapement of Pierre Le Roy, as used in the marine clocks of 1764-66; e is escape wheel (star) of brass with six long radial teeth, bent at the ends.
A - Detent (escape wheel lock), with two locking pallets p and pl at right angles, underneath at two other levels the arms bl and b, S-S-rim (Serge) of the balance. I - impulse taking pallet, attached to the rim of S-S.
rl, r2-turning pins (striking limits for p, pl). On the rim of the balance is the releasing apparatus, namely two pivots, i below and r above, each of which is contacted by a cylinder arc s-m above and k/u below; these form shoe-like panels and are safety measures to prevent A from tipping over, but b or bl will not normally touch them.
(Ditisheim)

Explanation of the operation according to Auricoste: assume one movement of the balance like x in clockwise direction. The escapement star E is locked by p on its arm 1. About in the middle position of the balance, u turns over bl on A, which releases arm 1. Immediately thereafter, the impulse pallet I *receives* an impulse from arm 2 of the escapement star; immediately after that, arm 1 is locked by lever pl.

When the balance swings back, arm L is released by i over b on A, so that arm 2 now reaches the former position of 1 and remains held fast there; after the balance swings back, the same operation is repeated.

contract from the Board of Longitude. This clock of Kendall's, usually called Kl (#51), proved itself splendidly on Captain Cook's second great expedition. Thus the new principle introduced by John Harrison finally attested to his assiduity, for the performance of Kl on Cook's long expedition, described in Chapter 5, provided proof that the initial success of H4 was not, as many people claimed, an unexpected success due to chance but the result of a well-conceived new type of clock.

Larcum Kendall was born in Charlbury, Oxford in 1721 and became an apprentice of John Jefferys of Holborn in 1735. It was Jefferys who built the famous predecessor to H4 for Harrison in 1753. After his journeyman years he founded his own business, and in 1765 he had already won such renown as a master of his trade that he was called by the Board to join the Harrison Committee. Larcum Kendall was in fact a very outstanding clockmaker with great talent for fine mechanics. His copy of H4 shows even finer workmanship than the original itself. Personally, though, Larcum Kendall was an extraordinarily cautious man. When he first accepted the contract to copy H4, he stressed that he would produce only a strict part-for-part copy of Harrison's clock without being responsible for its behavior as a unit, since he mistrusted Harrison's methods of adjustment. Kendall was not psychologically capable of accepting John Harrison's achievements for what they really were — a great technical, scientific accomplishment. After the completion of Kl, Kendall was asked by the Board under what conditions he would be willing to direct other clockmakers in the production of additional copies of H4. Kendall objected to such an idea on principle and told the Board they would be disappointed if they depended on clocks such as H4 for general use at sea.

As a result of further negotiations with the Board, Kendall received two more contracts for marine clocks, with permission to undertake such simplifications, improvements and economies as seemed appropriate to him. These clocks are known as K2 (#55a-c) and K3 (#57a-d). In making them, Larcum Kendall omitted the new ideas of Harrison that seemed inappropriate to him, such as the remontoire, and returned to a concept in the manner of the customary clockmaking practices of the time; the clocks are excellently made in terms of fine mechanics, but their results fell far short of H4 and Kl.

Thomas Mudge, who likewise belonged to the Board of Longitude's committee that judged Harrison's work in 1765, was a more interesting personality than Larcum Kendall. Mudge was born in Exeter in 1715 and, on account of his early interest in clocks, apprenticed by his father to George Graham at the age of 14; there he quickly became one of the most dependable colleagues of this great master. After Graham's death in 1751, Mudge took over Graham's business, taking his colleague William Dutton, another of Graham's students, as his partner.

Thomas Mudge participated in a modest way in two important advances in clockmaking. They were the invention of the detached lever escapement for clocks with balances, and the development of escapements with constant force (Echappements à force constante).

We have passed by the period in which the detached lever escapement was conceived, although the detached lever escapement prevailed throughout the world — long after Mudge, of course — as the only acceptable escapement for pocket watches, remaining so until the most recent times; it also achieved some significance in the construction of chronometers. The first pocket watch made by Mudge that was equipped with a detached lever escapement is known as the Queen Charlotte Clock. R. Good has described it in depth.[20]

The detached lever escapement allows a free, unhindered and wide swing of the balance, in which the relatively small transmissions of energy are symmetrical to the zero line, so that the system possesses remarkably good properties of oscillation. The detached lever escapement by Thomas Mudge is a brilliant further development of the dead-beat escapement originated by George Graham, which is used in pendulum clocks and regulators, and which increased their accuracy remarkably at one stroke. The dead-beat escapement is a further development of the recoil anchor escapement that William Clement seems to have invented by 1670.

Other than the Queen Charlotte Clock, there is a single, recently discovered, similarly constructed clock made by Mudge and originally fitted with the same detached lever escapement — known to specialists as the Flint Clock — whose dial bears the inscription: "THE ORIGINAL WAS INVENTED AND MADE BY THOMAS MUDGE FOR HER BRITANNIC MAJESTY A. D. 1770."

It is interesting to note that, despite the obviously great advantages of the detached lever escapement, it was hardly used at all in Mudge's day except in a few rare pieces by the well-known watchmakers Margetts and Emery. Not even for Count Brühl, the Saxon ambassador in London and a friend of his, was Thomas Mudge willing to build a clock similar to the one for Queen Charlotte. But at least he passed on the most essential instructions, so that Emery could build such a clock for Count Brühl.

The fact that introduction of the detached lever escapement took so long is surely attributable to Mudge himself, who took no steps to utilize his invention himself or to promote it, or even to make it known in the trade. According to several remarks in Mudge's letters, he was probably convinced personally of the inherent value of his invention but had no confidence in its economic potential. Thus it took decades before the detached lever escapement, which was naturally very fragile and costly to produce in the beginning, was introduced in precision pocket watches, but it was finally accepted throughout the world as the only escapement worth using, and remained so until the most recent times.

A very different case from that of the detached lever escapement was the development of the second technical principle, the constant-force escapement, through the use of which Mudge later hoped to compete for the marine chronometer prize.

Not long after his participation in the commission established by the Board in 1765, he submitted a memorandum[21] to the Royal Society of London, containing his suggestions for improving marine clocks. The essence of it was the introduction of a drive with constant force, acting directly on the balance arbor, so as to eliminate the recoil effect of the varying rate of the drive spring.

Until a short time ago it was believed that Mudge had invented the principle of the constant-force escapement. But during a meeting of the Antique Horological Society in the Science Museum of London on January 5, 1961, much attention was aroused by Dr. Geschwind of Basel, who displayed and described a clock built by Mudge in 1755 and having such an escapement with constant force; this clock is now at the Historical Museum of Basel. It could be documented at the same time that a young Swiss, J. J. Huber, then 22 years old and a pupil of Johann II and Daniel Bernoulli, had come to England with, among other things, the suggestion of utilizing the principle of constant force which he had originated, to solve the longitude problem. He had presented this to the well-known astronomer Bradley and, on his recommendation, then given a contract for its production to Mudge. Between June 9 and October 31, 1755 Thomas Mudge actually built this clock, which J. J. Huber then took home to Basel. It still exists and includes this fragile escapement today. The thorough description by T. P. Camerer Cuss[22] is complete proof of where Thomas Mudge got the idea for the *constant force escapement* that later influenced his life so greatly.

Mudge's first chronometer (#56a-c) of 1774 is made to run eight days; it originally gave excellent running results at Greenwich, but after several months, numerous mechanical difficulties appeared which Mudge attributed to inept handling by the Greenwich personnel. Probably, though, the sensitivity of the system was the ultimate cause. In the following years, Mudge built several other marine clocks in his manner, including the *Dresden Clock* (#60a-e) and the so-called *Blue* and *Green* (#61a-f) Marine Clocks, with running times of only 36 hours. These two were tested at Greenwich in 1779-80, 1783-84 and 1788-90; in the beginning they were very regular, but after a long time they developed some irregularities.

According to the available data it is understandable that the Mudge chronometers could not score any particular success in the Greenwich tests which had, in fact, been made much more stringent as of 1774. Thus they did not come into actual maritime use, and so Thomas Mudge's lifetime desire to win the chronometer prize could not be fulfilled, although his workmanship certainly ranked above that of Harrison and Larcum Kendall. It is not surprising that Mudge, annoyed as he was, engaged in unfriendly discussions and petitions. They are described in detail in his son's book.[23]

This book also contains 132 pages of copies of letters that Thomas Mudge Sr. wrote to his great patron Count

23 Thomas Mudge's arrangement of two serially operating springs, as used in his first chronometer (#56).

The lower barrel (5) is attached by three screws to the drum for the chain of the fusee. The lower spring (3) is hooked to (5) by its outer end; its inner end is hooked onto the center (4). The upper spring (8), coiled in the opposite direction of (3), hangs with its inner end of the center (9), which is linked to the arbor of center (4), but by its outer end to the upper barrel (6), which can be tensed lightly with fusee pinion (7), attached to the bottom plate; thus the two springs are wound simultaneously via the chain drum (1).

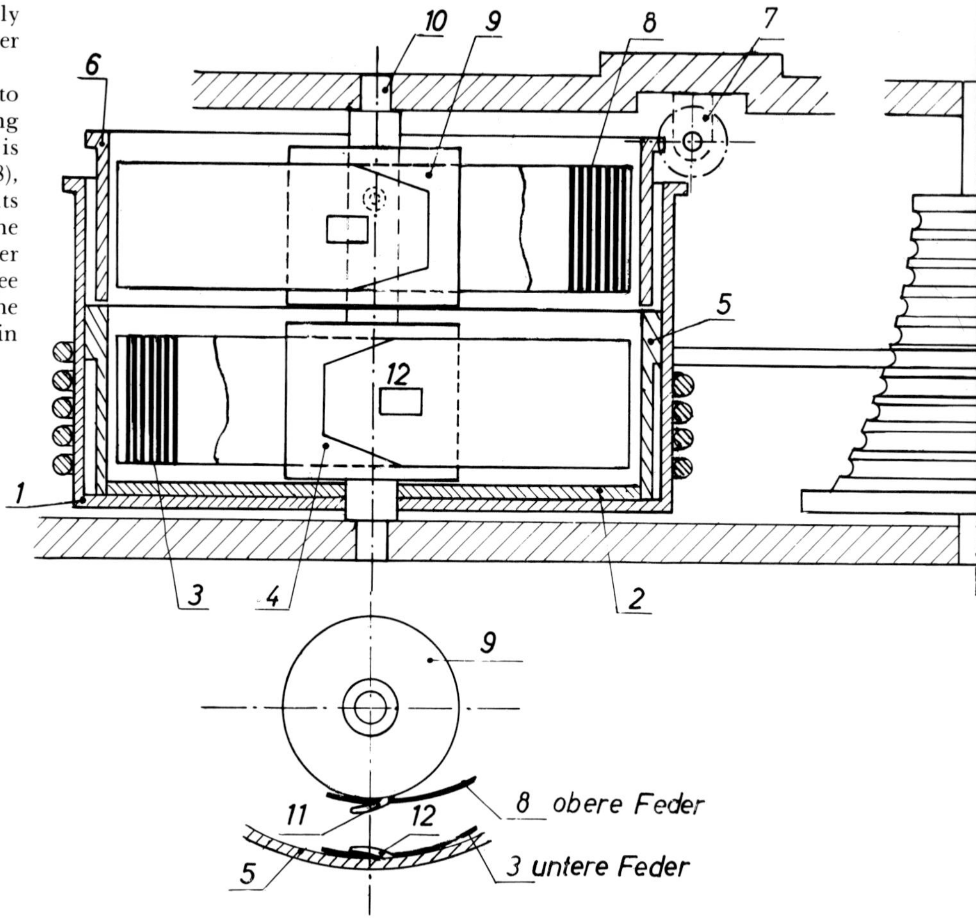

Brühl, the Saxon ambassador in London, in which many personal and technical details are discussed. Included as well are nine plates with numerous drawings of technical details of Mudge's designs, so that a study of them provides a good impression of the problems of highly developed clockmaking in the late 18th Century.

Among the many interesting technical particulars that Thomas Mudge developed and put into practice is a drive with two sequentially activated mainsprings, (#23) which he used in his first chronometer. It was an attempt to gain an even driving moment as the spring wound down, a design principle that was tested and utilized again much later but never really found acceptance because of the great difficulties involving breaking springs.

The final impression left by Mudge's work is based on the fact that despite his talent in conceiving and utilizing basically new and subsequently important technical ideas, and despite his superb technical ability, he never succeeded in producing marine chronometers that could take rough handling at sea, as the complicated construction and the sensitivity of his designs and the extraordinary difficulty of installation made them unsuitable for it.

After Thomas Mudge's death his son, Thomas Mudge Jr., continued to build marine chronometers according to his father's design principles (#110). He had the help of several colleagues — Pendleton, Pennington and Howells — but did not achieve any noteworthy success. Thomas Mudge Sr. died in 1794; he lived long enough to know that a marine chronometer made by his son ran almost as well as his own. But Mudge Jr. was not a very competent businessman and gave up the problematic work a few years later, having to take considerable material losses in doing so. In all, he built only twenty marine chronometers, some of which can still be seen in collections today.

John Arnold and Thomas Earnshaw

Both of these men stand on the boundary between the heroic pioneering age of marine chronometry, in which the foundation for later progress was built, and the era of economic utilization through fine mechanical methods and advances. The great importance of John Arnold (1736-1799) in the development of the marine chronometer can best be described in the words of Colonel Quill:

"And so Arnold actually succeeded in turning the extraordinarily expensive and complicated timepiece of John Harrison and the somewhat primitive-looking, not very compact Continental marine clocks of Le Roy and Berthoud into compact and robust box chromometers that are not essentially different from the instruments used today."

Arnold's undeniable pioneering achievement is the introduction of the pivoted detent with locking pallet in English chronometers on a broad, economically interesting basis. Exactly how and when this was achieved cannot be determined today. There is extensive literature on Arnold, but it lacks — especially for his early chronometer work — concrete contemporary data.

Here, on the basis of fairly trustworthy data, we shall try to present a chronologically reliable picture:

V. Mercer has developed the technology and tradition

linked with Arnold's name into an extensive biography[24] in recent years. It concerns not only John Arnold (1736-1799) but also his son John Roger Arnold. John Arnold Sr. had already attracted notice by about 1762 through his particular clockmaking ability and accomplishments (a repeating watch in a ring for George III in 1764, plus another ring watch with a ruby cylinder in 1769). Over the years the author has assembled a wealth of material. It is remarkable how many marine chronometers and pocket chronometers the Arnolds built; their numbers can scarcely be estimated. But it must not be overlooked that before official numbering began, a series of unnumbered chronometers of both types had already been built. Unfortunately, the numbering system does not afford a clear picture of the actual extent of production. In the appendix of Mercer's book there are lists of the pieces still in existence today, according to which more than 200 chronometers survive from John Arnold's lifetime.

Mercer's illustrations are remarkably instructive and show John Arnold's vivid imagination regarding variations of balance and balance-spring design as well as other components. The information is also valuable in showing how many later changes in the chronometer were made during the following decades. The workmanship and the extent of the use of Arnold's pocket chronometers are very instructive; he began to work in this interesting realm very early, for by 1780 his pocket chronometer No. 36 was being tested in Greenwich and remarkably good running results were achieved with rates of only 0.3 to 0.5 seconds per day.

The beginnings and chronological succession of progress in John Arnold's marine chronometer activities are somewhat obscure. Around 1768 a marine clock with an unusual further development of the cylinder escapement seems to have been built, but after the war it disappeared to the USA. In May of 1770 he presented a marine chronometer of his own design to the Board of Longitude and received a contract for a second specimen. Arnold delivered the second specimen to Admiral Sir Robert Harland (who had already obtained Arnold's 1770 watch) in Portsmouth at the beginning of 1771 for use on the ship *Northumberland* during a voyage to the East Indies. The 1771 delivery is important in that for the first time a marine clock was entrusted to a ship for a major voyage other than the Board's official testing expeditions. As for the success of John Arnold's clock, Harland wrote an extremely favorable account of the trip to the Earl of Sandwich, reporting on the performance of Arnold's chronometer and stressed that with its help, the determination of positions, formerly so unreliable, now made possible a safe landing in Madagascar.

How many of John Arnold's chronometers went to sea in the years just after that cannot be determined precisely. Harland seems to have had two of Arnold's clocks with him from March 1771 to 1775. It has been proved that Arnold supplied three marine clocks to Plymouth in 1772 for Captain Cook's second expedition, and two marine clocks in the summer of 1773 for the North Pole expedition of Captain Phipps and Ludwidge.

What kind of design these marine clocks of Arnold's had can be determined only in part. Of the three clocks

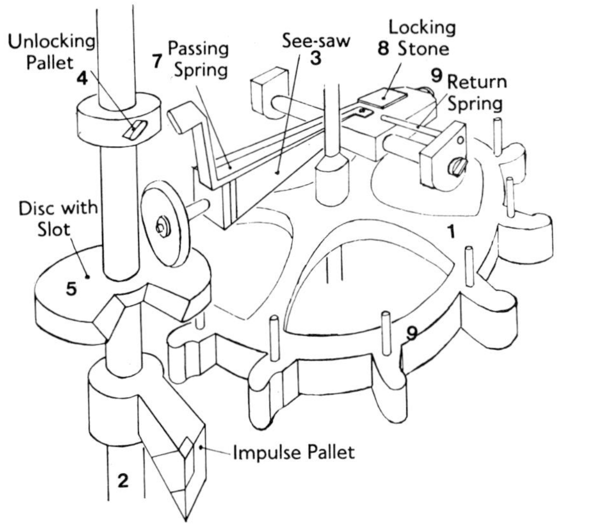

24a Axonometric portrayal of the critical parts of Arnold's first chronometer escapement.
1 Escape wheel
2 Balance arbor
3 Pivoted detent
4 Discharging pallet
5 Locking disc with slit
6 Impulse pallet
7 Passing spring
8 Locking stone
9 Passing spring
(after V. Mercer, Pl. 35)

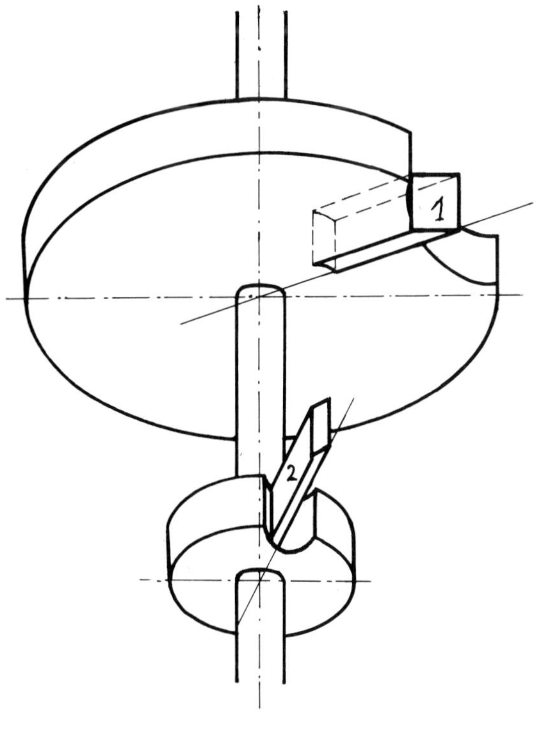

25 The basic arrangement of the two active elements of a chronometer escapement.
1 Impulse pallet
2 Discharging pallet

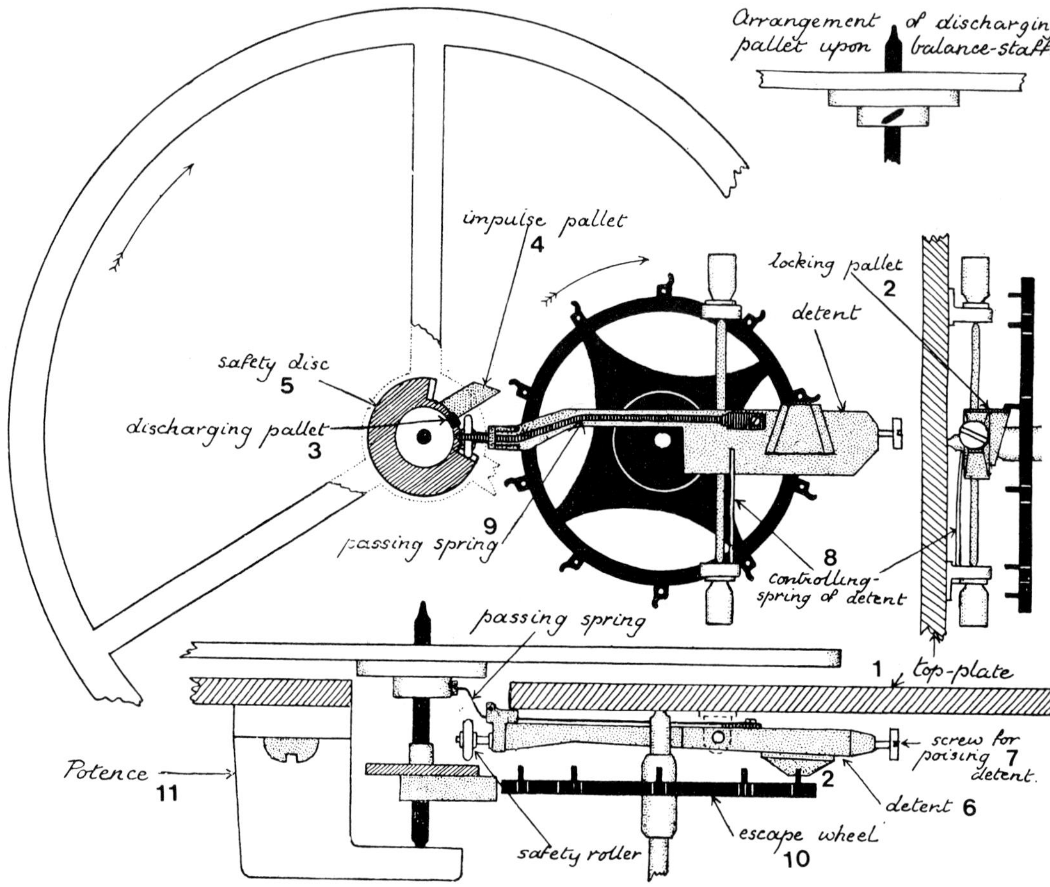

24b Arnold's first chronometer escapement of 1771, with a pivoted detent; the detent moves in a plane vertical to the top plate. Three essential components of the movement are provided:
for the impulse (impulse pallet)
for the locking (locking pallet)
for the release (discharge pallet)
Arnold's marine chronometer No. 2, now owned by the Royal Society of London, still has these historically important forerunners of the later chronometer escapement:
1 Top plate
2 Locking pallet
3 Discharge pallet (at upper right, details of its position on the balance arbor)
4 Impulse pallet
5 Safety disc
6 Pivoted detent: this moves by its own weight, is thus gravitationally activated
7 Adjusting screw for detent
8 Detent spring
9 Discharge spring (passing spring)
10 Escape wheel
11 Potence

delivered to Captain Cook, two are now owned by the Royal Society in London and were examined thoroughly some time ago by Commander Gould; as he describes in his book, they include Arnold's first known chronometer escapement on the rocker principle. V. Mercer assumes from a note in Captain Phipps' report on his North Pole expedition of 1774 that his marine clock was a pocket chronometer with a different, simplified chronometer escapement, later described as the chronometer escapement with passing spring.

John Arnold's first known chronometer escapement is shown in #24a, b. What appears to be of decisive importance here is the probably new simple arrangement of the two important active elements of a chronometer escapement on the balance spring, as shown in #25, namely the impulse pallet and the discharge pallet, on two different planes. Here a principle of the chronometer escapement that remained in use for more than 150 years was introduced.

Had Arnold heard of the similarly functioning French escape wheel lock of Pierre Le Roy and Ferdinand Berthoud and found in them an inspiration for his locking principle? We have no proof of this. But it can be assumed that a man with his breadth of vision must have heard at least rumors, if not more specific technical information, regarding the work done in France by predecessors involved in his own specialized area. Whatever the case may be, John Arnold's see-saw detent is a pioneering, independent, and great achievement.

Related to this is another very important improvement, the turning-pivoted detent (#27), the whole design of which already formed the basis for the spring escape-wheel lock (spring detent) that was in general use more than a century later.

How a final solution was achieved during the course of the next decade has never been explained, what with all the long polemics, spirited partisanship and resulting claims of priority in British specialist literature. To evaluate chronometer development overall, though, these details seem pointless and insignificant.

In his lifetime, John Arnold was a venturesome, forthright, technically and commercially very gifted entrepreneur. Thus he arranged for his first chronometers to be taken on such prestigious expeditions as those of Captain Cook.

Arnold's rocking cradle escapement (#24a, b) is practically identical in its operation to Berthoud's escape wheel (#26a). It is, to be sure, constructed more complexly in that it includes the element known as the *cradle*. In addition, its functioning requires the effect of weight; only when the roller of the rocking element (3) enters the cut in roller 5 is the locking pallet 4 released. Pressure in the direction of balance axis 2 can thus have a delayed effect, or none at all, at the moment of release. Aside from this, this arrangement (#24) already includes the truly simple and reliable setup of the active elements of a chronometer escapement (#25) so essential to all subsequent designs.

A clear chronological list of John Arnold's further development of the escapement scarcely seems possible.

In any case, a very important advance was the simple arrangement of the pivoted rocker (#27). One character-

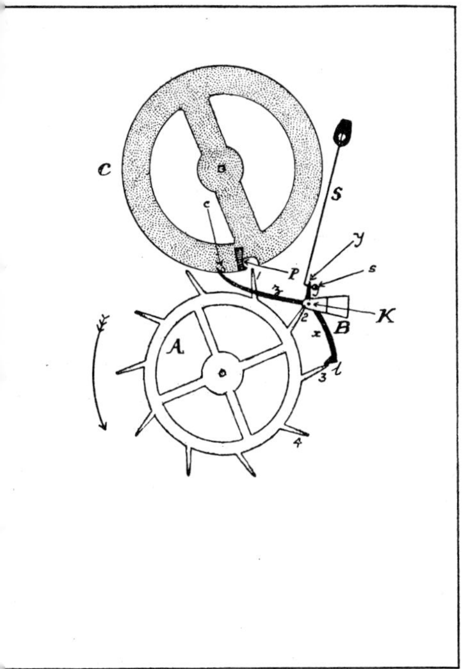

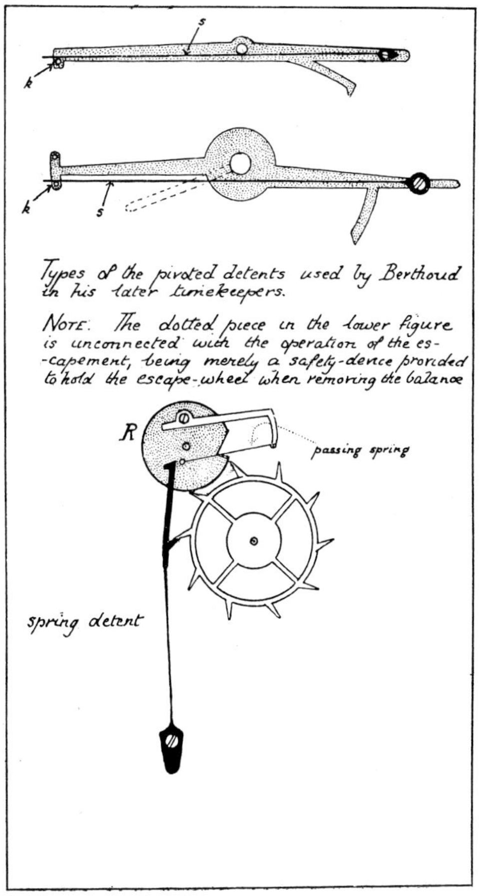

b

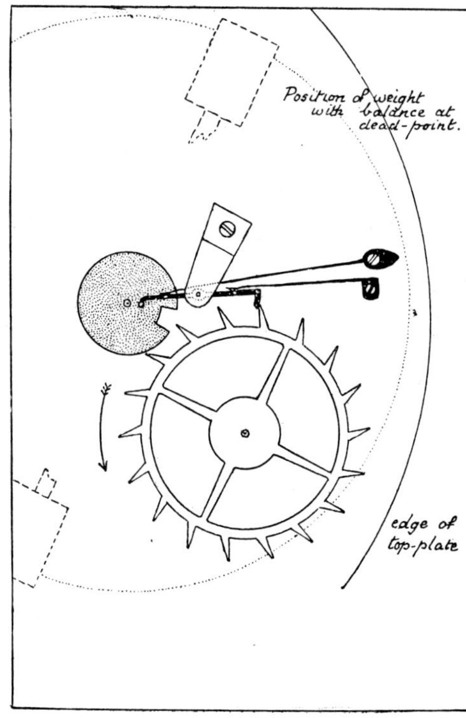

c

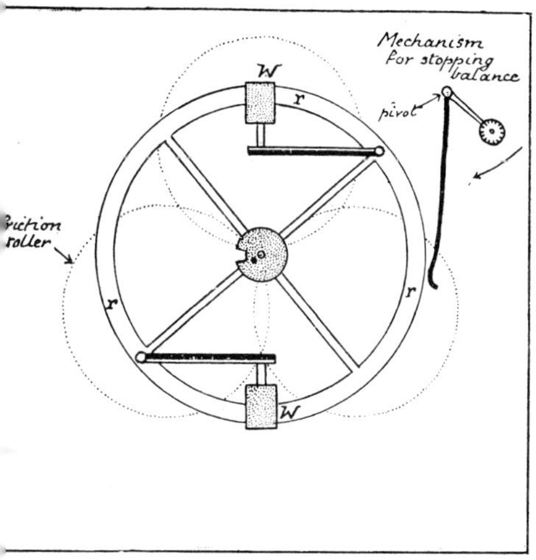

27 J. Arnold's later arrangement of the pivoted detent, according to Cmdr. Gould, in 1776; according to V. Mercer, already planned in 1772.

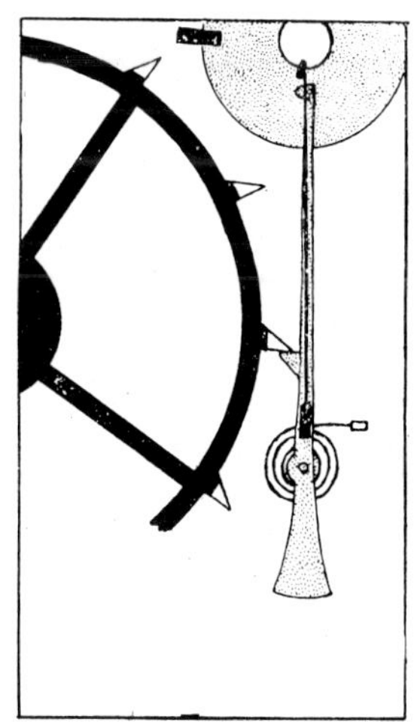

26 The most important detached escapement designs by Ferdinand Berthoud:
a) Berthoud's first escape wheel lock (pivoted detent of 1771). The escape wheel A enters the groove of the impulse wheel C; the latter is mounted on the balance arbor. Locking the escape wheel is done by the double lever B that is made to release A briefly by pin c on C. The use of two springs on the locking lever is regarded as a disadvantage because of the resulting complication of friction.
b) Berthoud's chronometer spring of 1782, already including a notable feature of all modern mechanical chronometers: the spring detent, as a further development of the pivoted detent that was used previously.
c) Chronometer escapement with detent, used by Berthoud in 1785 for marine clock no. 37.
d) Berthoud's bimetallic compensated balance, such as used in his marine clock no. 37. The adjustable weights WW move — more or less according to the temperature — from the inside out or vice versa. In marine clock no. 37 F. Berthoud also provided an attaching device at the upper left near the minute dial to avoid damage in transit.
(after Cmdr. Gould)

28 J. Arnold's chronometer spring (introduced ca. 1790 and used in Arnold's marine chronometer no. 26, according to Cmdr. Gould):

1 Balance arbor
2 Balance ring
3 Discharge roller
4 Ruby discharge pallet
5 Impulse pallet
6 Passing spring
7 Chronometer spring
8 Stud of spring
9 Adjusting screw for the spring
10 Locking stone
11 Locking wheel

After Gould

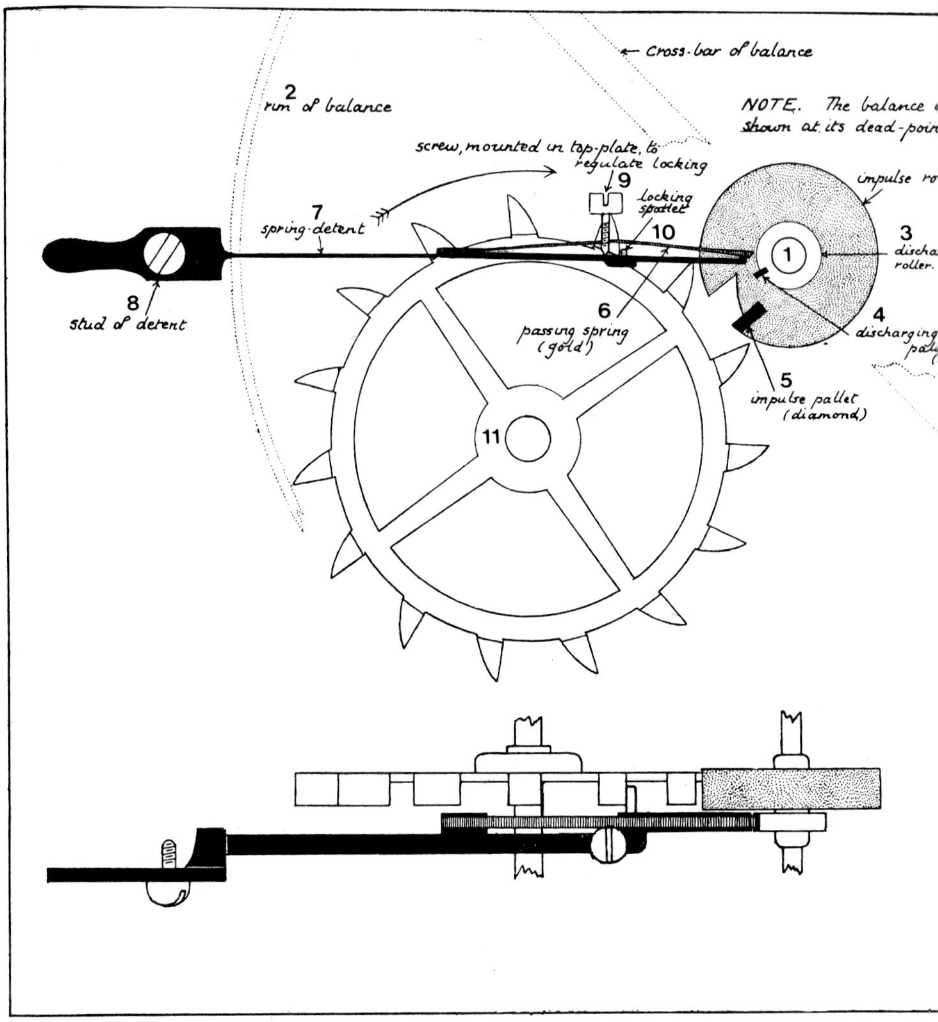

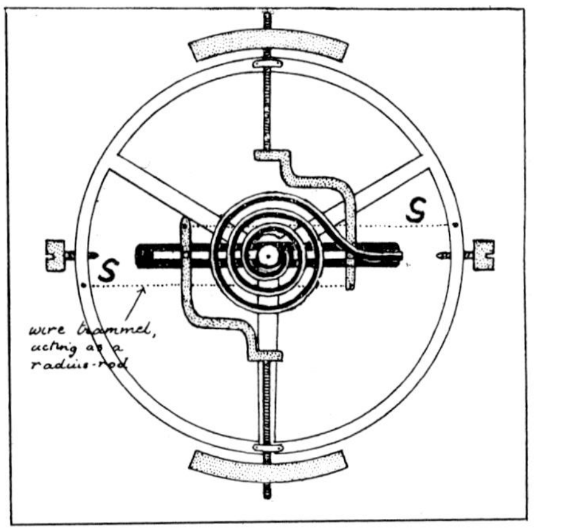

29 J. Arnold's first balance, of 1778, with built-in temperature compensation. The bimetallic hairspring in the middle turns the slit element and thus moves additional weights inward or outward. S S = tension wires.

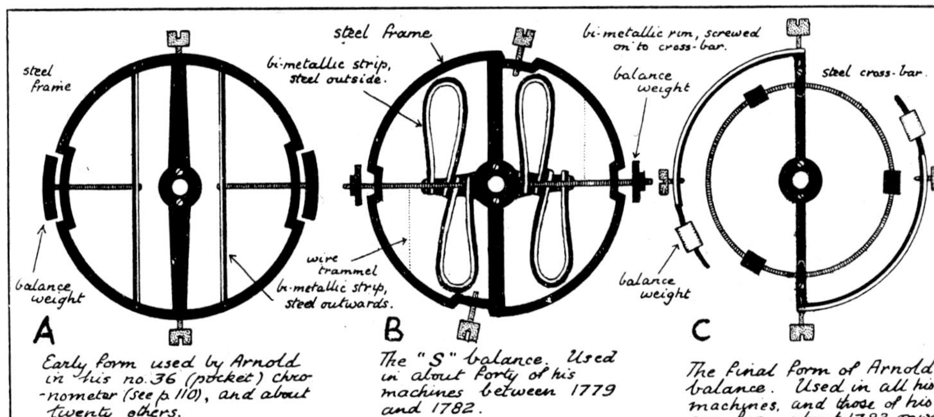

30 Various versions of temperature-compensated balances by J. Arnold (patent of 1782): (a) T-type, (b) S-type, (c) combined ring and Z-type (most often used later). The bimetallic strips are clearly marked; arms and rings are made of steel.
(After Gould)

istic of it is that the teeth of the escape wheel project vertically out of its plane. One after another, they are held by a ruby pallet attached to the lock (detent); the detent is mounted in bearings and held on the upper plate by a small hairspring. A small passing spring —projecting over the free end of the detent — moves it, through a turning motion of the balance, to release the escape wheel inwardly, but leaves it locked in the opposite direction, since the discharge pallet is let through by the passing spring. Impulse and discharge rollers (as shown in #25) with their pallets are firmly attached to the balance arbor. In the first versions, the pallets were made of ruby.

Later Arnold used a spring (#28) in place of the journal-bearing rocker but retained the original design tendencies of his turning pivoted detent. It is noteworthy that the teeth of the escape wheel in this new form are shortened to epicycloid form. This was not done to decrease friction between the tooth and the pallet, but to keep the effective radius on the running wheel the same through the entire impulse time. After John Arnold's death, his son went on using this design for many years.

The need for temperature compensation in the oscillating system that consisted of balance and spring had been recognized clearly from the previous work of Pierre Le Roy, John Harrison, and Thomas Mudge. John

Arnold had also worked on temperature compensation for the balance at an early date (1773). In his first versions he still used, as had Harrison, a bimetallic strip (curb) lying on the back plate and acting on the length of the hairspring. In isolated cases his bimetallic strip was bent like a hairpin.

Just becoming generally known Pierre Le Roy's recognition of the fact that using a bar to change the effective length of the hairspring externally, in the manner of the "temperature curb", influences the isochronism. Thus Arnold had soon (1773) located the compensation system — a combination of steel and brass — in the balance (#29) a step that suggests clear information on the ideas of Pierre Le Roy since John Harrison and others were still using compensators located outside the balance.

The principle illustrated in #29 allows a great number of constructive variations, which Arnold had included in his 1782 patent. Illustrations #30a-c show Arnold's later ideas of 1782 as to variations of his original temperature compensators located in the balance; #30c shows a thoroughly distinctive conception, dependent on the development of Pierre Le Roy's oldest idea.

In this original idea, bimetallic compensation always moves a pair of added weights inward or outward, according to the temperature, against the balance ring —changing the moment of inertia in this way. As for the construction of the bimetallic compensators originally used by Arnold, we do not know whether they were riveted or soldered.

The customary arrangement was that brass was outside and the steel inside. Because of the greater expansion of brass, the bimetallic strip bent toward the steel side when the temperature increased and toward the brass side when it decreased. Since the returning moment of the hairspring becomes greater at lower and smaller at higher temperatures, the oscillating characteristics of the system thus remain more or less unchanged. By changing the location of the weights by means of pushing them or changing screw settings, this can be influenced within certain limits.

V. Mercer's biography of John Arnold includes a great many detailed illustrations of the many designs and variations of Arnold's chronometers. It also includes a classification of the various forms of balance compensators used by Arnold, as follows:

Double T	Z	U
Double S	Y Z	U+O
OZ	Z+O	U+I

Subsequently, more information is included in Chapter 5 on John Arnold's activities and the further fate of his undertakings.

Thomas Earnshaw (1749-1829) made what may appear to be very modest improvements to the chronometer escapement, but all through the 19th Century they proved to be so effective that it seems appropriate to include Earnshaw among the pioneers in this field.

In general, the operation of his escapement is the same as those of Berthoud, and is superior to them in that the passing spring is less inclined to disturb the movement of the discharge pallet, (#26b). Compared to Arnold's system, (#28) in which the escape wheel is released in such a way that the detent is moved inward on the wheel, it has the following advantages: first, an easier release, because the locking is applied to the point of the tooth and not to its base, and second, an impulse is transmitted with less friction of the escape wheel on the impulse pallet. Thus it occurs that the individual impulses result from a direct hit on the tip of the tooth, which makes it possible to do without oil on the pallets, whereas oil is necessary for the epicycloidal flanks of the teeth in Arnold's system.

Earnshaw also made very significant technological advances in the construction of the temperature compensation for the balance. It was a brilliant solution that has been used widely to the most modern times. As #32 shows, it is simple in principle: As in Pierre Le Roy's version (#21), two bimetallic strips are attached to the balance arm, with the adjusting screws attached to the arm itself, while weights can also be moved on the bimetallic strips. When the temperature changes, the two strips bend inward or outward and the weights on them change the moment of inertia; by pushing the additional weights closer to or farther from the point of attachment, the compensatory effect can be changed as desired. Aside from this simplification of construction, Earnshaw made a much greater step. He used bimetallic strips that were not made as before — of a brass and a steel strip and at first riveted and later soldered together — but instead produced a metallic compound without solder or rivets.

Earnshaw's chronometer spring detent (#31 left) is an improvement of Arnold's see-saw principle. Which of the two (Arnold or Earnshaw) deserves credit for the decisive step in the successful introduction of it into practical use is still the subject of disagreements that still endure to this day. Whoever is interested in these polemics can find an outline of them —fair to both parties — in R. T. Gould book should not both of them, John Arnold and Thomas Earnshaw, be named with equal respect or a more partisan advocacy of Arnold in V. Mercer's book.

Earnshaw was a business competitor of Arnold and also his bitter personal enemy, which comes through clearly in Earnshaw's publications.[25] Surely his difficult rise above the most modest conditions and the difficulties caused him by Arnold's influential patrons and friends played a significant role.

The advantages of the chronometer escapement with spring detent are as follows: It has the least influence on the balance and hairspring. When the angles between the impulse pallets and the central position of the balance are chosen correctly, that is, when the system is adjusted correctly, the working impulses occur symmetrically to the central position of the balance, thus when it has the highest speed and is least disturbed by having to release the escapement and receives the impulse. Finally, the chronometer escapement with spring detent works without oil on the critical switching positions, thus giving an optimal working time until the next routine overhauling.

The invention of the chronometer escapement with a spring detent and a temperature-compensated balance

31 Thomas Earnshaw's chronometer escapement with spring, of 1784, compared to a modern chronometer escapement:
1 Escape wheel
2 Impulse roller
3 Discharge pallet
4 Passing spring
5 Locking pallet
6 Impulse pallet
7 Locking pallet carrier
8 Spring (chronometer spring)

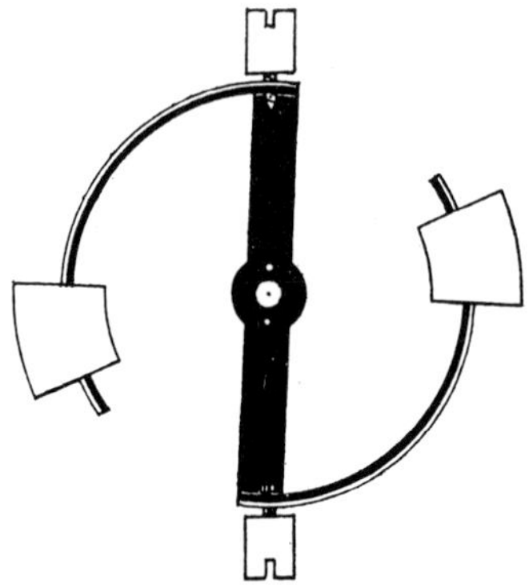

32 Thomas Earnshaw's temperature-compensated balance: The two metals of the ring segments are directly bonded to each other, not riveted or soldered. *(after Gould)*

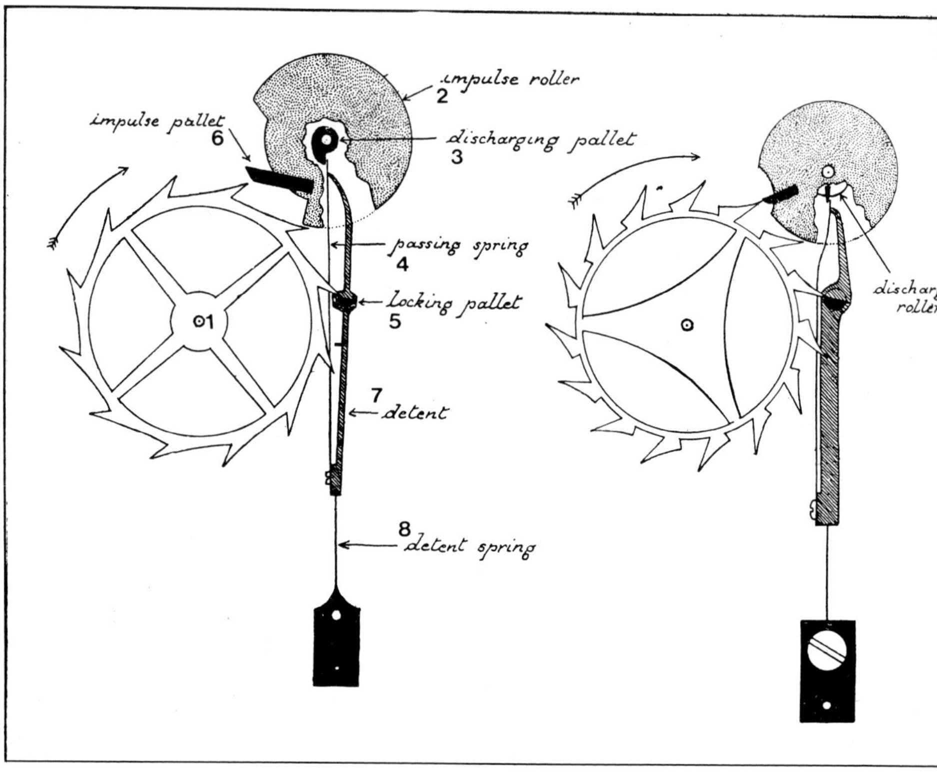

in Earnshaw's manner concluded the truly pioneering achievements that made it possible to establish seafaring on a safer basis, thereby increasing maritime commerce and political and naval power. For all of this it was necessary to bring the marine chronometer into as common use as possible, which required appropriate economy along with the greatest reliability. In this direction, Arnold and Earnshaw possessed not only the intellectual requirements, but also the fine-mechanical enterprise and talent that made their achievements the basis of the fine-mechanical era that followed, in which Britain attained leadership in chronometer production on a broad basis.

The heroic epoch of the longitude problem lasted something over fifty years. It can be said that the result of this half-century was as follows: It was shown that it was possible, to determine geographical distances by taking the time at the point of departure along through the use of dependable clocks built to withstand use at sea, even on voyages that lasted several weeks, and to determine these distances more precisely than was called for by Queen Anne's Act.

To finish the picture of this pioneering era, we should add a brief note on the varied but thoroughly unfortunate attempts of the Spanish government to build up the state-run manufacture of chronometers at the end of the 18th Century. Around 1800, the most important figures, Sanchez and E. Cruzado, died of the plague. Other men too, apparently quite talented and enterprising, could not achieve success either, hampered as they were by unfavorable external conditions. This unsuccessful development was described in detail by D.C. Fernandez Duro.[26]

Outsiders in the "Heroic Epoch"

All of the heroes named in this chapter, from Harrison and Pierre Le Roy to Earnshaw, have made great contributions to the history of time measurement, and of the chronometer in particular; their roles are generally recognized and no longer disputed today and they have been studied repeatedly in the literature since the 18th Century.

The systematic examination of collections great and small, plus a bit of luck in going through the literature and the catalogs of various auction houses, show that in this time period there were, above and beyond the great works of the master figures — also a series of well-conceived and very noteworthy examples of clocks that, by all their external characteristics, belong in the development of the chronometer for determining longitude but did not play a role in the great competition. Despite that, they should not be overlooked here.

Since there are hardly any technical analyses or historical documents concerning the men who built them, their intentions, and the testing they may have done, it is only possible here to mention the most important of these outsiders and hope that more information will be found in the future.

The first to be named here is M.V. Lomonosov,[27] born November 19, 1711, died April 15, 1765, Professor of Chemistry at the St. Petersburg Academy as of 1745. This scientist of European renown dealt with navigational problems from 1754 to 1759 and offered his own design for a marine chronometer driven by four springs.[28] A clock in a gilded copper case, with a strong resemblance to H4, and signed Johann Georg Thiel,[29] No. 1, 1768

Thos Earnshaw
INVT ET FECIT
No 509
ONE DAY

(Bremen), is shown in #49a, b; it has two oscillating systems of equal duration (full-oscillation duration: one-half second), two complete dials, each with hour, minute and second hands, both of which — independently of each other — can be turned on and off, as well as two escape wheels on one arbor with pallets set at unusual angles. The one system is driven normally by the mainspring and activates the remontoir for the second system, which also has an external bimetallic temperature compensation with pins on the hairspring.

Gould reported that this clock was tested by Maskelyne at Greenwich around 1777-78, and that its running was found to be only equivalent to that of the typical pocket watch. Unfortunately, further information on this interesting, inventive master from Bremen is lacking.

Two box-shaped chronometers were made by the Hollander Willem Snellen of Dortrecht, dating from 1770-80; one of them has a diamond pallet like Harrison's No. 4 and No. 5 (#52a, b) and is signed; the second (#53a-c) is very similar to the first in structure and was described exhaustively in 1959.[30] The author of this study ascribed it to Harrison's associates on account of the fine construction of the verge escapement. Without any proof, one might assume that Snellen had once worked with the Harrisons.

An unusual chronometer design by the Swiss Jean Facy in 1769 (#50a-c) shows a close kinship to the *Horloges Marines* of Ferdinand Berthoud, who was the only one to use weight drive with direction of the weights.

A big silver clock, dated 1775, resembling carriage clocks in style and signed "Les frères Goyffon à Paris, Nr. 837", (#58a, b), has an escapement of the same kind as is found in Pierre Le Roy's large marine clocks, in which the detached escapement is activated by the circumference of the balance, which was the first version of a detached escapement. For that reason, this second utilization of Pierre Le Roy's new principle is of particular importance, though nothing is known of the origins of this clock, and it would be interesting to know whether it was ever used at sea.

A very massive design by the Dane Armand (#64a-c), made in Copenhagen around 1780 and also clearly intended for use at sea, also shows French influence — especially of Pierre Le Roy's ideas — in the design of its movement. A pocket chronometer (#121) signed with the same name is to be found in the Aarhus Museum,[31] as is a marine chronometer movement similar to that shown in #64, doubly signed "Nr. 1 Bornholm" and, on a plate attached over it, "Nr. 11, Copenhagen." Dr. H. Henningen of the Danish Maritime Museum in Helsingör has recently learned from Danish archives that shortly after 1760 J. A. Armand was appointed "mechanikus" to the Danish Navy's chronometer builder, and built two chronometers to a special order of the King of Denmark; after factory testing, they were given sea testing that was requested and organized by the Royal Danish Scientific Society. This voyage took place between November 1782 and October 1793 on a Copenhagen-Le Havre-Lisbon-Madiera-Danish West Indies-Azores-Copenhagen route. The captain of the ship *Proewen*, which was used for the voyage, published a thorough report in the form of a book.[32] Details of Armand's chronometers and the results of this voyage is included in a dissertation by M. Passi-Drewes.[33]

1. See section: *A. Rees, Rees's clocks, watches and chronometers 1819-20*, London, 1970.
2. William Laycock, *The Lost Science of John 'Longitude' Harrison*, Ashford, England, 1976.
3. Humphrey Quill, *John Harrison, the Man who Found Longitude*, London, 1966.
4. Evans, D. W., *The First Hundred Years of the Marine Chronometer*, Hor. Journal, London, April 1953.
5. Johan Horrins (pseudonym for John Harrison, grandson of John Harrison), *Mémoirs of a Trait in the Character of George III*, London, 1853, p. 166.
6. Robert H. A. Miles, *Did Julien Le Roy Make the First Detached Lever Escapement?* A. H. III, September 1961.
7. Cecil Clutton, *Julien Le Roy's Lever Revived*, Ant. Hor. IV, Sept. 1963, and letters to A. H. of March and September 1964.
8. Three memoirs by John Harrison on the success of H4 on the voyage to Jamaica: a) Memorial concerning Mr. Harrison's invention for measuring the time at Sea, December 1762, no publisher, one page. b) Proposal for examining Mr. Harrison's Time-keeper at Sea, December 1762, no publisher, three pages. c) A calculation showing the Result of an Experiment made with Mr. Harrison's Time-keeper, in Consequence of the Instructions and Directions of the Commissioners of the Longitude, in a voyage from Portsmouth to Port Royal in Jamaica, in Order to Determine by that Time-keeper, the difference in Longitude between these two Places; and to find out how much Longitude found by the Time-keeper differs from the Longitudeformerly determined by Astronomical Observations. No publisher, one page.
9. James Short, F.R.S., presumed author of the anonymous pamphlet *An Account of the Proceedings, in Order to the Discovery ofthe Longitude: in a letter to the Right Honourable . . . ,Member of Parliament*, London, T. & J. W. Pasham, 1763, see*Clocks & Watches* (A Historical Bibliography), G. H. Baillie, London, 1951.
10. G. H. Baillie, *Clocks & Watches*, London, 1951.
11. R. T. Gould, *The Marine Chronometer*, London, 1923.
12. P. Ditisheim, R. Lallier, L. Reverchon, Vivielle, *Pierre LeRoy et La Chronometrie*, Paris, 1940.
13. R. T. Gould, *Early Marine Chronometers*, Hor. Journal.December 1923.
14. Pierre Le Roy, *Un nouvel échappement à repos dans lequella roue pose et est retenue à chaque demi vibration sur unepièce fixée à la platine et entièrement étrangèrebalancier*, (Mem. Ac. Sciences), Paris, 1748.
15. There is no corresponding German term for détent. In the Dictionnaire Professionne Illustré it is stated: Primarily used for striking apparatus, and under chronometer escapementwith resting jewel — a designation that does not fully describe the principle. The German "Gangsperre" (running barrier) seems more suitable.
16. Pierre Le Roy, *Recherches des longitudes en mer: Exposésuccint des travaux de MM. Harrison et Le Roy dans larecherche des longitudes en mer et des epreuves faites deleurs ouvrages*, Paris, 1768.
17. J. Auricoste, *Au Sujet d'une restauration de l'horloge marine de Pierre Le Roy* (Revue Chronometrique VIII), 3 pp, 1 illustration, Paris, 1939.
18. Robert H. A. Miles, *Did Julien Le Roy make the first detached Lever Escapement?* Pioneers of Precision Timekeeping,AH Monograph No. 3 of the Antiquarian Horological Society.
19. The "deposition" of new ideas, calculations or particular trains of thought to be opened at later points in time was a means — scarcely applicable in our time — of protecting new solutions to problems not yet to be made public.
20. R. Good, *The first Lever Watch made by Thomas Mudge*,Pioneers of Precision Timekeeping, Monograph 3 of the A.H., London.
21. Thomas Mudge, *Thoughts on the Means of Improving Watches and more particularly those for use at sea*, Chambers's Cyclopaedia, London, 1765.
22. T.P. Camerer Cuss, *The Huber/Mudge Timepiece with Constant Force Escapement*, Pioneers of Precision Timekeeping, AHMonograph No. 3.
23. Thomas Mudge, *A description with plates of the timekeeper invented by the late Mr. Thomas Mudge*, London, 1799.
24. Vaudrey Mercer, *John Arnold & Son*, London, 1972.
25. Thomas Earnshaw, *Explanation of time keepers constructed by Mr. Thomas Earnshaw for which a reward of three thousand pounds was awarded by the commissioners of longitude*, (from the communications made by him to the commissioners); London, 1805. "Statement as to the reward of three thousand pounds voted to him by the commissioners of Longitude", London 1806. "Longitude: an appeal to the public stating Mr. Thomas Earnshaw's claim to the original invention of the improvements in his timekeepers, their superior going in numerous voyages and also as tried by astronomer royal by orders of the commissioners of longitude and his consequent right to national reward", London, 1803.
26. D. C. Fernandez Duro, *Chronometrie*, Madrid, 1879.
27. Valentin L. Chenakal, *Watchmakers and Clockmakers in Russia (1400 to 1850)*, Antiquarian Horological Society MonographNo. 6, London, 1972.
28. Mikhail Vasilevich Lomonosov, *Meditationes de vi navis inmari certius de terminanda . . .*, St. Petersburg, 1759.
29. H. F. Tölke, *Das Chronometer des Bremer Uhrmachers Johann Georg Thiele*, Alte Uhren 4, October 1980.
30. Anonymous, *Ein sehr interessantes Zeitmesser mit Spindelhemmung*, Neue Uhrmacherzeitung XIII, 1959, pp. 6-8.
31. Robert H. A. Miles, *Tour of Denmark - Summer 1969*, (Part 3,Watches and Chronometers seen on the tour), A.H., March 1970.
32. Poul de Lóvenórn, *Beretning om en Reise foretaget . . . i Aarene 1782 og 1783 med Fregatten Próven*, forat untersóge de i Dannemark forfaerdigede Sóe-Lagende Uhrer . . ., Copenhagen 1786.
33. M. Passi-Drewes, *La Chronométrie et ses dévelopments dansle domaine hydrographique au XVIIIe Siécle*, dissertation, Sorbonne University of Paris, under the direction of M. Pierre Chaunu, 1981.

33 Harrison's marine chronometer no. 1, set up in the captain's cabin of the ship *Centurion.*
National Maritime Museum, Greenwich

34a Front view of Harrison's experimental marine clock H1 of 1735 with "grasshopper" escapement; with two opposed swinging levers coupled with four screw springs and linked with tensed steel wires. "Maintaining power" in the fusee; with two barrels, temperature compensation with grid system, some wheels of oak; rollers of the lantern pinions of lignum vitae; roller bearings for most arbor pivots, some with plates, some with segments, fastened on long double levers with counterweights and springs. Four dial rings, each with one hand; seconds above, minutes and hours in the middle, and the date below. Height ca. 76 cm, weight ca. 36 kg, construction time ca. six years.
National Maritime Museum, Greenwich

Opposite Page:
34b Side view of the movement of H1; the horizontal grid system for temperature compensation with effect on the coupling springs between the lever arms, the discs of the smaller roller bearing and the large lever system for the roller segments of the swinging lever are easy to see.

35b Side view of the movement of H2.

35a Harrison's experimental marine chronometer H2 of 1737, very similar in construction to H1, but somewhat smaller and simpler; all wheels of metal. Temperature compensation with grid consisting of only four bars, intermediate pinion of the 2nd wheel, works on the escape wheel arbor with the 3rd wheel and is pulled along 16 times per hour. Height ca. 66 mm, weight ca. 30 kg, construction time ca. 3 years.
National Maritime Museum, Greenwich

36a Movement of a very early marine chronometer (montre marine), apparently signed MM, No. 2, inventée et executée par Ferdinand Berthoud, 1755; free escapement, external grid for temperature compensation with 16 bars, roller bearing on the balance arbor, drive by chain and fusee on both sides; the dial has three rings for hours, minutes and seconds.
Arts et Métiers, Paris

36b Rear view of the movement of MM No. 2.

37a Harrison's experimental marine clock H3 of 1757, similarly built to H2, but instead of the second swinging lever, two opposed large heavy ring-shaped swinging balances; temperature compensation now by a bimetallic strip = compensation curb (riveted), metal wheels, only single pinions with lignum vitae rollers, intermediate drive directly on the escape wheel arbor, is pulled along every 1/2 minute; height ca. 53 cm, weight ca. 24 kg, construction time ca. 18 years.
National Maritime Museum, Greenwich

37b View of the movement and the back of H3.

38 Silver pocket watch with silver outer case (HM 1782-83); dial badly blackened by a fire. This historically important watch had been stored in an iron box when the building in which it was kept was set on fire and destroyed during a bomb attack in 1941. It has since been repaired as much as possible. Signed John Jefferys, London AD 1753. The movement includes many of the basic principles of the later so famous marine clock H4 of Harrison, namely a verge escapement with diamond pallets, an equalizing winding (Remontoir d'Egalité) and external temperature compensation — this is the first use of this principle in a pocket watch.

This watch made by Jefferys was used by John Harrison for the rest of his life; it passed out of the possession of Harrison's descendants and into that of a jeweler in Hull only in 1905.
Clockmakers Company Museum, Guildhall, London

39a Harrison's marine chronometer H4 of 1759, which achieved the solution of the longitude problem, as required by Queen Anne's Act of 1714, on two voyages to the West Indies in 1761 and 1764.
National Maritime Museum, Greenwich, London

39b The movement of H4, lifted out of the case.

b) movement

a) fully assembled

c) rear of the movement, potence removed

40a A marine clock similar to H4, signed "Ferdinand Berthoud invenit et fecit, 1763" (probably MM No. 3); dial with two separate partial circles (divided into 60 units), short hands, second hand out of the center, spring drive with chain and fusee, unusual detached escapement, temperature compensation with bimetallic strip, installed later in place of a grid system; this clock was used on several oceanic expeditions.
Arts et Métiers, Paris

40b Rear view of the movement of MM No. 3.

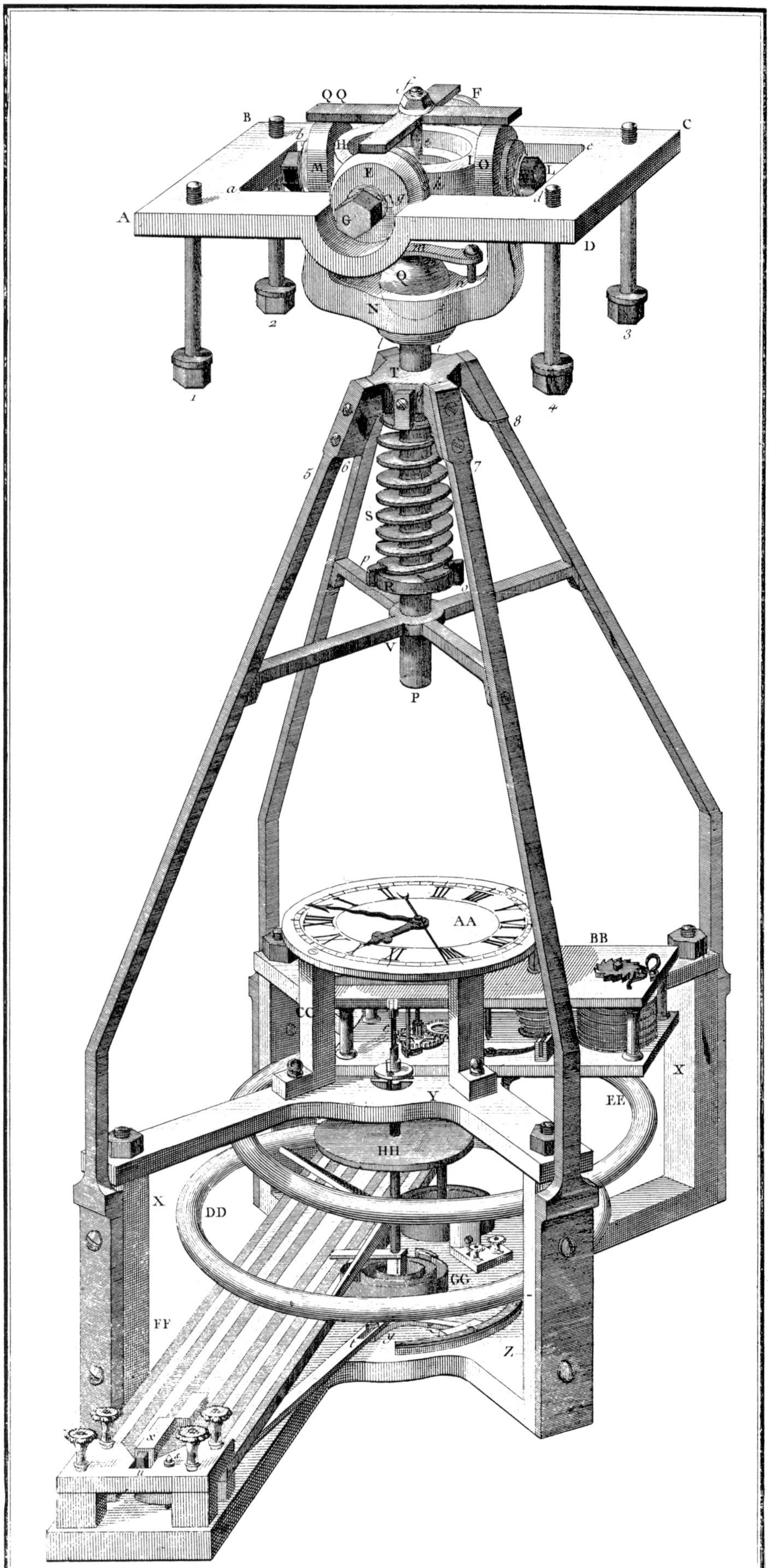
QQ
F
B
C
A
D
E
G
H
M
O
L
N
Q
T
S
R
V
P
AA
BB
CC
EE
DD
HH
GG
FF
X
Y
Z

41

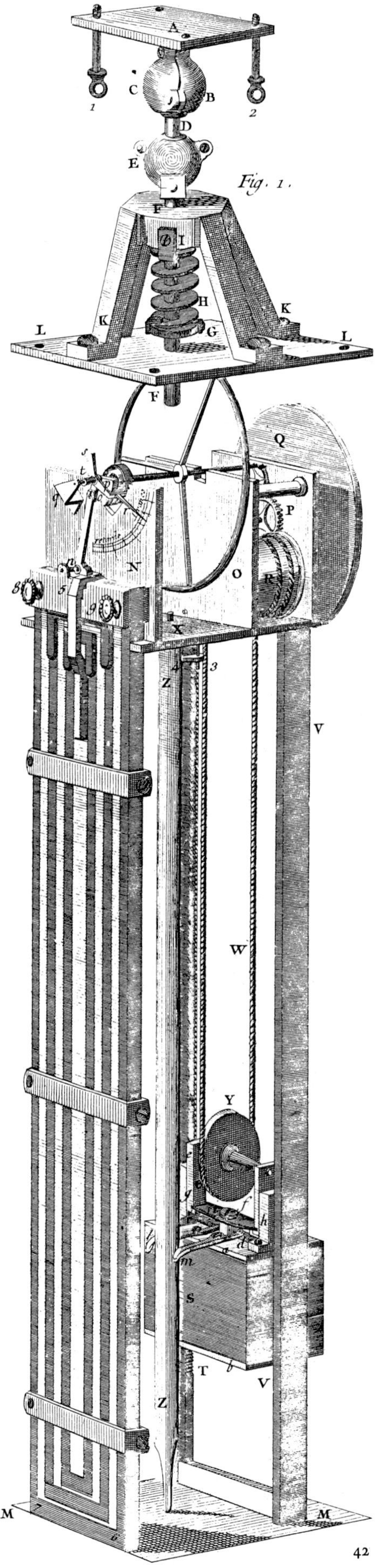
A
B
C
D
E
F
Fig. 1.
H
I
G
K
L
Q
P
N
O
R
X
Z
V
W
Y
S
T
M

42

41 Marine clock (Montre Marine) no. 1 by Ferdinand Berthoud, submitted to the French Academy of Science in 1763. Overall height about one meter, in a heavy iron frame with gimbal suspension and spring as well. Drive by spring, fusee and chain, two balances, each with a spiral hairspring, mutual external temperature compensation with a large grid made of eleven bars; undetached lever escapement.
Arts et Métiers, Paris

42 Marine clock no. 2 (Horloge Marine) by Ferdinand Berthoud, Paris, presented to the French Academy of Science in 1763. Height about one meter, in heavy iron frame with gimbal suspension and spring as well. Weight drive, one balance, external temperature compensation with a grid of eleven bars; cylinder escapement.
Arts et Métiers, Paris.

43a Movement of the famous (second) marine clock B by Pierre Le Roy, Paris, finished in 1764 (shown to the king in 1766); gimbal suspension and frame are considerably newer. Original case form unknown.

The three technical specialties of the basic principles for chronometers are realized and carried out here for the first time: Temperature compensation (with alcohol thermometers) in the balance, "detached" escapement (true chronometer principle), isochronism attained by two opposed spiral hairsprings; period of the balance 1/2 second.

For the escapement, see also 22a-b. It consists of various additional parts on the balance rim, a six-armed star with very thin arms, and a holder, which consists of a very complicated anchor with two resting arms and two releasing arms; control and impulse transmission both done by the parts of the balance ring; balance arbor suspended from a piano wire from above, set in roller bearings below.

This clock and a similarly built one, called marine clock A (no longer extant today), were tested successfully on three great marine expeditions; on the last voyage they were damaged when a heavy object stored nearby fell on them; clock A was unfortunately lost.

At the BrusselsWorld's Fair in 1935 the movement was displayed in action. When previously overhauled by M. J. Auricosta, the clock showed very good running results.
Arts et Métiers, Paris

43b View from above, showing the dial design with three separate rings for hours, minutes and seconds, and the winding box.

43c View from above showing the back of the movement; one can clearly see the big balance wheel.

43d Side view of the movement; the two thermometer tubes built into the balance for temperature compensation and the two big spiral hairsprings are easy to see.

44 Portrait of Pierre Le Roy with Montre Marine No. 1.
After an ivory miniature by Noemi Philastre

43e Large-scale view of the escapement of Pierre Le Roy's marine clock B with its balance, on the rim of which guiding elements for the holding piece and the six-armed star have been fastened.

45 Marine clock, signed Ferdinand Berthoud No. 8, Paris 1767; originally mounted in a gimbal system similar to that shown in #48, with central second, small minute dial and jumping dial for the hours. This chronometer and the identically built chronometer no. 6 proved themselves splendidly in sea trials.
Arts et Métiers, Paris

46 Side view of marine clock no. 6 (HM) of 1767, same design as HM no. 8, set up today in a glass cylinder for a better view of the movement; dial like that in #45.

47 Portrait of Ferdinand Berthoud.

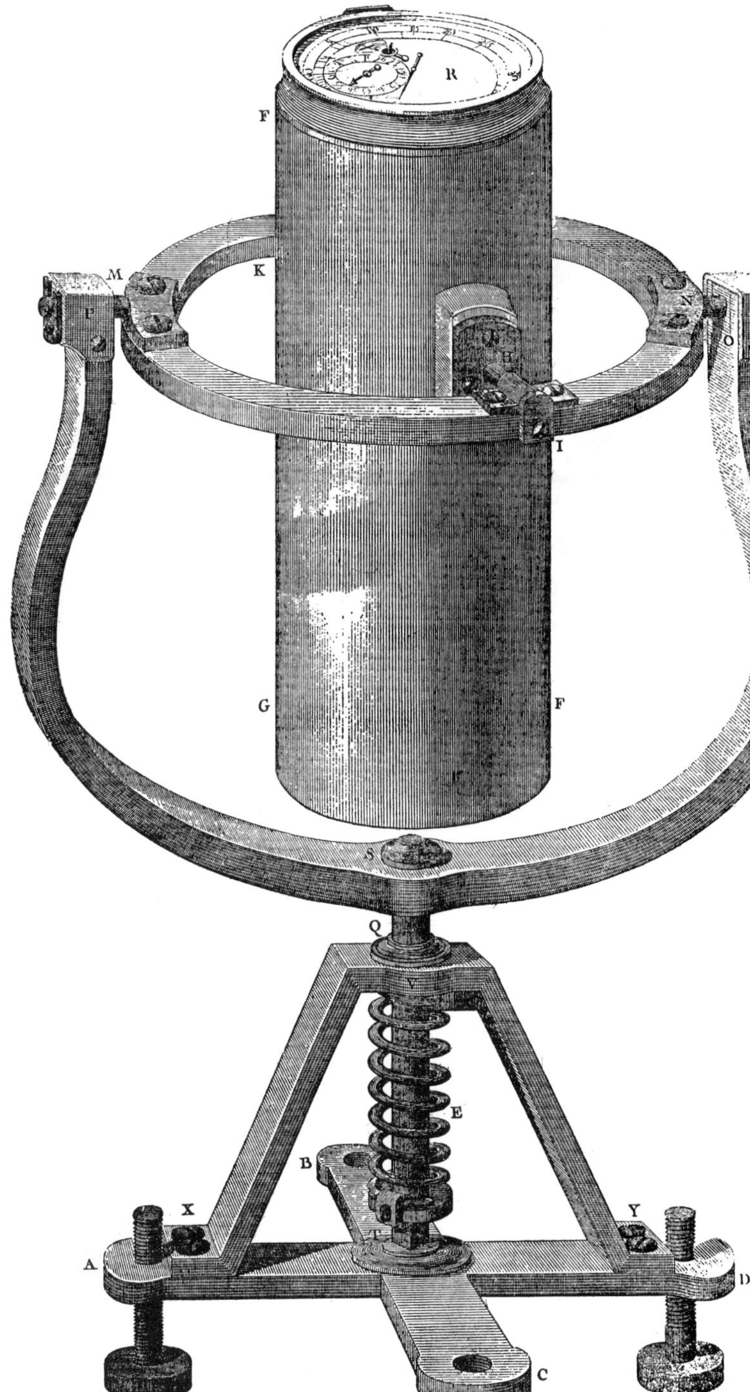

48 Overall view of the weight-driven marine chronometer (HM) by Ferdinand Berthoud.

49a Marine chronometer, signed Johann Georg Thiell, invenit et fecit No. 1, Octobr. 1768; case (diameter 96 mm) of fire-gilded copper; with two hour dials and two second hands outside the hour dials.
Württemberg, Landesmuseum, Stuttgart

49b Rear view of the movement of the chronometer by J. G. Thiell, Bremen; movement diameter 75 mm; with main and secondary movements functioning in the manner of a Remontoir d'Egalité. Two balances of equal frequency (2 Hz); temperature compensation by differential expansion of two metals affecting the active length of the hairspring of the constant-force balance; escapement after Flamenville; some ruby cap jewels but no whole jewels.

50a Marine chronometer, signed Jean Fazy de Genève, "l'élève de Horloger de la Marine Paris 1769." Strongly influenced by Berthoud's ideas; wooden frame with adjustable legs and two handholds on the sides. Dial diameter 195 mm, with five hands. Jean Fazy, born in Geneva in 1734, died in Moscow in 1811, worked first in Geneva and Paris, later in Stockholm and Moscow; from 1760 to 1795 he was the court clockmaker in St. Petersburg; he is known for this special work for Admiral Bellhausen, Governor of Kronstadt.
Antiquorum, Geneva

50b Side view of the escapement, a very early chronometer escapement with double arm with journal bearings, gold releasing spring, weight drive with three guiding rods and three rollers, movement diameter 155 mm.

50c Rear view of the movement.

51a Silver marine watch, Larcum Kendall's first copy of Harrison's marine clock H4; this watch, K1, was shown to the Board of Longitude in May of 1769 after a construction time of two years, and finally delivered in 1770. K1 is largely a copy of H4; only the running regulator, still "blind" (not used), in H4, and the corresponding adjustability were omitted. *National Maritime Museum, Greenwich, London*

51b Rear view of the movement with the signature "Larcum Kendall, London 1769."

52a Box-form ship's clock signed W. Snellen (probably Willem Snellen, Dortrecht, Holland), 1770; verge escapement with diamond pallets! Dial diameter 130 mm.
The Time Museum, Rockford, Illinois

52b Rear view of the movement, showing the temperature compensation, which acts on the hairspring with two screw pins (visible to the left of the balance verge).

53a Marine clock, unsigned (probably by Willem Snellen, Dortrecht), ca. 1770-80.
Stender, Sint Michelsgestel, Holland

53b This rear view of the movement shows the large balance and temperature compensation similar to that shown in #52b.

53c Side view of the movement.

△

53a △

53b △

52b ▽

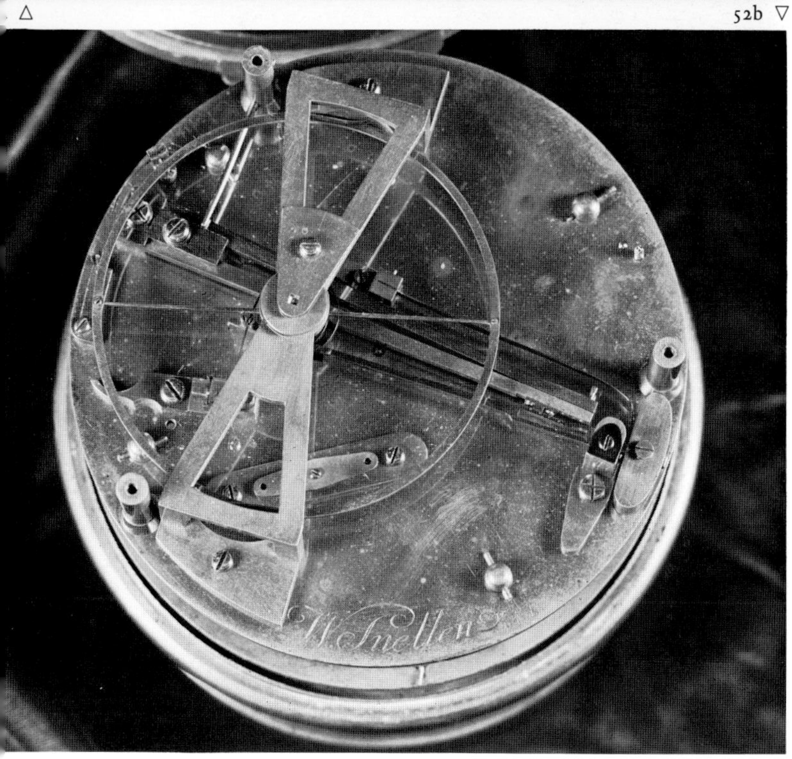

53c ▽

54aMarine chronometer, signed John Arnold London No. 1, ca. 1771, mounted in an octagonal wooden case, silver-plated dial, chronometer escapement with spring detent, two-armed compensated balance, drive by fusee and chain, additional winding with planetary gears. This historically important chronometer was rebuilt several times over the years by John Arnold himself, so its original condition can no longer be determined today.
Clockmakers Company Museum, Guildhall, London

54b The back of the movement shows the present conditions after being rebuilt several times by Arnold; typical Z balance by Arnold.

55a Silver marine clock, Larcum Kendall's K2 of 1771; the construction of the movement of this clock corresponds essentially to that of the typical pocket watches of the time; verge escapement similar to those of H4 and K1, but the pallets are made of ruby instead of diamond; in addition, Kendall did not use a Remontoir d'Egalité; power reserve during winding as handled by Harrison; for temperature compensation a spiral-shaped bimetallic device with a double lever was used; three separate dial rings for hours, minutes and seconds, the last out of the center.
National Maritime Museum, Greenwich, London

55b The movement of K2.

Larcum Kendall
London

55c The back of the movement of K2 shows the signature of Larcum Kendall, 1771, London, and the spiral temperature compensation built into a geared ring with fusee adjustment.

56a Marine clock, signed Thomas Mudge, of 1774 (No. 1); Mudge's escapement with constant force; eight-day running with separate dials for seconds, minutes and hours (the last with 360-degree calibration, plus a winding indicator sector).
British Museum, London, ex Ilbert collection

56b Mudge's marine clock no. 1 with its wooden outer protective case. The octagonal form was chosen by Mudge to be able to check the influence of various vertical positions during testing.

56c Side view of the movement of Mudge's marine clock no. 1; the escape wheel and the shortened balance arbor are clear to see.

57b Marine clock K3 of Larcum Kendall, lid closed.

57a Larcum Kendall's marine clock K3, 1774, with a movement very similar to K2 except that its verge escapement is fitted with two coaxial crown wheels and only a single ruby pallet; with three separate dials for hours, minutes and seconds; in an octagonal wooden case, lid opened.
National Maritime Museum, Greenwich, London

57c Movement case of Larcum Kendall's marine clock K3 removed from the wooden case.

57c Back of the movement of Larcum Kendall's marine clock K3.

58a Marine chronometer in silver case, signed "Les frères Goyffon à Paris", no. 37, detached escapement in Pierre Le Roy's manner; second hand out of the center, front winding through the dial.
ex Arckenbourg collection, Arts et Métiers, Paris

58b Rear view of the movement of the marine chronometer; balance with control elements on the ring and escapement components in Pierre Le Roy's manner easily seen, flat hairspring.

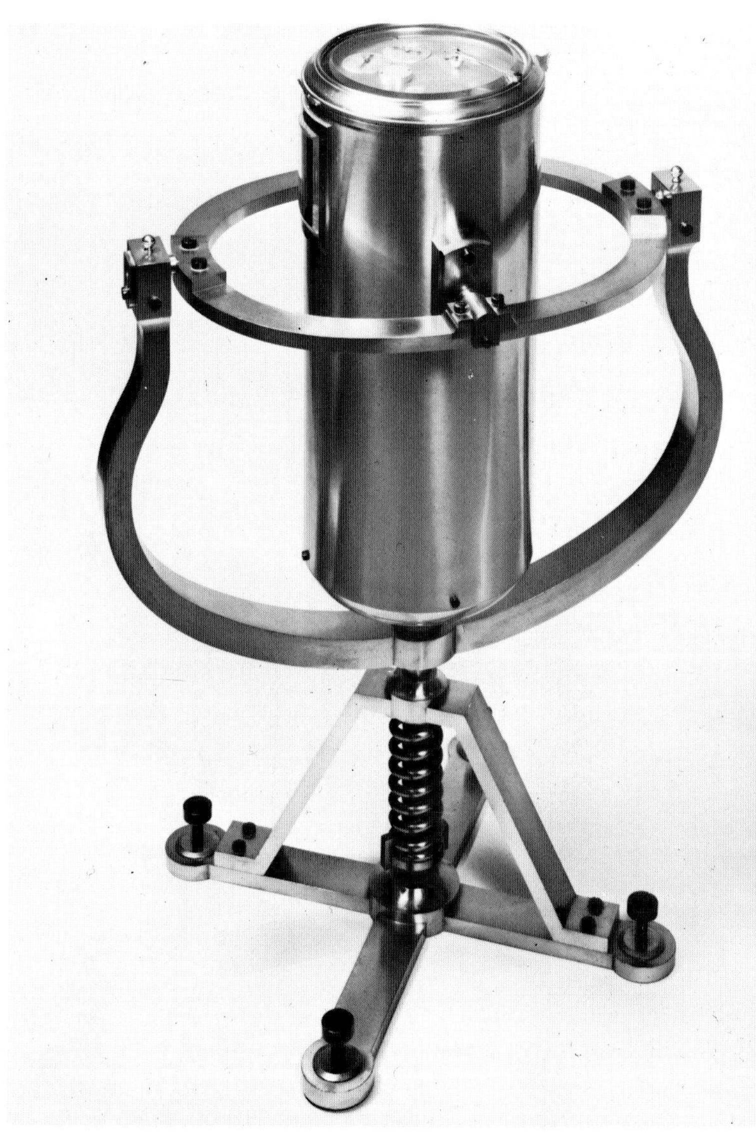

59a Large marine chronometer with weight drive by Ferdinand Berthoud, signed H. M. No. 16 Inventée et executée; Paris 1775; gimbal mounting with sprung original stand.
Th. Beyer Museum of Timekeeping, Zürich

59b Top view of marine chronometer H. M. No. 16 by F. Berthoud.

59c Side view of the movement removed from the housing, showing the grid system for temperature compensation, two drive weights with controls and roller bearings (both sides) for the suspended balance, beat two seconds.

59d View of the movement of the marine chronometer H. M. No. 16 by Ferdinand Berthoud, with dial removed.

59e Side view of the movement of Ferdinand Berthoud's marine chronometer.

59f Large scale view of the escapement of Berthoud's H. M. No. 16.

60a Marine chronometer, signed Thomas Mudge, 1777; shown complete, with two enamel dials, one for hours and minutes, the other for seconds.
Dresden, Math. Phys. Salon

60b Outside of the bottom plate, the four rollers on the bridge for the verge mounting are easily seen; signed on the bridge "Tho. Mudge 1777."

60c Side view of the movement, without the top plate.

60d A look into the movement, showing the inside of the bottom plate, the arrangement of the barrel (simple spring arrangement here) and special chain ("reversed fusee") arrangement.

60e Bottom plate of the movement after removal of bridge and balance, showing components of the temperature compensation with its lever.

a △ 60b ▽ 60d △ 60e ▽

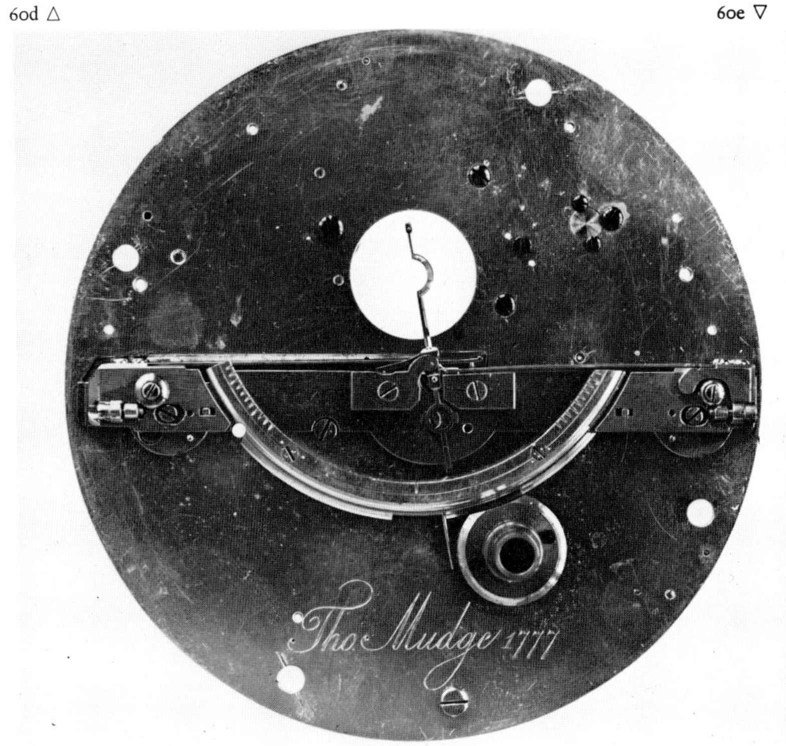

61a Marine chronometer by Mudge, the so-called "green marine clock" that was missing for years. Seen from above, along with a picture of the movement from the back.
The Time Museum, Rockford, Illinois

61b Parts of the movement of Mudge's marine chronometer.

61c Spindle bearings of Mudge's marine clock, enlarged.

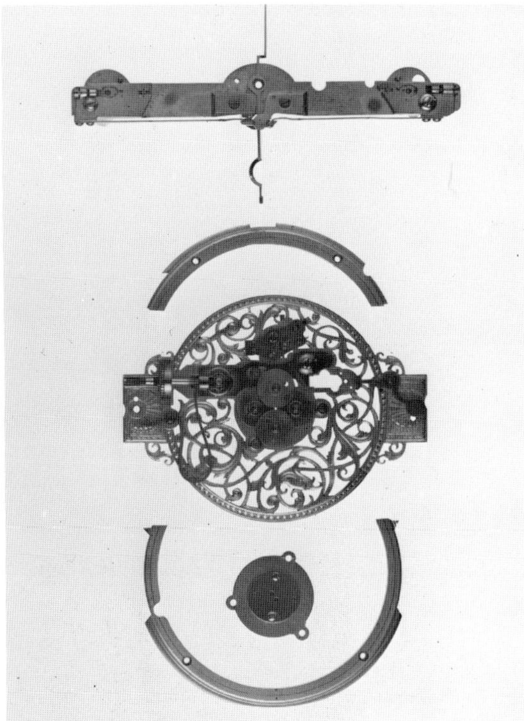

61d Balance bridge and temperature compensation.

61e Large-scale view of one of the impulse springs of the constant-force escapement.

61f Side view of the entire movement.

62a Marine chronometer by Ferdinand Berthoud, 1777, first piece of a new series for easy service and use in land vehicles as well as at sea. A further development of the type with spring drive; here are two barrels working together, chain and fusee. This clock was used on various expeditions.
Arts et Métiers, Paris

62b Enlarged view of the dial of F. Berthoud's marine chronometer with the signature: "A. At. M. inve et faite par Ferdinand Berthoud Nr. 1", seconds in the center.

62c Rear view of the movement of F. Berthoud's marine chronometer; balance arbor with roller bearings, detached escapement, originally with a grid system for temperature compensation, later replaced by a bimetallic strip.

63a Marine chronometer (montre marine) signed F. Berthoud Nr. 6, 1777, with gimbal mounting in polished wooden case; silver-plated dial, 145 mm diameter, 36-hour running.
The Time Museum, Rockford, Illinois

63b Rear view of the movement of Berthoud's montre marine, drive with chain and fusee, external grid system for temperature compensation, chronometer escapement with spring detent; balance arbor with roller bearings.

64a Marine chronometer signed Armand Kiobenhavn Nr. 7, circa 1780, especially well-made movement, influenced by the construction principles of F. Berthoud and Pierre Le Roy; drive with chain and fusee, compensated balance suspended and with "anti-friction roller", bearings on both sides; detached escapement of unusual design.
Th. Beyer Museum of Timekeeping, Zürich

64b Side view of the movement of Armand's marine chronometer.

64c Enlarged view of Armand's marine chronometer with "anti-friction roller" bearings.

IV: Sea Travel & Geographical Discovery

In describing the great significance of the longitude problem, several drastic examples of helplessness at sea were cited that were caused by the lack of knowledge of the ship's exact location at sea.

Many such cases can be found in old chronicles and newspapers, but the cases already noted will doubtless suffice to explain the various efforts made to determine longitude exactly and make clear the interest shown by all those involved.

The necessity of testing the new ideas and implements on ships at sea also contributed to the extent to which concrete suggestions and methods of determining longitude at sea were evolved and developed. Christiaan Huygens had undertaken such tests with his marine clocks, as had the French pioneers Sully, Pierre Le Roy and Ferdinand Berthoud, arranging for and carrying out practical tests that have been studied in depth in various publications. In the same way, Harrison immediately undertook a sea voyage to Lisbon[1] for testing, and the task imposed by the 1714 Act of Parliament resulted in H4's West Indian cruise from November 1761 to January 1762. Both were ordered and financed by the Board of Longitude; and the expenses of William Harrison, who went along in place of his elderly father, were paid as well.

The results of the second test voyage with H4, to Jamaica in 1764, as well as the French expeditions of 1768-69 (#65) with the marine clocks of Pierre Le Roy and Ferdinand Berthoud, were clear and undoubted proof of the outstanding suitability of mechanical clocks for determining longitude at sea — real revelations that the sought after solution to the longitude problem had now been found.

The contemporary publications, parliamentary discussions and books mentioned in the previous chapter, "The Heroic Epoch", have not resulted in an immediate

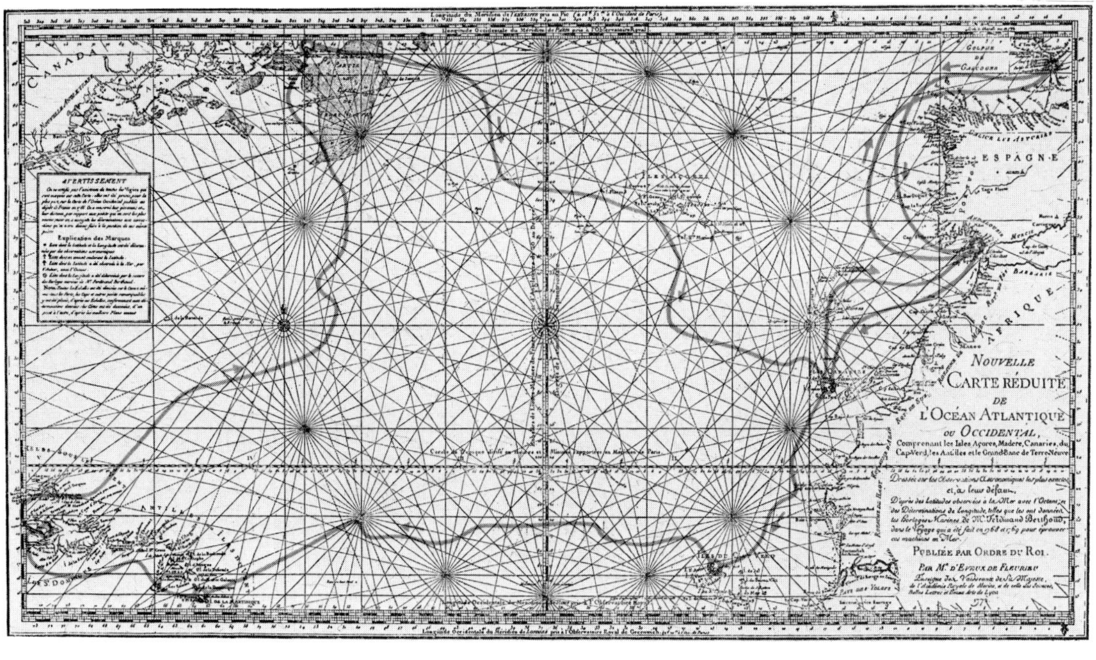

65 Map of the Atlantic Ocean with the route of the expedition under M. de Fleurieu on the frigate *Isis* to test marine clocks no. 6 and 8 of Ferdinand Berthoud, with the longitudinal values determined (both astronomically and by using Berthoud's marine clocks). *C. P. Fleurieu, "Voyage fait par ordre du Roi 1768 et 69 . . . pour éprouver les horloges marines par M. F. Berthoud, Paris Imprimerie Royale, 1773.*

or even speedy transition to practical seafaring through the use of the chronometer. These publications and public discussions surely helped the later spread of its use, but it is impossible to weigh the intensity of these effects today. Yet single incidents, particularly Captain Cook's second voyage of discovery, seem to have resulted in particular moments of awareness that contributed very effectively to the growth of interest in the use of chronometers among seafaring men and shipowners. Today, in the age of satellites and space travel, the adventurous voyages of Captain Cook are no longer as much a part of our lives as they were to earlier generations; it may be appropriate to summarize them on the basis of a report in the Viennese press[2] on the occasion of the 200th anniversary of Cook's death, which includes important data on the use of the chronometer on his voyages.

Cook was the son of a poor Scottish agricultural laborer, and was born in a small Yorkshire village in 1728. At the age of seventeen he left his home town and moved to a nearby seaport, where he sold imported goods. But the sea soon won the young Cook. At eighteen he developed a great passion for the sea and the unknown extent of the world, and it never left him until the day of his early death — 32 years later.

Cook's First Apprenticeship

In broad-beamed colliers Cook began a two-year apprenticeship. This loner devoted every free hour, as also during his next two years on a merchant ship, to the study of navigation. At 23 he was a helmsman, and four years later he was offered a command. But he declined to be a merchant-ship captain and began all over again in the Royal Navy. Did patriotism impel him, did he see a better chance of making progress in the navy, or both? It was June of 1755, and although the Seven Years' War was officially declared only the following spring, it had long since begun. The British blockaded the coast of France to prevent shipments of weapons to Canada. On one of the blockade ships, Cook advanced from First Mate to Chief Boatswain and, on his 29th birthday, October 27, 1757, became the helmsman of the *Pembroke*, a 64-gun ship of the line. He was now the highest non-commissioned officer on board.

He continued to study, mathematics, astronomy, and any subject relating to navigation; his knowledge must have been remarkable. His captain and admiral recognized his ability, named him helmsman on the flagship of the fleet, and employed him for three years chiefly as a cartographer for mapping the St. Lawrence River and the coasts of Newfoundland.

Expedition I

Only when the British Admiralty made secret plans that nobody but Cook could have carried out was he promoted to lieutenant. The undertaking was disguised as a research venture: Astronomers of the Royal Society, Britain's scientific center, were to study the transit of Venus between the sun and the earth on June 3, 1769 from the Pacific for the purpose of improving future determination of longitude on the high seas, and to prevent the French from annexing new territories there to make up for their last war losses (they had just been driven out of Canada and India). But the main goal of the expedition was to look for a presumed but not yet discovered continent in the south that had stimulated men's imagination for nearly two centuries: "The Great Southern Land."

They put out to sea from Plymouth in August of 1768. At Rio they had to endure 24 days of internment because the harbor authorities mistook the ship — which did not look much like a Royal Navy vessel — for a smuggling craft. Then they encountered icy weather off Tierra del Fuego, but rounded Cape Horn smoothly. Eight months after leaving England they reached Tahiti. In Tahiti Cook had to hold native chiefs hostage until two seamen, who had jumped ship and fled with Polynesian girls, were brought back and the voyage could proceed.

Off the east coast of New Zealand, Cook had no better luck with hostile Maoris than had his predecessor, the Hollander Tasman, who had discovered the western coast more than a century before. But during several brief landings the devoted scientist was able to collect many plants and animals unknown in Europe. Cook drew maps and named bays and mountains, including Queen Charlotte Sound, where he had his overworked ship careened on the beach for cleaning and caulking.

Cook annexed not only New Zealand for Britain, but also took possession of Australia for the crown. Then he sailed between the northern tip of Australia and New Guinea through the Torres Strait, the existence of which had only been rumored in Europe, and reached a shipyard at Batavia, the present-day Djakarta. The canals of this city, built by the Hollanders according to plans from home, were breeders of serious diseases in the tropics. Before the *Endeavour* reached Capetown in March, more than thirty members of the crew had died, including the ship's doctor, astronomer, and painter. On June 12, 1771, almost three years after they had left, the ship and half its original crew saw home again.

As a timepiece for measuring the transit of Venus, Captain Cook had taken along one of the regulators that had been bought from Shelton for the purpose. Plans had been made as early as 1761 for an expedition to observe a transit of Venus for the purpose of determining location precisely, and three astronomical regulators had been purchased from Shelton of London for the various expeditions to observe the transits of Venus.

Expedition II

Expedition II was intended to be a thorough exploration of the South Seas. The Board of Longitude played an important part in planning the details of this expedition. At its meeting of November 28, 1771 it was stated specifically that, second only to the intended geographical exploration, the most important task of the trip was to test the copy of Harrison's No. 4, namely K1, made by Kendall in 1770, and other chronometers obtained by the Board in 1771 — such as Arnold's No. 1.

An exhaustive report on the marine chronometers used by Captain Cook and his experiences with them was published by Comm. D. Howse in the September and December 1969 volumes of the A. H.[3] Many of these data taken from the Greenwich archives have been used in this chapter.

Before Cook's departure in July of 1772, two more marine chronometers made by John Arnold, No. 2 and No. 3,(#66), had been completed for sea testing.

The expedition consisted of two ships, the *Resolution*, of 462 tons, with 118 men as well as Cook, and the *Adventure*, of 350 tons, with 183 men, commanded by Captain Furneaux.

Cook's ship carried K1 and A3; that of Captain Furneaux had A1 and A2.

Extensive preparations and regulations for the treatment of the clocks on board had been made by the Admiralty and the Board of Longitude, so that the clocks would not be damaged during the sea voyage. Special cases with three separate keys had been built, and a complicated procedure for their daily winding had been worked out, using three different keys, for which the captain, the first lieutenant and the ship's astronomer had to be present. The written instructions for William Wales, the astronomer of the *Resolution*, read as follows:

"You must wind the clocks as early in the afternoon of every day as you can — you must compare the two with each other and record the indicated times on paper; you must also note the clock times when the sun's height in the morning or afternoon, or the distance of the moon from the sun or from fixed stars, have been observed; you must further calculate the geographical distance that is derived from comparing the clock time with the ship's local time based on the morning and afternoon heights of the sun; as often as you have the opportunity, you must compare one of the clocks in your care with those entrusted to Mr. Bagley on the other ship and record the times indicated by these clocks."

The chart in #67, which includes the main stations on the route, shows the behavior of K1 and A1 on the second expedition, as reconstructed from the recorded data. Wherever the ships lay at anchor for some time, the astronomers set up their observatories, as noted on the route of the voyage. With the help of regulators, a quadrant, and other instruments, they checked the running of the marine clocks every day, whenever possible, and thus determined the daily gain or loss.

At the Society Islands, Cook wrote in June of 1774:

"The clock K1 has lost 8 minutes and 34.5 seconds since leaving Queen Charlotte Sound, compared to its indication there; in these five months or more, extreme heat and cold have affected it. We have the impression that half the faulty indication occurred after we left Easter Island, so that it seems as if the clock worked more precisely in the cold than in the warm climate." At Christmas the *Resolution* was at Tierra del Fuego to begin searching the South Atlantic for the legendary "Southern Land." On this voyage Cook discovered the uninhabited South Georgia Islands, the glaciated, penguin-inhabited South Sandwich Islands, and South Thule, but could not find the slightest trace of a continent. Thereby the old belief that there had to be a mighty continent at the south end of the earth proved to

66 Arnold's marine chronometer no. 3 of 1773. This was taken on Captain Cook's second voyage around the world in 1773, on board the *Resolution*.

be untrue.

Cook's voyage home was made via Capetown, where he wrote to the secretary of the Admiralty: "The clock made my Mr. Kendall exceeded all expectations, even those of its enthusiastic partisans; we have checked it here and there by lunar observations too; it has been our loyal leader through all the difficulties of the various zones."

After rounding the cape, Cook made the following entry in his journal: "On our landing at the Cape, our error according to Mr. Kendall's clock was only 18 minutes of geographical distance . . . It would be very unjust to Harrison and Kendall if I did not add here that we gained a great deal of help from these very useful and valuable timepieces, as is also noted in other parts of my journal."

Finally, in 1775 Joseph Gilbert, Cook's helmsman on the *Resolution*, noted: "Kendall's clock is without doubt the finest mechanism that the world has created to this point."

On the trip home, Cook took a route to the island of St. Helena, trusting in the reliability of his chronometer instead of using the old seafarer's method of first reaching the latitude of the island in question and then, with more luck than understanding, sailing either eastward or westward. He was able to steer directly for the island, and he wrote about it:

"Trusting to the reliability of Mr. Kendall's clock, I made the decision to steer directly toward the island of St. Helena. Mr. Kendall's clock did not disappoint us, and at daybreak on May 15 it lay 14 miles west-northwest of us."

At the time when this direct course was taken for St. Helena, the ship *Dutton*, which belonged to the East India Company, sailed along with the *Resolution*. The successful course for St. Helena, steered with the help of the chronometer, was reported to the East India Company by the captain of the *Dutton* as a particularly noteworthy event and led to many of their ships being equipped with chronometers.

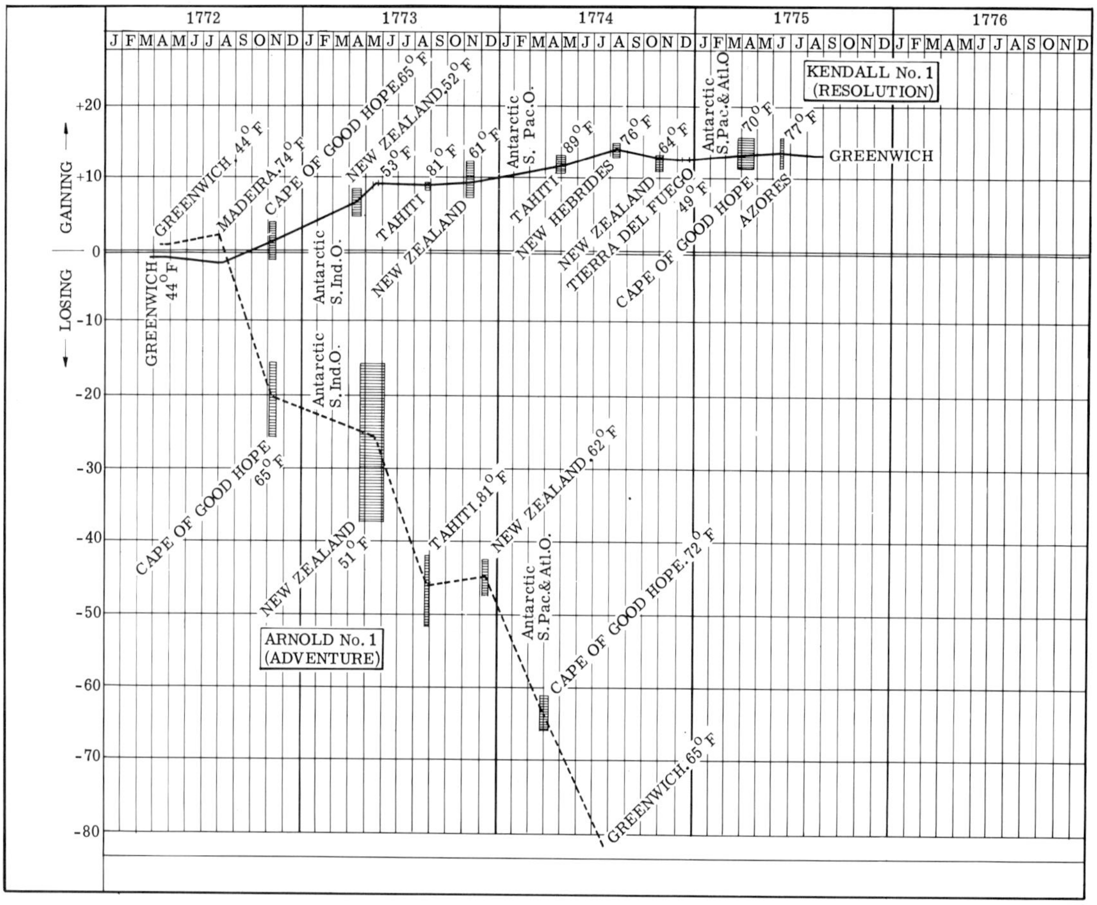

67 Voyage route data of Captain Cook's second expedition; the going rates (in seconds per day, mean time) of Kendall No. 1 (on the *Resolution)* and Arnold No. 1 (on the *Adventure*) were recorded as ordinates, and the mean daily temperatures in degrees Fahrenheit were noted; the darkened rectangles indicate control measurements taken on land, with the width of the rectangle indicating the number of days on which measurements were made and the height showing the distribution of the calculated rates; rate curves indicate the mean values of the rectangles according to how they are connected. *(Howse)*

On July 30, 1775, after being away for three years and eighteen days, Cook was able to anchor in England again, at Spithead. The last entry in his journal of this second expedition runs as follows:

"We landed near Plymouth Muker Church at 5:00 P.M. (North 10 degrees, west seven miles away); comparison with the time indicated on Kendall's clock showed that its error in geographical distance was only 7 minutes 45 seconds too far to the west."

This time the honors went directly to Cook. He was named a member of the Royal Society. During the entire unbelievable voyage, only one man had been lost. The king himself received Cook and promoted him to be commanding captain of a 74-gun cruiser.

Expedition III: Testing the Possibility of an Atlantic-Pacific Connection straight through America

On this third expedition too, Cook took Kendall's No. 1 clock in his own ship — again the *Resolution* — while Kendall's Clock No. 3, since completed, was taken on the escort ship *Discovery*,, with Lieutenant James King as astronomer. The performance of both clocks is shown in #68, in the same manner as the performance of the chronometers on Expedition II.

On December 7, 1777 the two ships sailed out of Tahiti, crossed the Equator and, on December 24, discovered an uninhabited island that Cook named Christmas Island. From there Cook steered northward again and, in April of 1778, reached Vancouver Island on the northwest coast of North America. He stayed in the Arctic some seven months, leaving via the Aleutians only at the end of 1788. From there on he headed southward again, and on January 18, 1779 he sighted his last important discovery: Hawaii. Did it happen quite by chance? Or had Cook's stay in Tahiti acquainted him with old legends of those early voyages of the Polynesians toward a land to the north? Cook's log books provide very exact details of the places and times of all the courses he steered, but why he chose this particular direction and no other remains his secret forever. He named the islands after his friend and patron — the First Lord of the Admiralty — the Sandwich Islands. The natives thought he was the returned harvest god Lono, as old legends had prophesied, and prepared an overwhelming reception for him.

The jubilant reception on the big island of Hawaii, after which the whole island chain is named today, is probably unique in the history of the entire Pacific. Thousands packed the bay. On some 1500 canoes, some 9000 islanders came toward the ship, offered fruit and women according to their tradition of hospitality; Cook noted without comment that the women granted their favors more willingly than any they had met before. The carefree Hawaiians slept on deck at night, leaving their canoes tied to the ship.

Early in February of 1779 the British set sail again. But fate allowed no easy departure. A storm off the north coast of Hawaii tore up both ships' sails and damaged the foremast of the *Resolution* so badly that Cook ordered them to turn back.

The First Sea Expeditions with Chronometer Testing after Cook

After the last two great voyages of Captain Cook, marine chronometers began to be used more often on expeditions. The concept gradually took hold that a reliable chronometer belonged among the equipment of every serious expedition. Of course there were still many seafaring men who, as Gould reports, had a deep distrust of chronometers. He tells of an admiral who, when he learned that a chronometer had been brought aboard his ship, ordered it taken back on land immediately with the comment that he wanted no such magic aboard his ship.

To answer such misconceptions and prejudices, the Board of Longitude made its chronometers available for important expeditions. Commander Howse reports the following in reference to the list of chronometers used on Cook's voyages:

1773: A North Polar expedition under Captain Phipps, for which the board made K2 and an Arnold chronometer available. K2 reached Spitzbergen aboard the *Racehorse,* and was then used in various ways until the unfortunate voyage of Captain Bligh on the *Bounty* in 1787, when a mutiny brought the voyage to an end.

The first fleet expedition to Australia (1786-92): This expedition with the frigate *Sirius* was led by Captain Arthur Phillip, who was to set up a prison camp in Australia and serve as the first governor of New South Wales. The ship set out, with K1, on May 13, 1787; Sydney was founded in January of 1788. In March of 1790 the *Sirius* was wrecked, but K2 returned to England intact early in 1792, was used there on sea duty until 1802 and then taken out of service.

A West Indian expedition under Captain John Elliot used K3 from 1782 to 1789.

1793-1802: another West Indian expedition under Sir John Jervis in which K1 again proved to be very useful; after this expedition the K1 was retired, having been in use for some thirty years.

A great Northwest Pacific expedition under Henry Roberts (1791-1795), with K3, plus Arnold's No, 82, 14 and 176, and a pocket chronometer by Earnshaw, which were tested here for the first time.

A further Australian expedition (1802-1804), with K3 and several Arnold clocks, which proved to be only partially useful.

Soon after 1800, though, the marine chronometer belonged to the standard equipment of British ships and also to those of other maritime countries, especially France, where a very significant manufacture of chronometers developed independently.

In the 19th Century, large areas of the earth's surface were still little or not explored, naturally giving rise to numerous expeditions, as much correcting of geographical positions was necessary. Thus most of these expeditions were equipped with appropriate astronomical instruments and chronometers. Their technical performance had quickly developed to the point where their operation during such a trip was no longer the object of special attention; instead, thoroughly reliable accuracy was a necessity, and thus in the records of most expeditions only details of the measurements taken are found. In spite of that, these scientific reports are of considerable interest to specialists in the study of chronometer history, for they often afford not only information on the numbers and types of chronometers taken along, but sometimes also describe unusual working conditions.

In the course of the chronometer studies on which this book was based, chance led the author to learn of several later expeditions that deserve mention:

The Austro-Hungarian North Polar Expedition of 1872-74, under the leadership of Julius Payer and Lieutenant Weyprecht on the ship *Tegetthof,* discovered Franz Josef's Land. The well-known Professor Oppolzer was in charge of equipping this expedition with its instruments, which returned to his possession after the expedition was over; they included marine chronometers by Onesime Dumas of St. Nicolas and a pocket chronometer by J. Vorauer of Vienna (#69a-e). The latter had an unusual fate; during an attack by polar bears it was swallowed by one of them. Later, when the bear was shot, it was found in its stomach, still running. Engraved on its case (#69d) are the expeditions on which the chronometer was used.

During World War I (1914-1918) Professor O. Osterhelt of the Technical College of Dresden traveled widely in China and calculated numerous positions.[5]

Alfred Wegener's Greenland expeditions[6] (1929-31): On the main expedition (1930-31), Dr. Georgi functioned as leader of the central glacial station, later called *Eismitte,* located some 400 kilometers from both the east and west coasts of Greenland. In a cave in the ice, Georgi carried on important scientific research at temperatures down to -60 degrees. The Eismitte Station was situated at about 71 degrees 56.2 minutes north latitude and 39 degrees 56.2 minutes west longitude. This longitude, along with various other geographical locations during this expedition, was calculated with a pocket chronometer (#70) that Dr. Georgi carried in a leather bag on his person throughout the entire expedition. Georgi's experiences at Eismitte Station are described impressively in his book.

The difficulties in determining time, and thus also position, that occur and must be dealt with even in modern times are shown by excerpts 4 and 6 from the reports of the aforementioned expedition:

Excerpt 4: "Determining Location (from Eismitte)"

The first temporary location determination by Dr.

Weiken on July 31, 1930 gave, with time indications by two clocks, the coordinates of 71 degrees, 8 minutes north, 39 degrees, 59.2 minutes west; a later review of the clocks' performance after time comparisons on the west coast gave the longitude as 40 degrees, 3.5 minutes. It was intended that, when trigonometric measurement of the height at Station Eismitte had been made in the summer of 1931 by using radio signals, final and exact measurements would result. This did not take place, though, and thus it was attempted to calculate longitude just before the departure by using the radio set aboard the propeller sleds, while latitude was determined on the basis of the sun's height before and after its midday height, as was done during the stay at Eismitte only to check the clocks for the meteorological measurements. I am particularly indebted to Ship's Officer Paul Nickstadt of Hamburg for the laborious evaluation of these data to derive reliable indications of locations.

The observations were taken with a naval sextant of about half the usual size (a so-called half-sextant) and a mercury horizon indicator in box form, both made in England and the property of the Imperial Central Colonial Office.

Excerpt 6: "Determining Time (at Eismitte)"

According to A. Wegener's daily temperature curves of the interior of Greenland in summer, and likewise the curves of Nansen and de Quervain, it was necessary, because of the extraordinarily fast temperature changes in the morning and evening, to limit the clock-time error as much as was possible without radio signals. Because of the special conditions of the first sled trip, we had reached Eismitte with comparatively imprecise clock time. This error could now be estimated after a determination of longitude based on radio time signals was achieved. The difference between Dr. Weiken's temporary determination on July 31, 1930 with the mean longitude calculations on August 4 and 5, 1931 amounted to 38.6 minutes, representing a time difference of 2 minutes 34 seconds, if the instrument errors and changes in position can be disregarded, by which calculations Dr. Weiken's clock must have run fast. For my main clock (a half-second pocket chronometer by Nardin of Le Locle), I had regarded Weiken's clock as correct and then calculated my correction as -24.5 seconds. According to the above, the actual correction was 2 minutes 58.5 seconds, practically -3 minutes. Later it also appeared as if the time brought to the inland ice from the coast was considerably inaccurate. The pocket chronometer was always hung in a bag and worn on the body, while an ordinary alarm clock served everyday needs. In 1929 it had already been found in calculating location that the chronometer showed an error (ran slowly) when it was directly exposed to the cold during these measurements. As a result, the time indication during the astronomical observations was provided by an ordinary, inexpensive pocket watch and an alarm clock; only for the last calculations of longitude in July and August of 1931 was the chronometer used, being removed from the for only a moment. Even so, an irregularity seems to have occurred here too; while the indicator clocks already mentioned, while exposed to the cold, strongly increased their running rate. For example, the alarm clock gained 22 seconds from 11:15 to 1:25 during the August 24, 1940 observation; the pocket watch gained 16 seconds from 3:02 to 4:10 on April 11, 1931. The main clocks too (pocket chronometer by Nardin and lever escapement clock M 1170 of Dr. Sorge), worked under unfavorable conditions; being always worn on the body to keep the same temperature they suffered some hard blows and movement.

Later the attempt was made to calculate the probable running of the Nardin pocket chronometer from all the time indications and transmissions, and then from this to calculate the running of the observation clock M 1170 (R. Sorge). As a result, the chronometer observations at Eismitte indicated at best an even running of -1 minute per month or -2 seconds per day. The irregularities in July and August of 1931 could have been caused by the use of the clock outside in determining location. The running of the M 1170 clock, based on the deviations of the running curve of the chronometer remained unexplained, even when the attempt was made to ascribe a part of the difference to a change in the running of the chronometer. It is probable, though, that the hard work in the building of the 16-meter shaft by Dr. Sorge somehow resulted in irregularities in his clock. Two time determinations were of particular importance. On August 24, 1930 a time error of -8 minutes was determined and the clock was set back this much. Through the previous and subsequent measurements on August 24, a failure or at most -5 minutes was shown to be likely. All the same, the extent of the deviation in the ensuing time was influenced favorably, since the clock constantly gained time. On the other hand, the time determination of December 30 worked in the opposite direction. Until then, a very regular error in the readings of +4 minutes (late) had been found. From these observations it appeared that the main clock, instead of gaining as had been assumed, actually ran slowly. This was applied to further running and every observation time was set earlier by the appropriate amount. Only the measurements in the spring showed the error in this indication, so that in March and the first third of April there was a considerable time error of up to 20 minutes. This time error was enlarged negatively by the earlier acceptance of a longitude of 40.0 degrees west, while this had now been worked out as 40 degrees 42 minutes, so that all observations had in reality been two minutes 42 seconds earlier than the actual local time."

The Filcher expedition (between the first and second World Wars) to central Asia, in which important determinations of location had been carried out, sometimes also on the basis of astronomy.[8]

Details of Chronometer Use on Shipboard

The practice of longitude calculation at sea has developed in principle as Gemma Frisius had predicted in 1530, but not so simply or smoothly.

Before a sea voyage began, the marine chronometer was first set to the exact time at the point of departure, for example, time at Greenwich meridian. To do this, the exact time had to be brought to the port from the nearest observatory where exact time measurements were made on an astronomical basis. The observatory itself and the

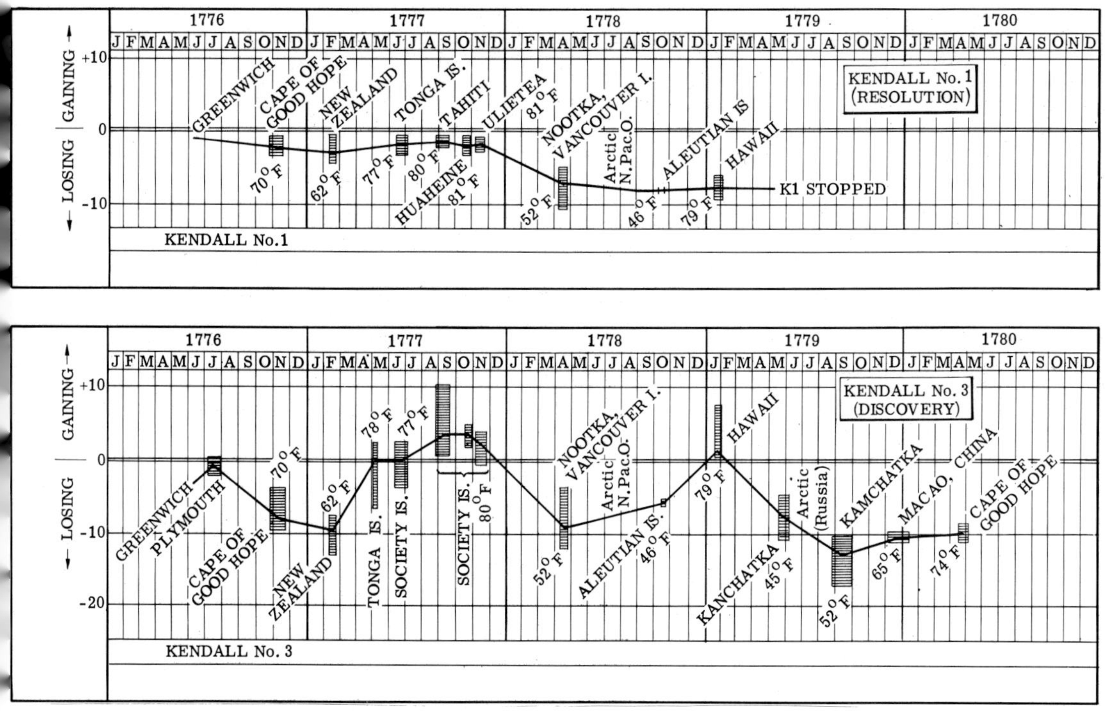

68 Voyage route data of Captain Cook's third expedition; the daily going rates (in seconds per day of mean time) of Kendall No. 1 (on the *Resolution)* and Kendall No. 3 (on the *Discovery)* are recorded as ordinates, with the mean daily temperatures noted in degrees Fahrenheit. The darkened rectangles indicate control measurements taken on land, with the width of the rectangle indicating the number of days on which measurements were made and the height showing the distribution of the calculated rates; rate curves indicate the mean values of the rectangles according to how they are connected.
(Howse)

regulator in operation there were naturally fixed at one location; thus a small, portable, precise clock was needed that could be set to the regulator's time at the observatory and then brought to the ship, a type of clock known as the deck watch or observation watch. It soon proved that setting the marine chronometer to the exact time and regulating its running in the manner in which a pocket watch is set in everyday life — by moving its hands forward or backward — is not allowable. The chronometer must run just as it comes from the workshop, and all time operations must be carried out in terms of reckoning and documentation. Thus it is necessary to record in a book, day after day, the difference between the marine chronometer and the time brought on board, and also to make the necessary corrections regularly. Since the chronometer has a definite going rate, (a daily rate or error), that depends on the external temperature, these data had to be kept in the form of regular records.

In calculating the daily rate for the application and use of the chronometer, one would assume that thorough information on the subject is to be found in the literature — at least for the better-known instruments. But this is not true, although the archives at Greenwich and also at Kew have much material on the subject.

The chronometer literature, though, has given little attention to this aspect. For example, if one tries to derive from the three key numbers of Greenwich — as were used for prize testing (see Chapter 6) — a key to the actual extent of the daily rate for certain instruments, one will be annoyed at the uselessness of the data.

In Chapter 6, several tables of running results over long periods of time are given for British, German, Swiss and American chronometer tests and competitions, which makes clear the general and decisive conclusions from the great body of material attained under constantly varying working conditions. In Chapter 6 it is also shown in more detail that some general problems of natural science cannot be solved in the classic manner.

To conclude this chapter, it seems appropriate to mention some details of practical longitude determination at sea. The process was as follows: First, appropriate instruments were used on the bridge to attain data for determining the true sun time of the ship's location, by determining the necessary sun or star altitudes and the appropriate times. But to compare these data with the ship's marine chronometer, a portable clock, reliable enough for short periods of time, was indispensable — the observation chronometer.

The data and the observation chronometer were now taken to the captain's cabin, the local time of the ship was calculated and compared with the time at the point of departure, as indicated by the ship's chronometer, to give the geographical longitude.

One sees that for every individual determination of longitude a considerable amount of calculation must be done, which requires a good deal of time, and that the presence of reliable running data to make corrections the marine chronometer is absolutely necessary.

1. Daniel A. Baugh, *The Sea-Trial of John Harrison's Chronometer, 1736*, The Mariners Mirror, August 1978, National Maritime Museum, London.
2. Conrad Anders, *Wer war eigentlich James Cook? Seefahrer und Entdecker aus Passion*, Die Presse-Spektrum, Vienna, 17/18 II,1979.
3. Derek Howse, *Captain Cook's Marine Timekeepers*, (A. H., VI, 1969) *Captain Cook's Minor Clocks and Watches* (A. H., VI, 1969) *Captain Cook's Pendulum Clocks* (A. H., VI. 1969) *Margetts' Watch*, (A. H., VI, 1979) *Regulator Clocks of the Royal Observatory Greenwich* (A. H.,V, 1968) Derek Howse & Beresford Hutchinson, *The Sage of the Shelton Clocks*, A. H., VI, 1969)
4. Julius Payer, *Die Ost. Ung. Nordpolexpedition 1876*, A. Hölder, Vienna. This interesting book also includes brief treatments of the second German North Pole expedition of 1869-70 and the German polar expedition of 1871.
5. O. Osterhelt, *Routenaufnahmen in West-Szetschuan*, Petermanns Geogr. Mitt., H 255, Gotha, 1958.
6. *Wissenschaftliche Ergebnisse der deutschen Grönland expedition Alfred Wegeners 1929 und 1930-31*. Published by Kurt Wegener for the Union of German Sciences, F.A. Brockhaus, Leipzig, 1940.
7. Johannes Georgi, *Im Eis vergraben*, Verlag des Blodigschen Alpenkalenders, Paul Müller, Munich.
8. W. Filchner, W. Prybyllok & L. Hagen, *Route Mapping in Unexplored Regions*, Basel, 1957.

a

b

c

d

e

69a-e Silver pocket chronometer, signed J. Vorauer, Wien 1864, that was taken on several historic expeditions, particularly the Austro-Hungarian North Pole expedition of 1872-74.
a. Overall view,
b. View of the movement with signature: Vorauer in Wien,
c. Inner lid engraved F. I. Vorauer in Wien, Nr. 70,
d. Outer lid with engraved list of all expeditions on which the chronometer was used,
e. Outer lid with Viennese silver punched 1864, master's sign MM u. (Wagner), and the engraved name of owner Prof. Theodor von Oppolzer, the renowned astronomer who prepared the instruments for the North Pole expedition.

70 Silver observation chronometer, signed Ulysse Nardin, Locle & Geneve, used at the central ice station "Eismitte" on A. Wegener's Greenland expedition (1929-31).
a. Front view,
b. Rear view with opened lid.
(Photo Abeler, Wuppertal)

a) Eight-day chronometer by John Arnold (ca. 1775), unknown until the end of 1980. Enamel dial signed "Arnold London", diameter 121 mm. Segmented dial circles for up-and-down indication (left) and regulation (right), front winding through the center of the up-and-down hand. The small opening under the "VI" allows the chronometer to be started when stopped by using a small key, via a lever moved by a spring working on a pin in one of the balance arms (see b).

Under-the-dial view (above): Monometallic brass balance between dial and front plate, with 7200 half-beats per hour, so that the second hand jumps by the second. Chronometer escapement with spring, flat balance hairspring, temperature compensation by two bimetallic strips on the regulator. The lever and its spring to swing the balance can be seen clearly in the "VI" position.

b) Side view of the movement. Drive by chain and fusee with large barrel, to produce enough energy for eight-day running. Balance between the dial and the front plate. The most noteworthy detail is that the chronometer spring detent escapement (activated by pressure, not pulling) is used. Since no references to the rebuilding of a chronometer can be found, new light is cast on the struggle for priority between Arnold and Earnshaw by the discovery of this escapement.

◁ a

b

V: The Epoch of Fine Mechanics

The Basis

The majority of the chronometers with which collectors and admirers come into contact come from the period that stretches approximately from 1785 to 1900. In this period of a little more than a hundred years, thousands and thousands of marine chronometers and observation watches (pocket chronometers) were produced to equip many ships and supply ships' officers. They show good workmanship, but they differ considerably in important technical details and thus deserve the attention of all who are involved in horology. In this period of time, covering some four generations, it was the chronometer builders — above all the fine mechanics — who strove to improve the reliability of their products through the quality of their workmanship. That is why this epoch has been given its name, although in the latter half of the 19th Century there were a number of theoreticians in France whose work came into use only toward the end of the 19th Century and in the 20th.

Many marine and pocket chronometers have been preserved from this era of fine mechanics because the chronometers used have always been replaced by new ones after a time, when their running results no longer lived up fully to expectations. In general, the chronometers used in seafaring were thoroughly overhauled every three to four years, but after two or three overhaulings they were no longer in their original condition. Some of the chronometers taken out of service were used as reserves in depots, while others were preserved privately as mementos.

The pioneers of the heroic epoch did much more than merely showing that Gemma Frisius' vision into the future could be fulfilled, that the time of the point of departure could be taken along and, even on a ship tossed by stormy seas for several weeks, still be accurate enough to be used in calculating longitude. They surely proved, as did John Harrison with H4, that with a lot of work and expense a usable marine clock could be built. At first it was only a unique device that could serve on one single ship; but then Arnold, Earnshaw and Berthoud, chiefly through brilliant simplifications, made less costly manufacturing possible. The vital problems involved were publicized partially through the competition fostered by the Board of Longitude and partly through various publications. What resulted was, of course, not optimal construction directions in the form of a handbook for engineers, but rather only descriptions or discussions of such problems that seemed particularly important to the involved chronometer builders, including hints as to the directions in which solutions were to be found and how improvements might be attained. Included were the 1762 memoranda of John Harrison,[1] the 1763 brochures of J. Short,[2] and the many further publications — whether positive or negative —inspired by Harrison's great success, such as Baillie lists.[3] But technically important information was provided by the publications of Pierre Le Roy (1770) and Ferdinand Berthoud (1773) noted in the previous chapter.

A thorough study of the many writings available today, beginning with the documents of the Board of Longitude and including the numerous monographs and biographies concerning important chronometers and their builders that have appeared in the most recent years, plus the many comprehensive reports, makes clear the fast spread of chronometers in seafaring at the very beginning of the fine mechanical era. This is undoubtedly attributable to the Board of Longitude's activities, which may well have contributed more than the publications already mentioned. Both the assured lack of bias on the part of the Board and its establishment of valid testing regulations for the chronometer as well as the test procedures, plus its granting of prizes for special achievements all had a favorable effect. So did their concern for the long-term operation of chronometers in maritime service, and their constant contact with the various chronometer builders was a steady stimulus to good workmanship and progress.

To go more deeply into the technical details and activities of the fine mechanical era, and in order to be able to understand and evaluate correctly the motivations that led to the progress made all through the 19th Century, it is useful to start from the problems of chronometer building already known in the heroic epoch and the first solutions of them.

The situation at the end of the heroic epoch can be evaluated as such:

a) It was shown to posterity by Harrison's H4 that it was possible for a spring-driven clock to be taken on a six-week voyage to America, keeping the time at the point of departure so precisely that even after six weeks a sufficiently accurate determination of longitude could be made.

b) Thanks to the invention and technical development of an escapement for seagoing clocks — the chronometer escapement —it had been possible to replace the extraordinarily refined arrangement of diamond pallets used by Harrison with robust, relatively easily built and economical systems such as Arnold's pivoted detent and Earnshaw's improvement — the chronometer spring-detent — that afforded better release of the locking jewel.

c) The further existence of the Board of Longitude as a permanent institution with the job of being a test center for new designs and a driving force in the use of chronometers in shipping in general.

d) The recognition that two things were essential: automatic temperature compensation devices for the chronometer balance and the development of simple balance designs.

Earnshaw's invention that replaced the original riveted bimetallic strip with a fused iron and brass strip was a decisive step.

The introduction of a unified adjusting system for the chronometer, based on the need for it to work as accurately as possible with minimal running errors, at both high and low temperatures. The adjustments were thus made in winter below the freezing point and in summer when it was very hot.

f) The recognition of the great importance of the configuration of the balance spring, based on Le Roy's determination that only one single spring length can afford isochronism for a given spring profile at varying amplitudes, a discovery that inspired numerous subsequent discussions, experiments and developmental work of a metallurgical as well as a theoretical nature.

g) The fact that in the 18th century British clockmaking, especially in the realm of the pocket watch, had attained a height well above those of all other clockmaking nations. The reason for this was not only the wealth of ideas in terms of design, such as the invention of the lever or cylinder escapement, not to mention capable workmanship, but also the high quality of work in the assembly and component factories. This includes the brass, the steel of the pivots, and especially the escapement and balance springs — particularly important components — which could be produced at constantly high quality only by the British manufacturers in the 18th Century. Finally, it was also the unique experience in the production of jewel bearings, invented by Facio in Britain, which was a British monopoly for a long time. Another important factor was the good training of the watchmakers. We know little of such details as where the good materials, especially the good springs, were produced; it must have been particularly in the workshops of central England. We likewise lack information on the methods and formulas used in the manufacturing and testing of the springs.

Because of the advances made in the navy as well as merchant shipping at that time, the need for marine chronometers was remarkable. The good conditions in Britain already portrayed also led to the great increase in British chronometer production. Two basic necessities were decisive in the process: first, keeping the price of producing a chronometer so low that even small ships could be equipped with them, and second, making the chronometer as reliable and durable as possible. It had been recognized very early that use in ships exposed to rough seas tends to decrease the quality of chronometer performance. It was soon seen that balance and escapement springs wore out, lubricants thickened and solidified in the bearings, and that mechanical changes of other parts took place as well — including loosening or displacement, as well as metal fatigue.

Attempts to maintain the original quality of operation and to improve the precision of the chronometer by making the daily errors and their fluctuations smaller, motivated the men to make fine mechanical measurements and thus establish the importance of fine mechanics in this era.

In order to bring the detailed problems recognized in the heroic epoch nearer to a solution, the clockmakers now proceeded along predominantly fine mechanical, empirical and intuitive lines. After relatively primitive considerations, the designs of the critical parts were varied, based on the results of performance in practice, first in testing and then at sea, and compared to expectations. The consequent observation over long periods of months and years, and the possession of actual long-term records, were necessary prerequisites.

How much interest the general public took in chronometer problematics at the beginning of the 19th Century in Britain is shown by an 1806 report,[4] or can be seen in the number of pages devoted to this subject in the Rees lexicon.[5] The latter is a collection of excerpts concerning problems of handicrafts, and the section on clocks covers 295 pages in all, including 40 pages on chronometers in general, ten pages on temperature compensation, 34 pages on escapements and twelve pages on remontoirs. There are also 55 good technical drawings that make the designs of those times easier to understand. Illustrations include those of refined and very complex fine mechanical designs.

It is not easy to evaluate fine mechanical work realistically. To the layman and the beginner, the quality of fine mechanical work will consist primarily of the neatness and high polish of the eye-catching surfaces of plates, bridges and mountings. This is, in fact, important to the quality of the actual functioning of the chronometer, since the problem of the lubricant's tendency to spread or stay on the lubricated parts of the clock plays a role. The effect of surface polish on the spreading of the oil, and thus its importance in maintaining effective lubrication, was recognized early.

Among other factors, the precise centering of the wheels, pivots and arbors, their degree of polish, the evenness of the gearing and gear form in the wheels and pinions and their fitting together, the saving of weight in all moving parts, and the evenness and steadiness of the springs (for both the power and the balance system), are all qualities whose evaluation is scarcely possible from outside and without special measuring procedures. But it was just these qualities that constituted the worth of products made by individual masters and their workshops; they also provided the basis for evaluating certain products according to their maker's reputation. The reinforcement subsequently provided by the demanding strictures of the official test centers and observatories generally agrees with the traditional means of evaluation.

To achieve progress in terms of durability and precision, it was necessary to proceed on a number of lines of fine mechanical development. These will be treated below in terms of their main tendencies.

The Middle Temperature Error of Bimetallic Compensated Balances and Efforts to Eliminate or Decrease It

In the heroic epoch it had already been recognized clearly that the system of *balance with spring* was the heart of the marine chronometer, but that the precision of this system was very much influenced by temperature changes. Temperature compensation of the oscilating system by external elements soon proved to be unfavorable, showing the need to move the temperature compensation to the balance itself. The production of the balance ring of bimetallic strips according to Earnshaw's method immediately proved to be an excellent improvement in lessening the sensitivity of the rate to temperature.

After the crudest effects of temperature had thus been eliminated, a further effect of temperature made itself known. Even when the marine chronometer was set up in the captain's cabin the temperature on a cold winter night in the 18th Century, without any particular means of producing heat, could sink well below the freezing mark. In tropical waters in midsummer it could rise to and even above 40 degrees Celsius. This temperature variation in the environment in which the chronometer had to work thus went well beyond the two boundaries to which the chronometers were subjected in the testing process and brought to light another, unexpected characteristic of them. The rate slowed strikingly in both areas, both over and under the temperatures for which they were adjusted for optimal running, and as the temperature went farther beyond the borderline, the rate decreased more rapidly, while it increased between the boundaries. This phenomenon, known today as the middle temperature error, was found particularly in expensive chronometers in which chance variations were largely avoided. Depending on the temperature variation of the chronometer's environment, it amounts to changes in rates in terms of one to several seconds per day.

When and how the phenomenon of middle temperature error was first recognized is not clear, despite numerous studies and controversies. It probably became known first to those chronometer makers whose chronometers did not run satisfactorily at varying temperatures that they were dealing with a basic, typical error that was related to the way chronometers were made.

Thus the earlier cases cited in this regard by Commander Gould can probably be understood in these terms. For example B. W. Hardy's "permanent compensation balance" (#71) of 1804, (which can be regarded as the first treatment of the middle error, as the weights on their carriers parallel to the balance pivot move inward faster at high temperatures than they move outward at low temperatures), and also those of the complicated middle grid system of J. G. Ulrich, made in 1814. The middle error was definitely known by about 1820, but the final understanding of the situation was complicated and unclear. Around 1828 the British chronometer maker John Sweetman Eiffe developed special compensation devices for the balance and had them tested at Greenwich. The director there, Astronomer Royal Sir G. B. Airy, then published Eiffe's descriptive manuscript in 1842, including his technical considerations and the practical application of them. But Airy also included in this publication a patent document by Robert Molyneux that proceeded in the same direction. The latter described what were probably independent but extremely similar developments in content to those of Eiffe, and showed measures taken (#72) to eliminate the middle temperature error. An essential part of both elements is the inclusion of additional components to the ring-shaped balance with bimetallic strips of the Earnshaw type, which affect the bimetallic strips when a certain temperature is exceeded and thus cause a change in the moment of inertia of the balance. This official publication by Airy attracted the attention of many other chronometer makers and inspired them to action in the same direction, so that soon numerous new suggestions for special balances with additional compensation devices to deal with the middle temperature error came forth, most notably a patent by E. J. Dent.[7] In it he refers to his work with temperature-compensated balances since 1829.

The middle temperature error is of great importance for the whole fine mechanical epoch to the extent that it inspired the great majority of chronometer makers to seek new special solutions for the balance and brought forth solutions of very different types, sometimes brilliant ones whose conception and practical application include much to inspire one to become acquainted with the chronometers of the fine mechanical epoch.

The principle of all fine mechanical designs to decrease the middle temperature error is as follows: No matter in which direction the borderline temperatures are exceeded, the moment of inertia of the balance must be decreased to make the clock run faster. The application of additional bimetallic systems or other temperature-dependent mechanisms to carry out constant or momentary changes in the moment of inertia is obvious.

Illustration 73 shows the numerous special designs illustrated and described by Commander Gould; the caption includes brief references to their particular details, while #74 supplements them with a series of special designs for eliminating the middle temperature error. In retrospect it can be seen that they scarcely fulfilled the hopes pinned on them, and that thus only a few of them, such at that of Kullberg, have been used in large numbers over a number of years since they all involved considerable difficulties in terms of manufacture and maintenance. In descriptions of the chronometers displayed at Greenwich, the comment is often found: "An imaginative design, but very difficult and only adjusted correctly by a few specialists."

A basic error of all added mechanical components is,

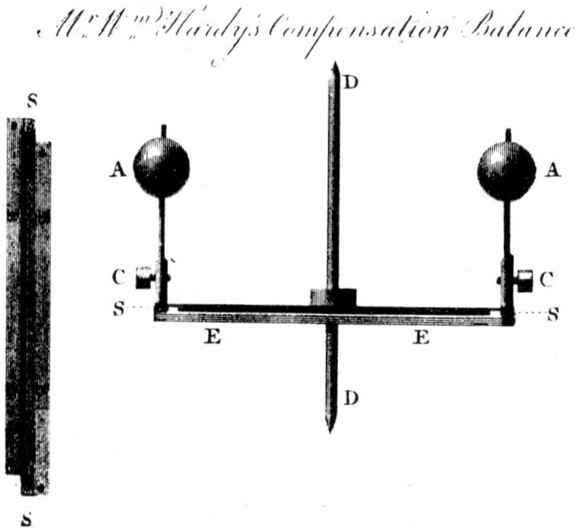

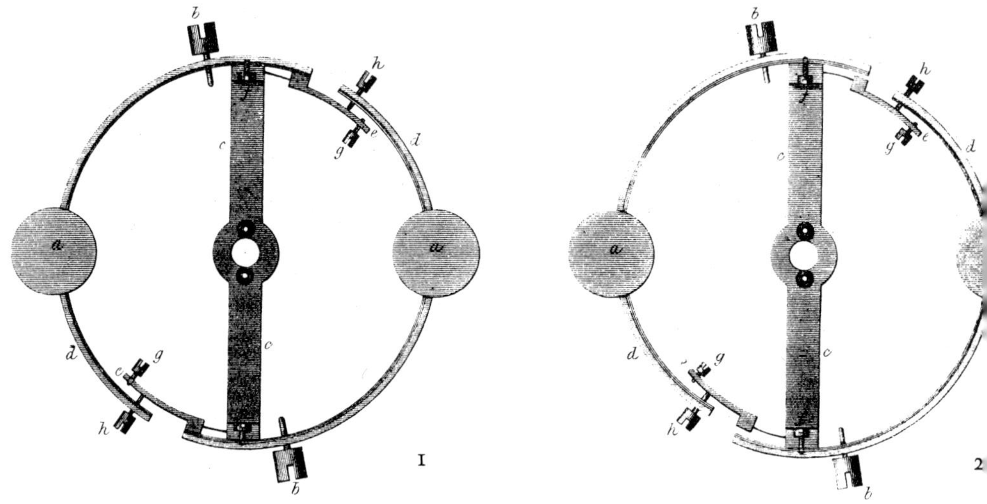

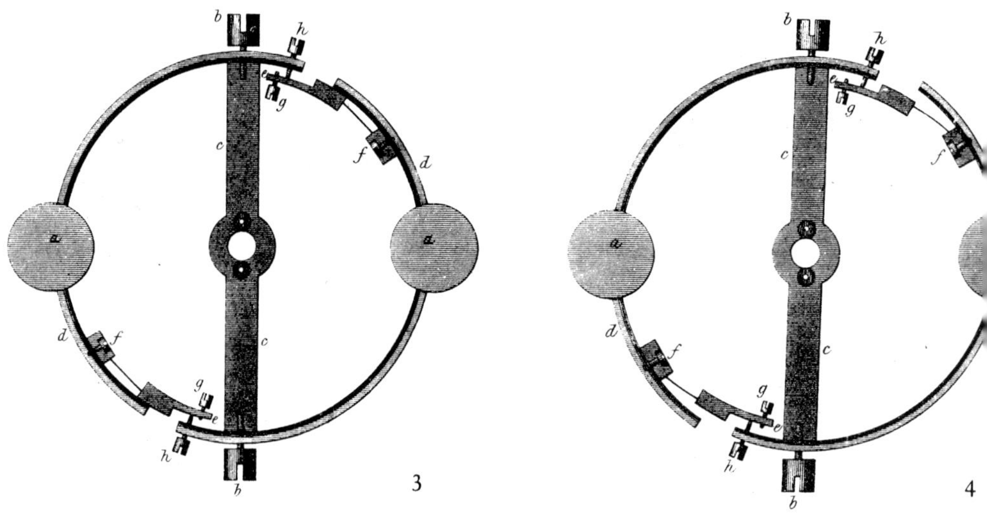

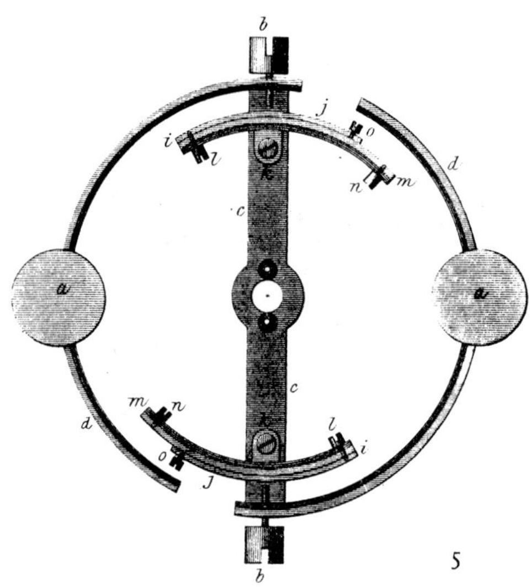

71 W. Hardy's balance principle (1804) to decrease the middle temperature error: The two weights A are carried by the lower part of the sprung crossarm EE; the upper crossarm part SS lies loose on EE and, depending on the temperature, the arms carrying the balls are bent more or less in the plane of balance arbor DD and crossarm EE SS.
After Glasgow: Watch and Clock Making, London, 1885

72 Molyneux' balance designs — from a British Admiralty publication of 1842 on the reduction of the secondary temperature error with additional components: No. 1 and 2 show the added elements on the short, no. 3 and 4 on the long sections of the ring; in no. 5 the elements are attached to the crossarm of the balance. No. 2 shows the influence of an inward bending, no. 4 an outward one; with sharp inward bending of the bimetallic arms (at higher temperatures), more weight is moved toward the center and the moment of inertia is reduced more strongly than through the mere moving of the bimetallic arms.

in fact, a limitation of the balances' strength and stability over long periods of time.

The usual conception of the cause of the middle temperature error that prevailed well into the 20th Century is shown in #75 and described in its caption — the varying course of the moment of elasticity in the balance spring relates to the moment of inertia of the balance. By shortening the latter, very precise functioning can be expected in only two points — the descending points of intersection of the two characteristics of the balance spring and balance. Between the two points fast running results, and beyond them slow running is unavoidable. This is the middle temperature error. Although the diagram of #75 agrees with the practical experience that only working points are possible for a zero running rate, it is still an artificial diagram, because the basic conditions for the actual behavior of the balance spring and balance in temperature changes do not apply, according to the later research of C. E. Guillaume.[8]

In Chapter 7 *(Mechanical Perfection via Modern Metallurgy)* the metallurgically controlled influences of temperature on the expansion and elasticity coefficients of the balance and balance spring materials are treated specifically. When he accepted the Nobel Prize for Physics in Stockholm in 1920, Guillaume spoke of attaining a metallurgically correct explanation for the cause of the middle temperature error on this basis. His diagram (#76) anticipates the discussions of this subject in Chapter 7, as it and its caption show that the functional behavior of the two critical parts — the balance spring and the balance — occurs, because the quadratic behavior of the material affects the balance spring, while the bimetallic system ultimately reacts in a linear manner on temperature changes. This is the exact opposite of what was assumed formerly.

It is not surprising that the lack of insight into the thermo-mechanical behavior of materials used in the components of the oscilating system led the chronometer makers of the fine mechanical epoch to combat the middle temperature error with the means at their disposal, as can be seen in the designs of many 19th-Century chronometers. A report by Corvette Captain Rottok[9] can be cited, who notes particularly in a study of the practices of chronometer testing in several navies that at the time at which he made his study, almost all English chronometers had balances with additional compensatory devices.

In connection with the variations of balance and chronometer construction, it should not be forgotten that various adjustable screws and weights were used in attempts to attain the correct balance and moment of inertia. Thus there are many variations for the purpose of simplifying the regulating and adjusting problem of these elements and, at times, of decreasing the air friction via these elements.

Equalizing the Moment of Power with Spring Drive

Deriving the power of a clock movement from a wound-up coiled spring seems to be a simple, problem-free solution to the layman. But if one asks how even is the power that a spring provides, one immediately faces a remarkably difficult problem, insoluble through analysis. The turning moment that a spring provides is not at all constant and does not even have a simple relationship to the number of windings that can be expressed in a law. The characteristics of the spring steel used, its surface condition, and its state of lubrication at any given time, play important roles. In addition, the behavior of springs that operate in a barrel (spring housing) is inherently different from the behavior of a coiled spring that is not enclosed. It is appropriate at this point to remember that the spring, when fully wound, forms a practically closed ring around its center (#77a) and, after running down, forms a similar closed ring in the barrel (#77b). Thus a change of power or moment occurs such as is basically shown in #78: At the beginning of winding, the turning moment increases strongly, but then increases only slightly after several windings, and at the end of winding it shows another strong increase.

It is particularly important that the characteristics are influenced, within considerable limits, not only by the material but also by the way it has been worked. If the oscillating system is driven directly from the spring, or from the barrel in which it is located, then varying amplitudes, corresponding to the varying moments, must be expected. This is in fact the case, and since the period of very many oscillators depends on their amplitude — the oscillator is thus not isochronic — then the variations in the power of a spring drive are one of the major causes of imprecise time-keeping in spring-driven clocks. Attempts to eliminate or at least limit this disturbance are as old as spring drive itself. One obvious way was to conduct the power from the barrel via a chain and fusee, whereby the curve of the fusee must be dimensioned so that it corresponds to some extent to the spring characteristics (#78). This means that for every individual spring there must be an equalization of the fusee with the spring (or vice versa), and it is not possible to use any spring that happens to be available as a replacement. Although the use of the fusee made the movement a good deal more expensive, systems with fusee and chain have been used preferentially, so that good springs could remain in service for many years.

It is appropriate at this point to mention the other ways that are or were employed to act against the change in spring power during its running down:

A) Spring brakes (stackfreed), in which a curved cam affected a fully wound spring more strongly than later. This primitive equalizing apparatus was originally developed in southern Germany for use on watches worn on the body, and was used frequently until the beginning of the 17th Century.

B) Attachments (partial or complete stops) to the barrel that allow only a limited part of the total possible length of the spring to be effective. According to the number of windings and the limitations caused by the stopwork, one can limit the moment to a greater or lesser extent over the effective extent of its functioning, as a comparison of the diagrams in #78 shows. This simple method is still used today, generally in expensive watches in which the oscillator is made as isochronic as possible, so that the period remains as constant as possible despite the changes in amplitude that occur.

73 Various attempts to eliminate or reduce the middle temperature error with fine mechanical designs; each balance design is accompanied by its inventor's name and the approximate date of publication.

The designs shown in 73a-f work by replacement, that is, additions, after going beyond certain bending changes, cause a different bending relationship and thus a different degree of change in the compensation.

The additions with adjusting screws are usually attached to the crossarm of the balance, sometimes (in the case of Kullberg) only to the encircling rim of the balance. According to whether the additions come into play when they bend inward (at high temperatures) or outward (at low temperatures), the rate is slowed at high or at low temperatures and hopefully evened out. An analytical description scarcely seems possible, though, all the more so as the stiffness or springiness of the added systems with the screws could have further influence. In retrospect, one cannot help but doubt that any of the inventors could imagine a quantitative effect of his design at different temperatures. The idea seems to have been a very intuitive one, that the effect of unlinearity would counteract the disturbing temperature influences within certain limits and approach an optimum through the use of adjusting screws.

The designs shown in 73g-o give constant influence in additional compensation, above and beyond the primary bending changes of the bimetallic balance.

In the constantly working additional systems too, the difficulty of a quantitative description of the effect of additional compensation, as noted above, appears to apply, so that here too the achievement of a greater or lesser success must be regarded as mainly a result of the maker's talent.

The mechanical concepts of the inventors on which the individual designs are based were described thoroughly by Commander Gould in 36 pages of his book. Only the balances of Kullberg's later design, as of 1860, were successful in practice. See #306.

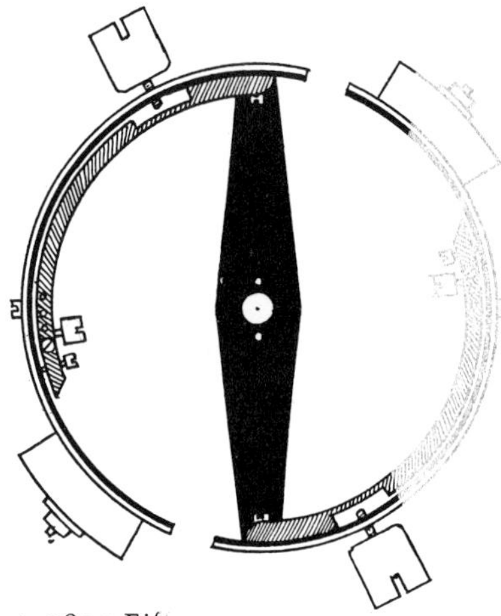

a 1835 Eifie

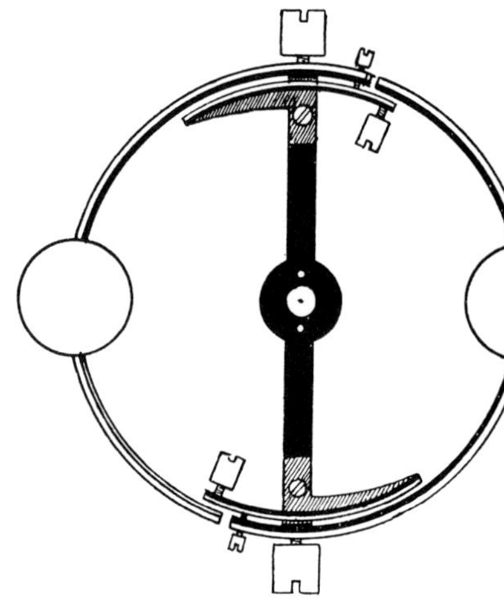

d 1860 Mercer

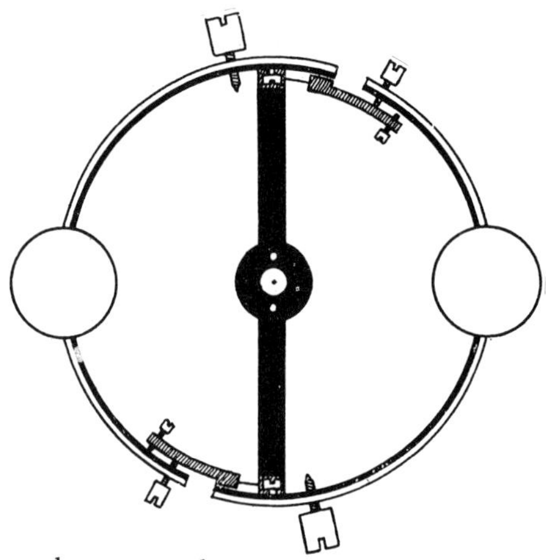

b 1840 Molyneux

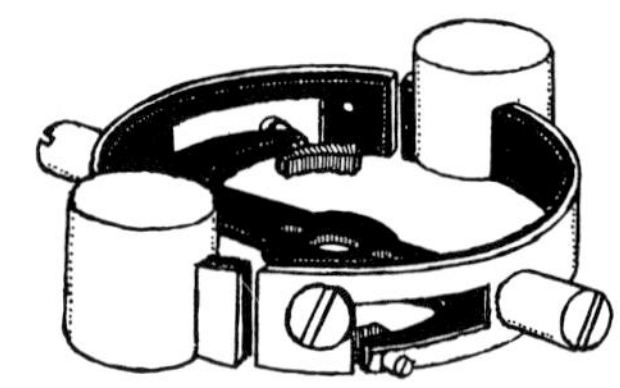

e 1860 Kullberg

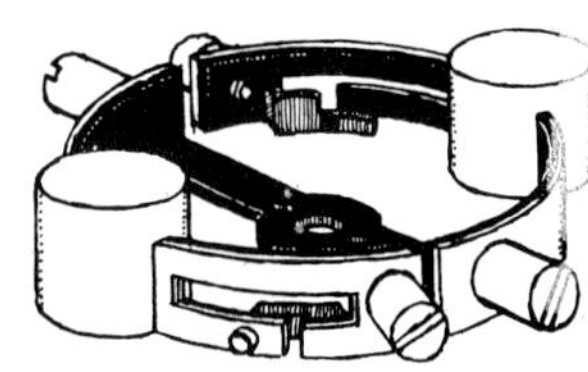

f 1870 Kullberg

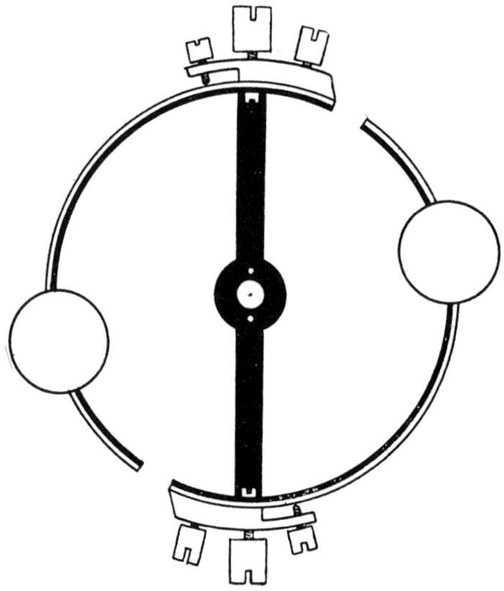

c 1845 Poole

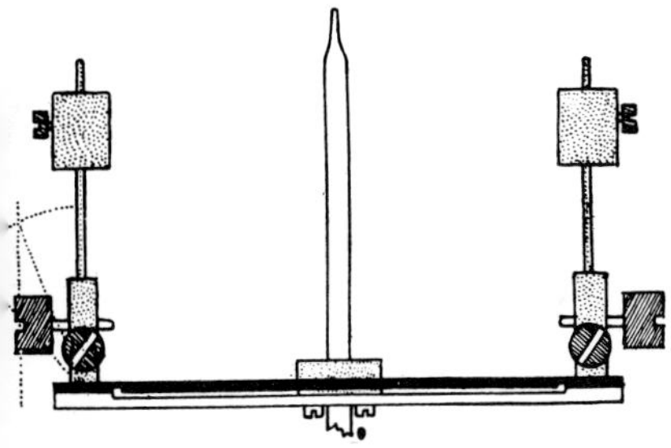

1804 Hardy

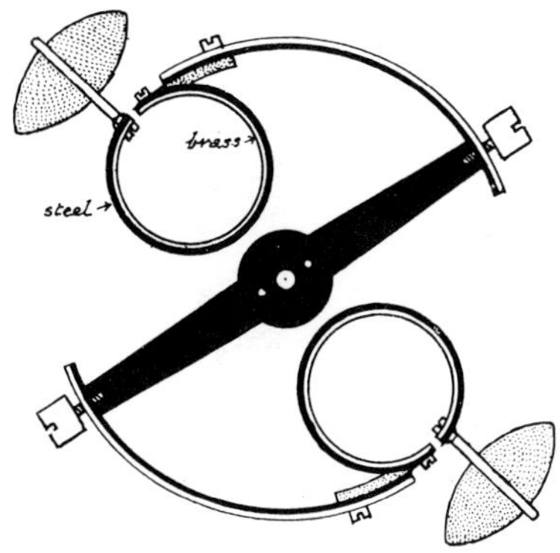

j 1840 Ulrich

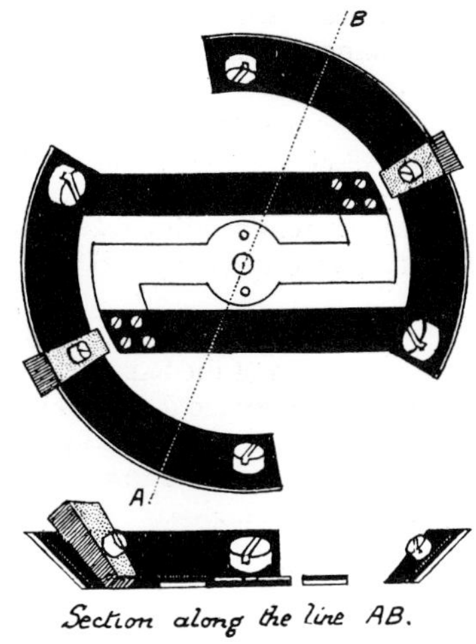

m 1845 Hartnup

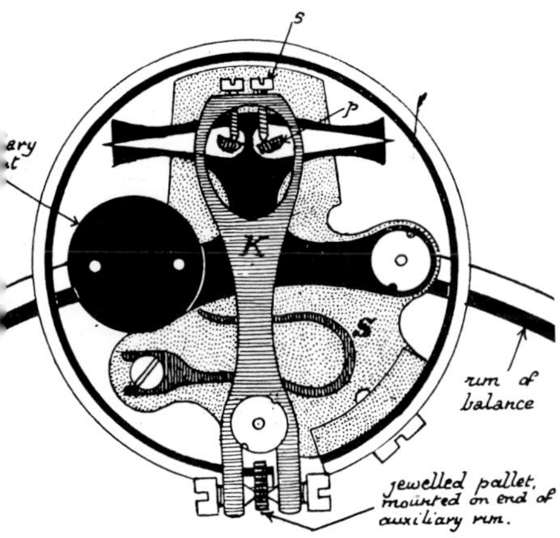

1810 Barraud

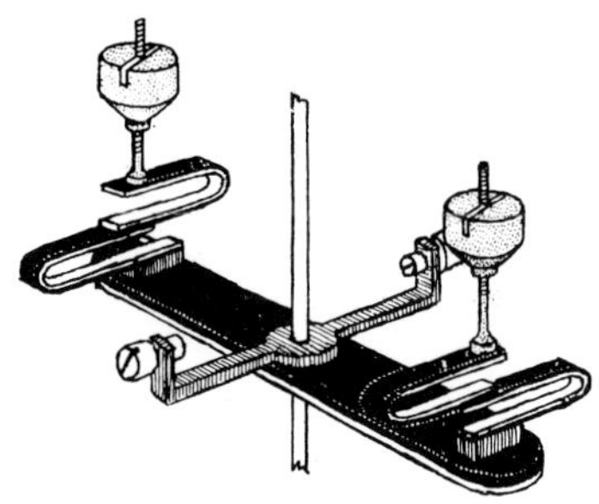

k 1842 Dent

n 1850 Dent (Querschnitt des Unruhreifs)

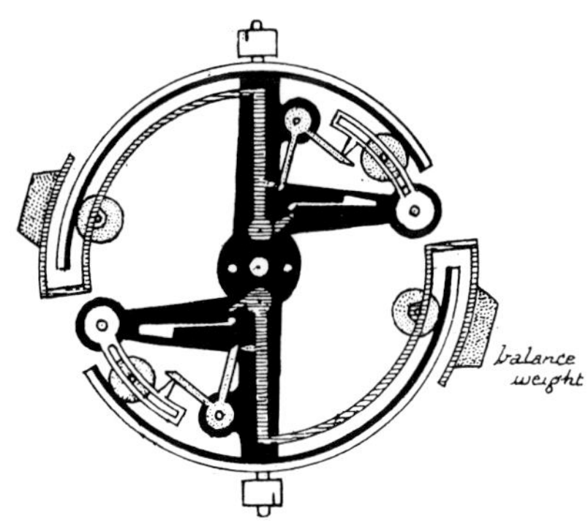

1815 Ulrich

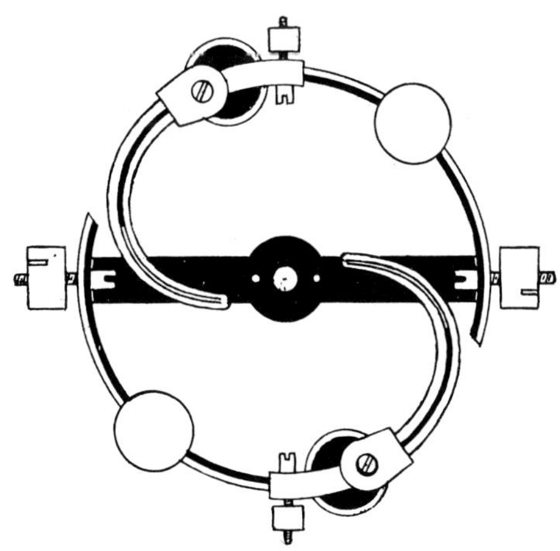

l 1843 Loseby

o 1855 Kullberg

74 Pictures of particular balance designs to compensate for the middle temperature error of bimetallic balances.*

a) Compensated balance after William Hardy, ca. 1821; the two horizontal bands consist of two overlapping loose strips of brass and iron, linked with the weights above so as to move them in or out according to the temperature. Arnold also used the same principle along with circular bimetallic balances.

b) Balance with U-shaped bimetallic strips (staple balance) to eliminate the middle temperature error — Dent's patent 9302 of 1842, intended to offer a particularly precise change in the moment of inertia.

c) Compensated balance developed by Hartnup, Superintendent of the Liverpool Observatory (ca. 1845). This arrangement is supposed to move the compensating weights inward more quickly with rising temperatures and outward more slowly with lowering temperatures. In addition, circular elements that could easily be turned on a lathe were regarded as a practical advantage. a- view from above; b- from below.

d) Compensated balance by John Gottlieb Ulrich, London. Ulrich's principle was developed within five decades and often varied. Here is a later version in which the equalizing weights on the balance move farther away from the axis of the system at high temperatures but nearer at low temperatures. This theoretically correct idea is hard to put into practice, though, is not easy to adjust, and has proved to be not stable enough in practice.

e) Experimental balance with 8 spaced radial arms extending from the balance ring; probably a stable design but surely involving the disturbing influence of air resistance. *(T. Mercer)*

f) Bimetallic balance with secondary error compensation after E. Pike (ca. 1860), a further development of the Eiffe-Molyneux principle.
(T. Mercer)

g) Bimetallic balance with secondary temperature error compensation after Thomas Mercer (ca. 1887).
(T. Mercer)

h) Balance design by Frodsham (ca. 1865), round balance ring with two bimetallic strips over it, which bend according to temperature and thus swing the weights attached to their ends inward or outward.
(T. Mercer)

i) Compensated balance by Frodsham, London (ca. 1871), with two bimetallic strips as in (h) but without round rims.
(T. Mercer)

k) Compensated balance (incomplete), further development of the types shown in h and i; the added weights are missing from the two screws that project upward.

l) Balance with simple adjustment, known as Airy's bar (ca. 1871), clever design for fine adjustment by moving the two sectors. This design was required by the British Navy for some time but did not prevail for lack of stability.

m) Compensated balance by Winnerl, Paris (ca. 1876), essentially a combination of Arnold's and Hartnup's ideas. The hoped-for theoretical advantages did not work out thoroughly because it was not possible to make the structure strong enough.

* All pictures whose origins are not stated are of models now owned by M. Paul Rudolf, Triel-Sur-Seine. They were submitted by the chronometer suppliers to the French Navy for testing in connection with governmental longitude measurements.
(Photos by G. von Voithenberg, Munich)

a

b

c

c

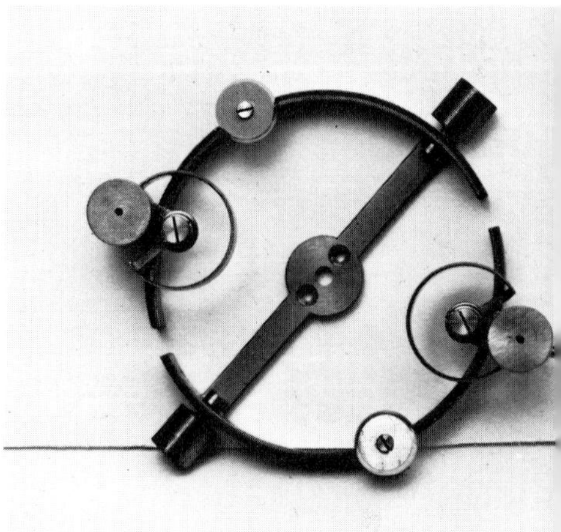

d

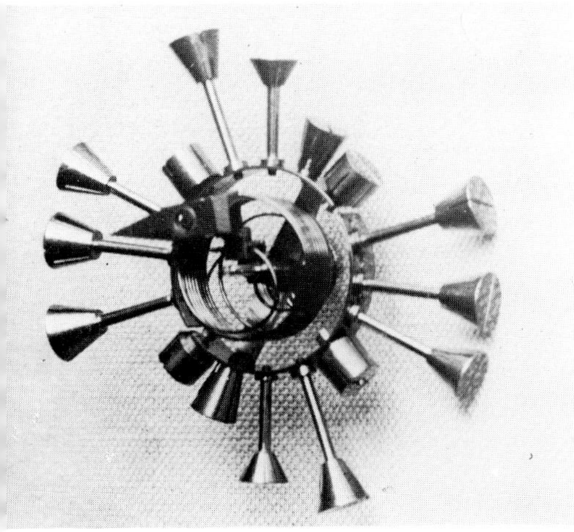

h

k

i

l

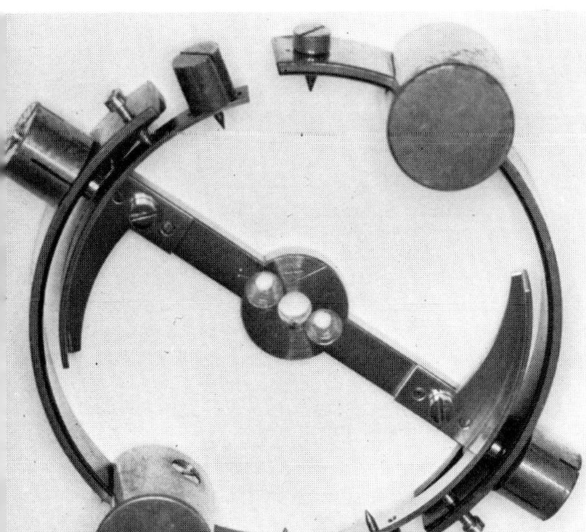

i

m

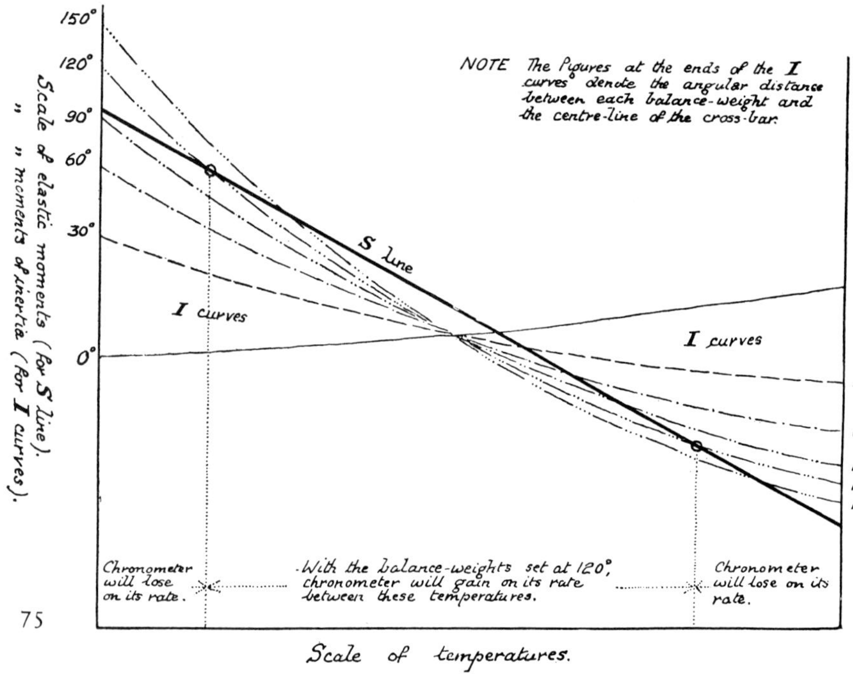

75

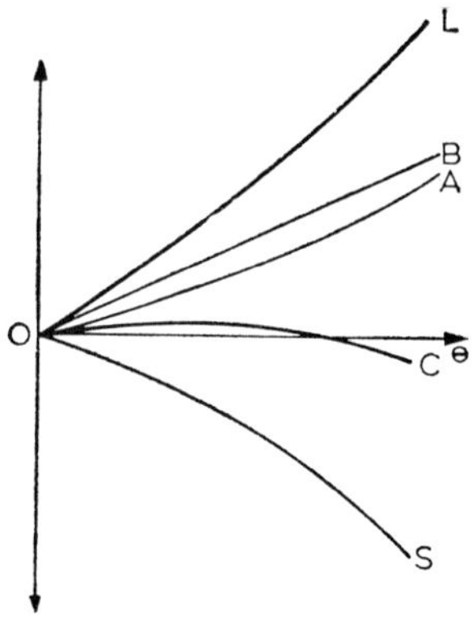

76

75 Drawing illustrating the concept of the origin of the middle temperature error prevailing into the first quarter of the 20th Century. Above the surrounding temperature ⊖ are both the extent of the elastic spring moments S and that of the moment of inertia I of a bimetallic compensated balance; the latter correspond to the curves and can be determined by the angles of the weights to the balance arm, stated in degrees.

The two crossing points between S and a certain one of the I curves represent the two temperatures at which the balance assembly should work without error. If, for example, the weights of the balance are set at 120 degrees, then the chronometer should run fast in the area between the two dotted vertical lines X and Y, but slowly outside them.
(after Commander Gould, p. 178).

76 Guillaume's diagram to explain the middle temperature error of chronometers with bimetallic compensated balances after Earnshaw. S, the extent of the modulus of elasticity $\Delta\varepsilon$ of the steel of the balance spring, increases with ⊖ and includes a component that increases with $\ominus^2$. The resulting loss of rate is thus weak at first and increases faster only later. The expansion coefficients Δ of brass (L) or steel (A) in the bimetallic ring of the compensated balance include both linear and quadratic parts, but the later are practically the same in brass as in steel and thus cancel each other out; thus there results a linear loss of temperature for the moment of inertia of the bimetallic balance corresponding to the line B and a corresponding gain in rate.

At O it is assumed that the moment of inertia and elasticity coefficient of the springs provide a correct rate. Since with increasing ⊖ , at first $\Delta\varepsilon$ decreases more slowly than ΔL increases, therefore at first the rate increases, but later this changes to a rate decrease.
Guillaume, Nobel address, 1920

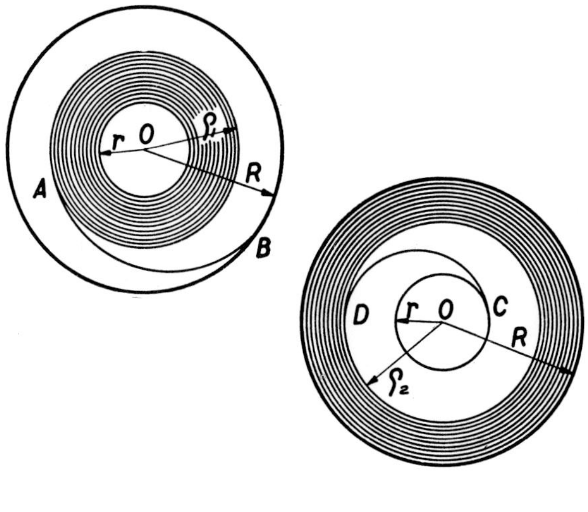

77 The positions of the drive spring in the barrel: a) fully wound, ρ_1 ..radius of barrel arbor + spring; b) unwound to radius of ρ_2; o axis of barrel, r radius of barrel arbor (chosen as 1/3 of R on basis of experience) B, C attachment points of spring to barrel and barrel arbor. The best spring development is attained when the volume included by the wound spring equals half the free volume between the barrel and arbor.
(after "Theorie Generale de L'Horlogerie" by L. Defossez, Vol. I)

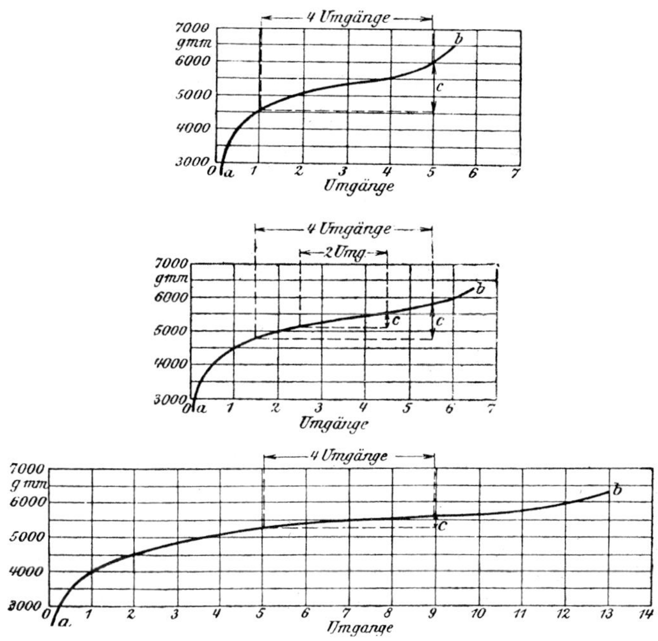

78 Extent of turning moment of main springs housed in barrels; full winding and unwinding extend between a and b.
1 Spring with 5.5 turns, fully wound; when limited to 4 turns, a change in moment of +/-14.3% is to be expected.
2 Spring with 6.5 turns, fully wound; when limited to 4 turns, a change in moment of +/-9.4% is to be expected, when limited to 2 turns, one of +/-3.7%.
3 Two springs of type (2) in series: when limited to 4 turns, a change in moment of +/-2.7% is to be expected.
(from A. Helwig, Drehganguhren, pp. 56-57)

C) Two or more main springs are made to work together, which can be done by either parallel or serial linkage — causing a problem that, despite its importance in chronometer development, has been studied relatively little to date.

In the interaction of two or more springs, one can attain a greater evenness of power and limit the danger of spring shocks to the movement by brief stops between individual windings of the spring, and the low cost of thinner springs can also be made use of.

By dividing the stored power among several barrels and the simultaneous use of longer and thinner springs, the whole path of the spring is probably increased considerably; and the possibility of using a barrel arbor of larger diameter is also provided, thus lessening the danger of spring breakage. But to attain very even power one must be able to make adjustments by appropriately locating the beginning and end positions of the spring characteristics, as shown in #78; here too, only a part of the entire path of the spring can be utilized. Of course the space and work needed for multiple springs are also larger and more expensive than for a single spring.

All in all, though, the use of two or more springs has considerable advantages:
1. Reduction of the high point of power by dividing the total power. 2. Reduction of disturbances in power caused by short delays between individual windings. 3. Reduction of the danger of spring breakage by using thinner springs and a greater arbor diameter. 4. Reducing friction. 5. The possibility of producing flatter movements.

The great majority of all multiple spring drives utilize parallel spring functioning. Only in England did Thomas Mudge, as noted in Chapter 3, use two springs working serially (one after the other) in his 1774 marine chronometer — probably for the first time. Leading French clockmakers such as A. Janvier and F. Berthoud had already used two parallel-working springs in regulators at about the same time. After the French Revolution and particularly in the first half of the 19th Century, two parallel springs were often used, especially in chronometers, as shown in the designs of A. L. Breguet, L. Berthoud and H. Motel in the illustrations.

The fact that these outstanding clockmakers often used two spring barrels leads to the conclusion that the French spring manufacturers of the time made remarkably good springs, whose performance when running down was more favorable than those made in many other lands. This complex question has scarcely been dealt with to date, since many of the old springs have been lost or weakened and replaced by modern springs that were not made in the original manner. It appears that the original French springs provided a particularly smooth and even running, and thus made particularly quiet functioning possible. The author's own experiences with a year clock made by Janvier, that requires an unusually high number of windings, and in which the springs had been replaced long before the clock came into the author's possession, indicate that the original springs from the early 19th Century gave a more even performance.

In all clocks with spring drive, the springs themselves — generally coiled strips — are of major importance in terms of durability and precision timekeeping. The problems of drive springs have been discussed remarkably little in the literature. The reason for this is probably the lack of elementary, easily described relationships between geometrical dimensions such as cross-section form and spring length and the operating potential of a specific spring between being fully wound and fully run down, the material (usually steel, though of very varying composition), which possesses characteristics of elasticity that are hard to determine because of the variety of preparations.

A very worthwhile approach to the complicated relationships between the qualities of drive springs already noted and the operating potential theoretically attainable from them was taken in a clock-technology colloquium at the Technical University of Stuttgart in 1966 by Professor Glaser, who later expanded it into a book.[10] Some additional factors noted (some of which have considerable retroactive effects): the varying behavior of open springs as opposed to those in barrels (spring housings), the convexity of the spring cross-sections, the presence of inner tension, the influence of the rounding of the cross-section in the course of the bending that changes as the spring works, the type of spring attachment at its ends, and the very important friction as the spring runs down.

An exact measurement of spring dimensions is not at all easy, and is to be regarded primarily as a scientific special task for special cases; but with the numerous diagrams provided by Glaser that were based on practical experiments, it is not difficult. One must proceed from the quantities that were formerly taken into consideration exclusively for determining spring dimensions: length, width, strength, plus arbor and barrel diameters, to calculate the attainable turning moment, its decrease as well as the spring energy and the usable number of coils. Certain physical-technical values of the spring material and the friction to be expected must be taken into consideration. The latter in particular introduce a considerable scattering into the data, as the friction can easily change, depending on the type of lubricant, the surface conditions and the degree of winding of the spring at any given time, at a ratio of one to four. The information gained, though, is of considerable value in evaluating a mechanical system, as it shows, for example, how the uneven changes of the turning moment come to exist according to the degree of winding.

The complicated problematics indicated here make clear that over the centuries clockmakers, and especially chronometer makers have depended on obtaining good springs from spring specialists without setting particular testing criteria. In this respect, the use of two or more barrels in the fine mechanical period seems especially significant.

Since relying on unknown spring sizes for clocks of high value means a high degree of uncertainty for the maker and user alike, it is understandable that other ways of becoming independent of the unevenness of spring quality were sought. But all of them resulted in complications and were thus used only to meet very high demands. With proper adjustment, they did provide good results, but the added complications in construction prevented the proliferation of these technically interesting systems in greater numbers.

Structure of the Movement and Spring Barrel

The movement of a chronometer is the constructive solution of the challenge of combining the cooperating elements of the movement in one unit. As a development of the mechanical clock, its constructive principles were the point of departure for the development of the various chronometers. On the one hand, those components (as the individual inventors and designers conceived of them), were of decisive influence on the high degree of precision to be attained. On the other hand, adaptations were added to meet the special demands of use at sea. Since there was no experience in this area in the beginning, the test clocks of the preliminary period and the beginning of the heroic epoch show varying, individual constructions, as shown by the illustrations in Chapters 2 and 3. By the fulfilling of the requirement of Queen Anne's Act, Harrison's marine chronometer H4 had a decisive influence on the direction taken by the development of the movement. One can understand that the structure of the traveling or carriage clocks of the day, like those of large pocket watches, had movements whose designs showed the trend that was being followed.

Housing a chronometer movement between two full plates, as in a pocket watch, was not possible at first since greater demands had to be met in terms of operation, servicing and adjustment. Some of these demands counteracted each other. Withstanding all forces requires a firm opposed position for the individual bearings and reliable access to them. But it is not only the static pressure of the pivots against the bearings during the individual stages of power transition, but also the dynamic pressure resulting from individual impulses via the escapement onto the balance, and the way the whole power system moves and then stops at every impulse.

The necessity of precise adjustment of the balance to achieve the correct period and the adjustment of the escapement require easy access to the balance and the escapement through correspondingly long distances. But the attempt to keep the wheels and arbors light demands thin arbors, as short as possible, which makes short distances between the bearings necessary, so that these thin arbors do not bend; only thus can one keep the acceleration functioning of the works small. The fulfillment of these opposed requirements of design is attainable through the introduction of separate sets of components, for example, separating the balance with the escapement from the rest of the movement and forming a special set of components with its own plates or special partial plates, a tendency that also proved useful in series construction.

Something similar can also be achieved with cocks or bridges on a main plate.

In this way many variations in construction are possible, and fulfilling the various requirements becomes a more or less architectural, artistic activity. Thus the various structures that make up the movement, with full and partial plates, multi-stage constructions with bridges and cocks and the like, have come into being; all the structures that give the chronometer movement its particular fascination.

Abraham-Louis Breguet was especially active in engineering this structure. His movements show not only a richness of artistic imagination in the utilization of available technology but also reveal his efforts to fulfill certain technical demands by decreasing or eliminating unfavorable mechanical effects or affording the best possible access to the critical components (#156).

In his chronometers, A. L. Breguet often used two parallel barrels, doubtless inspiring many outstanding chronometer builders to begin to use the same two-spring system, especially in the first half of the 19th Century. Watchwords of design such as simplicity and economy in construction could not have advocated the use of two mainsprings. There were important artistic and technical experiences that impelled Breguet to turn away from economic and traditional thinking, and through his imagination and sense of practicality, he developed a very personal style, as shown in #115, 137, 141, 142, 149, 156, 165, 179, 181, 182, 185, 187, 188, 189, 246, 260 and 377. Breguet's imagination in the realm of structural design of the movement led to the great variety of his creations.

Breguet was not exclusively devoted to using two mainsprings in chronometers. There are many marine and pocket chronometers, especially with eight-day movements, in which he used one barrel with chain and fusee. The illustrations show how Breguet strove again and again to form an optimal structure in his clocks and movements, weighing the advantages and disadvantages of the possible solutions against each other, and these efforts had an important effect on the subsequent notable French masters in the first half of the 19th Century, as well as on numerous British chronometer mnakers. Only in the second half of the 19th Century did an increasing simplification in chronometer construction appear, aiming at a simple two-plate structure.

The housing and attachment of chronometers on shipboard was originally done in glass cases or, as with Harrison's H4, by being set on a cushion. Brass cylinders with weight drive were also used for chronometers, such as those of F. Berthoud. To the extent, though, to which flat spring-driven movements became established, six- and eight-sided wooden cases that set on a table were used at first, but soon the typical chronometer box with gimbal mountings came into use. The British chronometer cases usually had a rectangular cross-section, the French often a square type with unusual elliptical gimbals. The workmanship of the cases was always of very high quality, and expensive woods such as mahogany, ebony, and at times oak, were used. To strengthen the corners, metal bands were often inlaid. The use of cubic wooden cases was the rule for maritime use throughout the 19th Century and to this day; sometimes wall or table clocks were fitted with chronometer movements for home or office use, but only in very rare cases were costly cases made, like those made of nephrite jade by the famous jeweler Fabergé. The great majority of marine chronometers had to be wound daily; but there are also a good number of chronometers with eight-day movements. In general, the chronometer movements mounted on gimbals are turned around for winding and then wound with a ratchet key inserted into the chronometer casing. But there are also a number of interesting makeshift arrangements, in which winding is done without tipping the movement, through openings in the case or through the glass over the dial. Finally, some chronometer makers also provided built-in winding systems that could be wound without an additional key.

In pocket chronometers, tourbillon and carousel watches, though, the normal round form of customary pocket watches was taken for granted. During the first half of the 19th Century they were wound with a separate key, while in the second half of the 19th Century, crown winding as in pocket watches soon prevailed.

The Chronometer's Positional Error and its Elimination

In the heroic epoch it had already been seen while testing Harrison's chronometers that the position the clock was in while being tested had a considerable influence on its rate. Such changes in the rate were called the positional error. It is understandable that the position error was reviewed quite early, and the desire was expressed to keep a chronometer's running rate as constant as possible, whether it was at the top or at the left or right side, regardless of whether the movement was standing or lying or of how nearly vertical the position of the XII on the dial was — since observation watches could take a variety of positions in a pocket, the problem of the positional error was of great importance for them.

The decisive step in eliminating the positional error was made by A. L. Breguet, as is described in detail in Chapter 8, "Tourbillons and Carousels."

The Isochronism of the Oscillator

The isochronism of an oscillator is its ability to keep the time of an oscillator always equally long, regardless of the distance of the oscillation. It is important for the chronometer maker to strive for this as much as possible. The reason for the occurrence of varying amplitude is mainly the changing power from the mainspring at different degrees of winding, which is never really constant even with well-tuned fusees; in direct barrel drive, the amplitude of the balance often returns halfway by the end of the period. Other factors that influence the period are changes in friction on account of oil thickening, changes in damping by air resistance at various barometric pressures and in moist air, magnetic influence, etc.

It is particularly the balance spring in the balance assenbly that has a considerable influence on the period. Therefore much thought and care were devoted to the construction of these springs. Probably in connection with Pierre Le Roy's finding that every balance spring shows an isochronic behavior at only one particular length, the chronometer makers and clockmakers of the fine mechanical era devoted a lot of thought to the design of the balance spring. Illustration 79 shows a number of such springs that were suggested for chronometers and precision pocket watches. In the very first tests using spring-powered oscillators as the standard, a great variety of spring types were put to use: R. Hooke is said to have used a leaf spring in 1659; Christian Huygens used a flat coil spring in 1663, for which he can be given priority; in 1676 the Abbé Hautefeuille suggested a cylindrical spring turning in the direction of the cylinder axis. The flat coil spring (#79, 1-3) later became almost the only imaginable type for pocket watches and many other clocks, while helical springs like that in #79, 4 were used above all others in chronometers, though sometimes conical or spherical types (#79, 5-14) were used. The use of conical or spherical springs was an attempt to avoid the danger of two subsequent windings causing blockage through external shocks, which occurred now and then with tightly wound cylindrical springs. The various spring shapes in #79 are, after all, just the results of wondering whether one or another shape would provide better isochronism for the balance. Which particular geometrical factors resulted in good isochronism could only be determined much later on the basis of thorough analytical study, which showed that the transitional area between the actual balance spring and the attachment points on the balance and plate made the difference.

Numerous clockmakers and chronometer makers also worked out the possibility of using other materials than steel for the balance spring, and occasionally used them. The following table provides a list of various materials considered or used around 1880 for the production of balance springs, along with their elasticity coefficients; some of them were also considered for use in bimetallic compensated balances. In seeking and testing various materials for balance springs, J. Dent, for example, used glass springs (#80) as well asgold-plated steel springs (#81).

79 Various designs for balance springs used in the balance assembly of chronometers.
1. Classical, simple form.
2. After Breguet (overcoil).
3. After Cole (double overcoil).
4. Helical form, after Arnold.
Special forms: 5. Berthoud cone, 6. F. Houriet ball, 7. Helix and spiral, 8. Helix and two spirals.
9 through 14: Special forms used by H. Motel, but only the conical forms 12 and 14 gave the desired lasting precision. 1-8 after Gould, 9-14 after Sabrier.

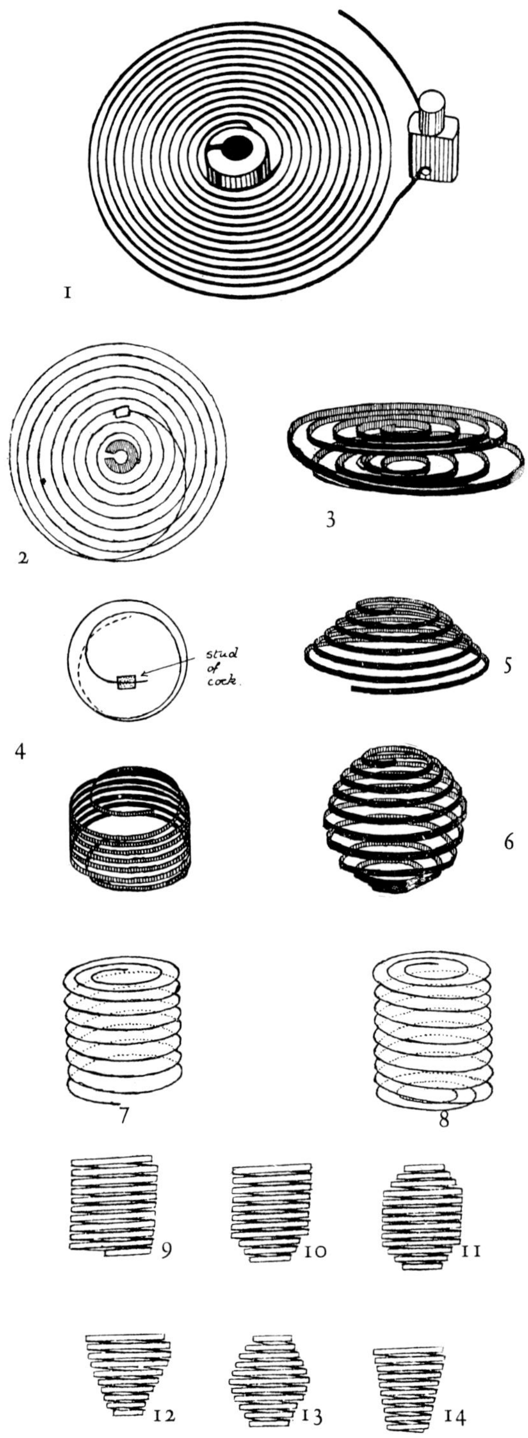

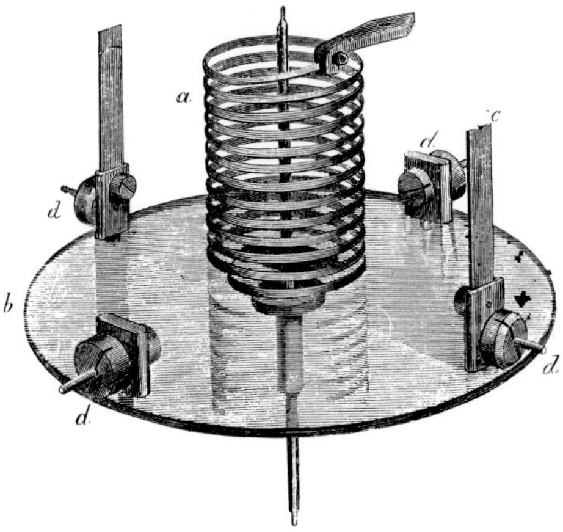

80 Dent's balance of glass with glass spring (ca. 1835-40); the adjusting screws d and two compensating strips c have been attached to the glass plate.
(after V. Mercer)

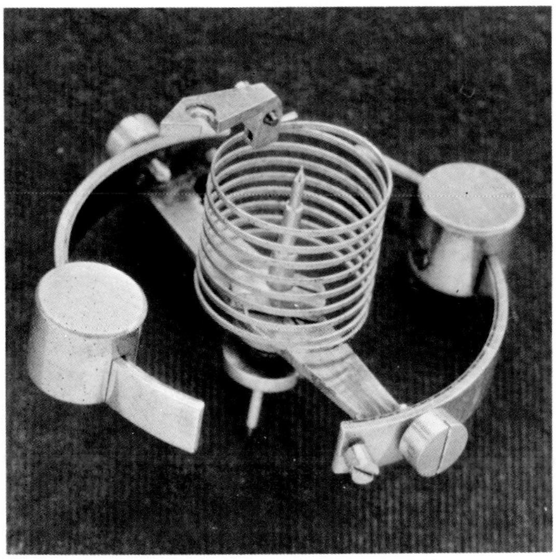

81 Two-armed bimetallic compensated balance with gold-plated helical steel spring, after Dent (without middle temperature error compensation).

Table

Material	Elasticity	Coefficient
Steel	20.5	10^9 kp/m^2
Palladium	14.4	,,
"Geneva Palladium" (alloy)	13.8	,,
Platinum	19.1	,,
Nickel	23.2	,,
Cobalt	21.6	,,
Brass	11.4	10^9 kp/m^2
Gold	8.4	,,
Silver	7.5	,,
Zinc	9.7	,,
Aluminum	7.8	,,
Alu-bronze	13.6	,,

(After Ed Phillips)[11]

Along with the shape of the balance spring, the choice of the material of which it is made and the thermal treatment in its formation play a major role — factors that are discussed only rarely in the literature. The heating, hammering and such were regarded by the masters and workshops as the keys to their quality and usually kept secret.

One of the rare instructions on this subject was provided by H. Motel.[12] Motel used round steel of the needed diameter as a raw material to make it half-flat; he passed it through a small home-built rolling mill, then the steel was cut to the right length and stretched stiff, blued with a red-hot iron and stretched once again. Then it was rolled on a template representing the desired form and finally given its desired spiral form in an iron tube that was filled with a special mixture of about 3/4 animal carbon and 1/4 charred shoe leather. This was heated until red-hot, and the contents were then plunged very quickly into cold water. This process, which is described in greater detail, was supposed to give the steel a particularly enduring hardness.

For the production of spherical springs he used templates made of ten to twelve plates of copper.

The Most Important Masters of the Era of Fine Mechanics

Publishing a complete list of all chronometer makers would go far beyond the boundaries of general information on chronometers, for the number of masters is very great. Many of them worked independently for only a short time, and a great number of names to be found on chronometers are those of dealers or middlemen. For some years, scarcely an auction of old clocks has taken place without the names of several chronometer makers appearing in the catalog, often including some whose names scarcely appear in the literature, or not at all. Valuable sources of names of chronometer makers are the test registers from Greenwich and Kew, hundreds of pages long. In addition, there are the numerous competitions of the test centers listed in Chapter 6. There are numerous research reports of a geographical nature in which chronometers and their use have been cited, but which can barely be mentioned here.

A list of chronometer makers and dealers, including many of the latter who signed their names on dials, is found in T. Mercer's monograph.[13] This monograph shows the great extent to which the use of foreign chronometer parts, and even entire movements, had become the rule before the turn of the century —even by formerly independent masters. In addition to a large number of renowned British chronometer names, on 81 pages of the appendix of T. Mercer's book one finds many well-known chronometer suppliers from all over the world; the USA, Belgium, Canada, Denmark, Germany, Norway, and Sweden etc., who collaborated with Mercer in specified ways. This scarcely makes it possible to organize the subsequent fine mechanical era on the basis of masters or schools, which may help the reader to understand the following, somewhat arbitrary, division into three groups as an attempt at a geographical organization.

The Great European Chronometer Makers

Here are gathered all those names that chance has presented to the author, which may excuse the subjective choice of names. They are mainly names from Britain or France, but a few important names from other European countries have been added, having come to the author's attention during preparations for this book —because they have often been mentioned in the literature, either in their own time or later, because of either their noteworthy achievements or particular publicity.

The dates are those of their birth and death when available; when incomplete data was at hand the time of their professional training is indicated. For the names of some chronometer makers who stand out because of particular achievements, a short biography has been added to the list of their best works, while for most names only the most important dates and literary citations, if any, are noted.

Armand: Father (Jean Abraham, 1732-1819) and son (Johann Wilhelm, 1762-1806) both worked originally at Bornholm and then in Copenhagen, and made some noteworthy marine chronometers. Jean Abraham Armand came from Lausanne; his work in Denmark began at Reynsborg in 1764 as a mechanic for the Royal Danish Navy. After good initial test results on a West Indian voyage, Armand settled in Copenhagen and was to provide two chronometers a year to the government (#64). His son, Johann Wilhelm Armand, one of whose chronometers (#121) is preserved at Aarhus, worked with him.

Several marine chronometers were made at the Armand workshop later, but could not match the competition from the further developments of other masters, especially Jürgensen.

John Arnold and his Successors

The works of John Arnold, still in the pioneering era, were the basis for a successful enterprise for decades thereafter. The biography by V. Mercer, already mentioned, shows that John Arnold took his son John Roger as an apprentice at age fourteen and made him his partner in 1787. Three years before his death (1799), John Arnold seems to have withdrawn from the business. John Roger Arnold died in 1843; in 1830 he had taken E. J. Dent as his partner. John Roger Arnold was able to maintain the high standards of his father's production; he built numerous marine and pocket chronometers. The archives of Herstmonceux show that some ninety marine and thirty-six pocket chronometers were sent to Greenwich for testing. The entry of E. J. Dent resulted in a considerable increase in the company's production. In the next ten years, some 600 marine chronometers and 1200 pocket chronometers were manufactured. The majority of the latter were equipped with a crown winding invented by a colleague, Th. Prest. Interesting experiments as to the usefulness of various materials for the balance assembly took place in this era; gold balance springs and balances of platinum, silver and brass were made. In 1836 balances and balance springs of glass were even made for testing. In 1840 the partnership between J. R. Arnold and Dent was dissolved, and the latter opened his own business in the neighborhood. In his last years, J. R. Arnold still produced a number of pocket chronometers and pocket watches, but only a few marine chronometers from this era are known. But since his parting with Dent, J. R. Arnold is said to have sold some 300 marine chronometers and 400 pocket watches.

Soon after J. R. Arnold died in 1843, the business was bought by Charles Frodsham. It continued as Arnold & Frodsham, Chronometers Makers, Strand 84 until 1858 when the name of Arnold was dropped from the business name.

Of the illustrations in this book, #54, 92, 93 and 95 are by John Arnold, #94, 106, 107, 108, 109 and 124 by him and his son, and #119, 122, 133, 145, 177 and 178 by John Roger Arnold.

Auch, Jakob (father, 1765-1842, later with his son), worked first in Echterdingen, later in Vaihingen and Seeberg, and was court mechanic at Weimar as of 1798. J. Auch was one of Ph. M. Hahn's most capable students and built several interesting chronometers.

Barraud, Paul Philipp, London. At the beginning of the 19th Century he was very successful in the chronometer business. He was born to a Huguenot clockmaking family in 1752, worked with his father at first, where they suffered considerable financial difficulties in 1788. After his father's death (1795), he included marine chronometers in his production.

At first he worked with William Howells and George Jamison and used Thomas Mudge's designs, but could gain no commercial success in this endeavor; soon Barraud turned to a simplified design much like those of Arnold and Earnshaw. On this basis, P. P. Barraud soon developed a successful business that employed numerous chronometer makers, mechanics and watchmakers. From 1806 on, P. P. Barraud also held various administrative and honorary positions in the Clockmakers Company of London. Within two years he was first Guild Master and then a member of the corporation of the City of London.

About ten thousand pocket watches and over a thousand marine and pocket chronometers were probably made by Barraud's company, as his biographers[14, 15] report. Although Barraud himself built scarcely any of the watches, and particularly the chronometers, that bear his name, they always enjoyed a good reputation. The marine chronometers in particular are more durable than most other chronometers of his time; they ran at the same rate for up to five years without needing any repairs. The reason for the high quality of his products was the very dependable workmanship plus well-planned and well-executed details of construction. Barraud was a gifted manager to his numerous colleagues and suppliers. His legal action against Earnshaw in 1805 shows, though, that his relationships with them were not always free of friction. Here Barraud showed himself to be very experienced in using official procedures for the best results in his many personal relationships with other experienced chronometer makers such as Arnold or Brockbank; in this way he often pushed the factually brilliant but personality lacking Earnshaw into a corner and inspired some of the latter's many foolish and hostile statements that have often led to a negative assessment of Earnshaw's personality in the literature.

P. P. Barraud died in the summer of 1820. Two of his sons, Frederick and John, carried on the business, with the former taking a leading role. In 1838 I. A, Lund became a partner, and the business lived on under the name of Barraud & Lund. The last Barraud successor was Hilton Paul (1818-1880), but the firm carried on until 1929. The Lunds stood out particularly as pioneers in the transmission of electric time signals, which was carried on until 1963 by a separate corporation, Standard Time Co. See illustrations 117, 159, 160, 161, 162, 163, 174 and 220.

Barwise, John (1780-1842), London, is known for numerous pocket chronometers (#157, 221, 233).

Basnet, James, circa 1825, Liverpool.

Benoît, Achille Hubert (1804-1895), a well-known watch and chronometer maker of Versailles, later the director of the Royal Clock Factory there, and finally the director of the watchmaking school in Cluses; he developed numerous improvements to clocks and also worked with chronometers and tourbillons.

Berthoud, Pierre Louis (1754-1813), was a nephew of the famous Ferdinand Berthoud, probably learned from him, but did not enjoy the best of relations with him later. P. L. Berthoud worked for the French Admiralty almost all his life and devoted all his energy to the production of the best marine chronometers and high-precision pocket chronometers. In chronometer making, Berthoud retained the construction styles of his uncle to a great degree, never giving up the principle of the detent as an escape wheel lock. The workmanship of his marine chronometers is of remarkably high quality, to the extent that reports on them exist; all of his marine clocks performed extraordinarily well at sea. Still in all, relatively little on Pierre Louis Berthoud is to be found in the literature. Unpublished material such as working notebooks are privately owned in Italy, and their possessor intends to make a special study of them.

Only a few of his products are around today. According to the judgment of many specialists, scarcely any other chronometer maker created similarly high-quality and elegant chronometers. Berthoud developed, among others, a very elegant detent design (#86), a type that both A. L. Breguet and Henri Motel used. At the express desire of the French Navy, Berthoud took on four students, including Henri Motel and Pierre Saunier. At the end of their training period (1812) he recorded his most important experiences[16] to leave to his students. See illustrations 104, 114, 132, 144.

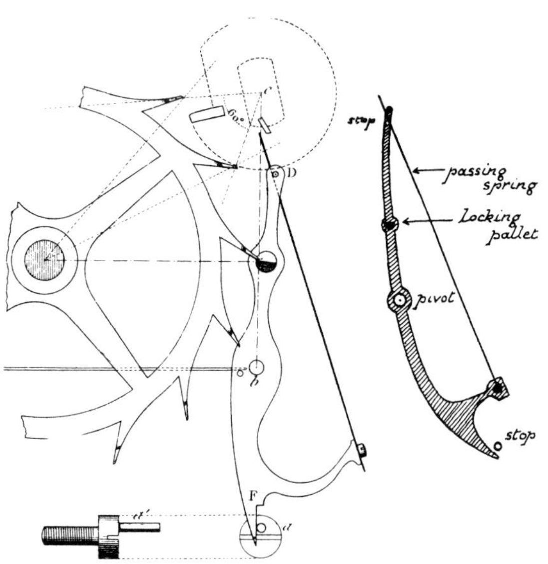

86 Drawing showing principle of pivoted detent developed by Louis Berthoud.
c Balance arbor with releasing and impulse pallets,
b Pivot of the detent,
d Banking point of detent, finely adjusted with pin d',
r Return spring for detent (steel), pivot of gold (lubricated)!
D Passing spring — locking point of detent after Saunier (right: Motel's version)

Abraham-Louis Breguet

A. L. Breguet was a very remarkable European personality at the end of the 18th and beginning of the 19th Centuries: an outstanding watchmaker, a brilliant designer and inventor of new constructions, a man of artistic sensitivity to the indications of his watches and instruments, and a very capable businessman.

Born in Neuchâtel, Switzerland, in 1747, he went to France in 1765, settled first at Versailles — then a center of arts and sciences — and then worked in Paris, probably with F. Berthoud and Etienne Gidem. He also studied physics from M. de Prony and mathematics from Abbé Marie (Collège Mazarin).

In 1775 he went into business with Xavier Gide, and thanks to his outstanding precision watches he was soon able to win the distinguished society of France and England as his customers. In 1793 he came under the suspicion of the French revolutionaries, but was able - with Marat's support - to obtain a visa to Switzerland for his son Antoine-Louis from the Comité de la Republique.

After the Revolution, A. L. Breguet's workshops, which had been confiscated and destroyed during that time, were restored at government expense. By the beginning of the Empire he had 100 employees and a yearly income of half a million francs. Under Napoleon his fame grew even more. In 1811 he became a supplier of the Tsar, and he died in September of 1823 after attaining much more technical and commercial success. His son Antoine-Louis became his successor. He soon formed a partnership with his younger brother Louis-Clement François. Antoine-Louis died in 1858. In 1855 Edward Brown of London had entered the firm of Breguet as a director, followed in 1878 by Edward Brown Jr., in 1912 by his brother Henry, and finally the latter's son George. A few years ago the well-known Parisian firm of Chaumet took over the old Breguet firm.

Abraham-Louis Breguet's field of activity was remarkably diverse. His particularly beautiful and interesting pocket watches, with a great variety of forms and complications, were particularly renowned. He also produced many traveling clocks and small standing clocks. Pocket chronometers (which he called "garde temps") were included in his manufacturing program very early, circa 1777, but he essentially followed the examples of Arnold and Earnshaw. Breguet took up the

87 Tourbillon pocket watch in gold case, signed (A. L.) Breguet, No. 1918. A: Front view; machined silver dial with blued steel hands, in the center is the hour ring with secret signature, with minutes outside just at the rim, two small second dials, left as stopwatch, below a sector for winding indication.
b. Rear view of the movement, escapement "a levees naturelles", according to original certificate, sold to Count d'Archinto on February 13, 1822.
Auktion Ineichen, Zürich, 1978

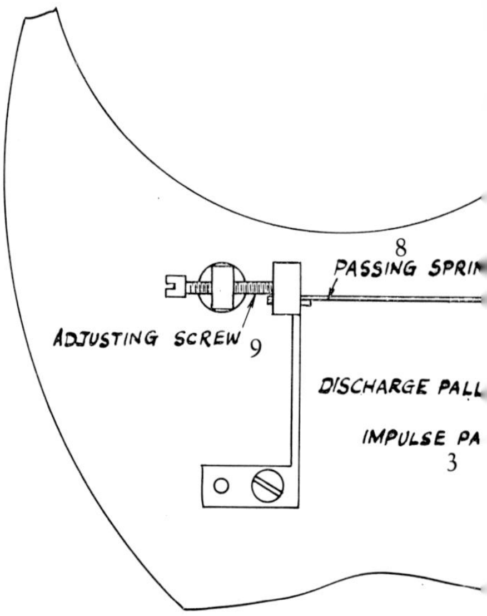

88 Chronometer escapement by Peto, so-called cross-detent, developed circa 1785. *(after E. L. Christmas Ant. Hor. 1963)*

construction of marine chronometers relatively late; only after about 1806 do notes and references begin to appear in the Breguet archives in Paris. As of 1818 he began to produce great quantities of marine chronometers of particular beauty and fine workmanship. Both these and his "garde temps" basically follow the already well-established traditions of the British. But what makes Breguet's work especially interesting all the same is that the wealth of constructive details that made not only assembly and adjustment but also repair work easier. Breguet worked out and employed outstanding engineering advances and developments in this area.

The external appearance of the marine chronometers made by Breguet is characterized by beautiful cases of rectangular cross-section with original types of gimbal mountings.

Proceeding from the general problems of movement structure treated in this chapter, A. L. Breguet put into practice various details of movement arrangement, as Egli[17] has recently described. Stressed there among others is the fact that, to have a good effect on friction conditions in the wheel train, Breguet connected the centers of the barrel and the first wheel. In addition, he put to good use the fact known to all clockmakers that higher numbers of teeth on the pinions decrease friction losses but cause more work and greater expense despite their favorable technical effects. In Breguet's chronometers one finds nine instead of the usual eight teeth on the first pinion. Thus he achieved an alternating functioning of the teeth of both barrels and thus a further decrease of peak power, though it made for a less favorable equalization of power.

The use of two barrels to make the movement as flat as possible is problematic in chronometers, as the outward appearance is less important than precise running. In marine chronometers, where the depth of the movement affords better stability in the gimbal mounting, one could even speak of less oscillating stability in the mounting as a disadvantage of a flat movement. But it is noteworthy that Breguet liked to use two barrels in this very type of clock.

A main reason for using two barrels instead of a fusee was, to Breguet, obviously the improvement of friction characteristics, which are clearly less favorable with a chain and fusee. In addition, the equalization of power is not as ideal in practice as one might conclude in theory from neglecting friction.

The great numbers of Breguet chronometers — and the even greater numbers of his pocket chronometers —

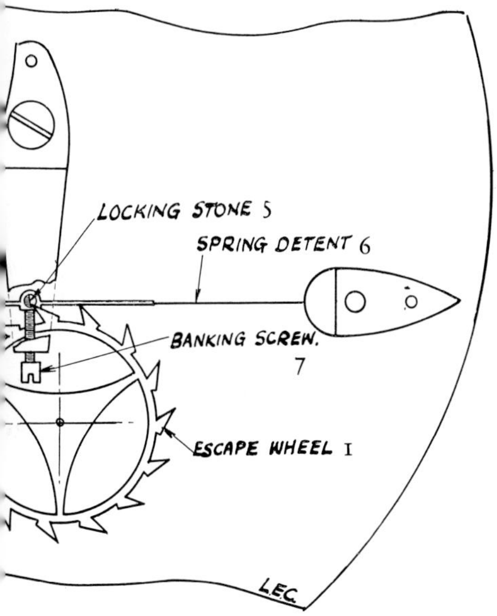

indicate an extraordinary technical versatility that cannot be defined within the limits indicated above. Daniels has analyzed a great deal of the technical particulars in A. L. Breguet's clocks and explained them with the help of drawings in an outstanding, illustrated, large-format book, thus presenting an almost lifelong study of the work of this great master.[18]

In general — probably because of his close friendship with Arnold in London — Breguet liked to use detents in his chronometers. Escapements with constant force, which Breguet used extensively otherwise, appeared only rarely in his marine chronometers, but he chose them for fixed-place chronometers just as he developed them for use in pocket watches (pendules sympathiques). Breguet also came up with variations on chronometer escapements, such as his "échappement naturel" or a type very much like Robin's escapement, for use in his chronometers.

One of Breguet's very special inventions should be mentioned which he developed to decrease the positional error that can be very disturbing for pocket chronometers when they are carried vertically or at an angle in a pocket.

The positional error comes about because the weight distribution in the balance is usually not completely equable, as a little more weight is present at some spot, and thus in different positions the clock takes on a greater or smaller additional pendulum effect that affects the action of the balance and its spring. To eliminate resulting changes in rate, Breguet got the idea of having the balance and escapement rotate around a horizontal axis so that in each complete revolution the different weight influences would cancel each other out. This is the tourbillon principle, a brilliant solution whose construction demands the highest degree of handiwork and fine mechanics. Illustration 87 shows a particularly beautiful tourbillon pocket chronometer by Breguet, that brought a previously unequaled record price at a 1979 auction in Zürich.

Watches with tourbillons (called Drehganguhren in German), like the *carousel watches* that seek to eliminate the positional error in the same way, are of particular interest to collectors and have always been highly prized by their admirers. Their most important features of design and construction, along with several masterpieces, will be described in Chapter 8.

The number of clocks and watches made by Abraham-Louis Breguet is remarkably great. There may be eighteen to twenty thousand of them whose existence can be proved by the existing record books — which include the dates of manufacture, the selling price and the customer. With such a great volume of production and such a remarkable personality as A. L. Breguet, it is impossible to include all of the highly interesting technical and artistic details of his work in a book devoted to the general development of the chronometer.

A. L. Breguet is undoubtedly one of the most important clock and watch producers of all time. The many thousands of valuable pocket and traveling watches, every one a jewel of fine workmanship, as well as the much less numerous standing clocks and regulators, are convincing evidence thereof. But A. L. Breguet did not merely recognize and understand what makes for high-quality fine mechanical work and carry it out; he also found many talented colleagues, most of them from his native Jura, and that he employed and instructed. By criticizing bad work but praising and exalting good achievements, he kept his workers in a good mood, and they were always receptive to his assignments. The fact that Breguet also had a supportive interest in the ability of others is shown by his friendly relationship with John Arnold in England, his closeness to Houriet in the Jura, and his readiness to train especially capable young masters and keep in touch with them afterward — as was particularly the case with the Dane, Urban Jürgensen.

A number of characteristic chronometers made by A. L. Breguet can be seen in #115, 137, 141, 142, 149, 156, 165, 179, 181, 182, 185, 187, 188, 189, 246, 260 and 377; the captions give concise information on the particulars of the individual pieces.

Brockbank, John (London, 1761-1806), became a master in 1777. He was an outstanding early chronometer maker with whom many later renowned clockmakers studied, while he made good use of their talents for his own purposes. The most important of his colleagues was Thomas Earnshaw, whose new ideas were utilized by him and other masters. Another well-known colleague was Peto, who developed the variation on the chronometer escapement named after him (#88), which can be regarded essentially as a way to circumvent the patent on Earnshaw's escapement. See #123, 126, 127, 139, 140, 147, 148.

Cole, James Ferguson (London,[19] 1799-1880). According to P. M. Chamberlain, J. F. Cole was an outstanding clockmaker and chronometer maker who developed numerous independent and interesting designs,[20] which rarely appear in present-day literature. Chamberlain described his ideas as a direct development of those of Mudge, Kendall, and also A. L. Breguet.

Delépine, Boromée. He worked in St. Nicolas d'Aliermont at the beginning of the 19th Century and founded a clock factory later acquired by Pons de Paul. The firm produced chronometers until 1910. See #288.

Dent, Edward John, London. This name still enjoys a good reputation today, after 150 years, for chronometer making as well as that of other valuable clocks and instruments. Edward John Dent (1790-1853) was a capable and considerably talented man who already had a good reputation as a watchmaker and chronometer maker. From 1830 on, he worked with John Roger Arnold, son of the famous John Arnold, as his partner, becoming independent in 1840. E. J. Dent died childless in 1853; of the four Rippon stepchildren of his wife's first marriage, two became chronometer makers. The younger, Richard Rippon, changed his name to Richard Edward Dent in 1850 and, after Dent's death in 1853, inherited a part of his business. After the latter's death in 1856, his widow Marianne directed the business under the name of M. F. Dent & Co.; it remained in operation under successors until 1920.

The older son, Frederick William Rippon, was apprenticed to the well-known master B. L. Vulliamy, also took the name of Dent and directed the part of the firm he inherited from his stepfather until 1860. After a troublesome court case involving the inheritance, his mother, Elizabeth, took over the shop in Pall Mall and ran it until 1865; then two of her daughters, named Bukney and Gardner, acquired it, keeping it until 1868. In that year Bukney decided to go abroad and sold the firm of E. Dent to other proprietors.

Vaudrey Mercer[21] has written a detailed history of the varying events, successes, and technical achievements of the two firms to the present day. In reading through it, one not only gets a vivid impression of the industrial importance of chronometer making in England, but also a conception of the role that one family and their relationships played. This includes collaboration with Professor Airy, with Lord Grimthorpe or noteworthy chronometer makers such as Kessels of Altona, Germany, and several others. The problems dealt with help to afford a view deep into the making of chronometers and regulators. Technically important advances of the Dent group concern the middle temperature error and how to eliminate or reduce it, for which Dent developed a series of typically fine balance designs. In addition, Dent dealt with magnetic influences on the balance assembly, and used glass and gold balance springs in these efforts. In the realm of large clocks and tower clocks, the firm of E. Dent & Co. built the famous tower clock of Big Ben in Westminster Abbey in London as well as that of the Stock Exchange and three notable regulators, including that of the Russian observatory in Bulkova; other production lines included marine compasses, aneroid barometers, chronographs, etc.

The economic volume of the two firms was great. Almost sixty pages of Mercer's biography are filled with the certificates for chronometers — from Kew (1885-1923) and Greenwich (1829-1933). Finally, Appendix IX includes a thorough listing of the names of workers and works masters, how they produced movements, escapements, balances and spring adjustments, and the dates of production. This listing shows how the section of the business devoted to chronometer making had progressed at various times. See #169, 218, 236, 275, 284, 298, 304.

Earnshaw, Thomas (London, 1749-1829). Over and above his role in the heroic epoch, he continued to work successfully at chronometer making until his death, as shown by his many surviving products. See #102, 111, 112, 118, 136, 146.

Eiffe, John Sweetman (London, 1800-1880). He was a capable chronometer maker, who successfully worked on special balance designs to eliminate the middle temperature error.[22] He holds the honor of being the first to recognize the middle temperature error in chronometers with bimetallic balances as a phenomenon inherently linked to these designs. He not only developed various improvements in his production, but also made all of England's chronometer makers aware of this important problem in chronometer making through his memorandum on the subject, which was published by Sir G. Airy, the Astronomer Royal, in 1842. A somewhat disturbing feature of Eiffe's writing is his naive denigration of Harrison's work. See #235.

Emery, Josiah (London, 1725-1797), was an outstanding clockmaker but achieved no noteworthy success with his chronometers in the tests of the Board of Longitude at Greenwich. See #103.

Frodsham, Charles & Co. For more than 150 years, numerous marine and pocket chronometers were signed with this name. The founder of the firm was William J. Frodsham (1778-1850), whose mother was a granddaughter of the famous John Harrison. W. J. Frodsham bought J. R. Arnold's business in 1843, after Arnold's death. His successor was his son Charles (1810-1871), who also acquired the firm of J. B. Vulliamy in 1854. In the course of his activities, he won many notable prizes for particularly good chronometers; for example, his Chronometer No. 1 showed the minimal variation of only 0.57 seconds in a year of testing at Greenwich.

Charles Frodsham was succeeded by his son Harrison Miles (1855-1927), who directed the firm along with E. E. Rutherford from 1871 to 1927. The consistent, meticulous workmanship and great reliability of the Frodsham firm's marine and pocket chronometers are well known. See #143, 205, 232, 248, 256.

Gannery, Victor (St. Nicolas d'Aliermont, 1820-1851), was a student of Perrelet and later of Winnerl in Paris. Gannery produced large numbers of raw movements (ébauches) for French chronometer makers (until about the middle of the 19th Century, "ébauches" referred to the bottom plate with pillars and bridges, the barrel, fusee, locking click and its rim, plus some screws, all roughly turned out with file and grinder). Finishing was done by the individual firms, though. After Gannery's death, Onésime Dumas took over his firm and continued to supply the most important firms in France, plus others (Le Roy, Jacob, Ekegren, etc.) with his ébauches. See #228, 229.

Gardner, Robert (London, 1851-1931), was an important chronometer maker from about 1885 on. He promoted W. G. Schoof's lever escapement for chronometers, among others. He also built up a very significant horological library, which later became the basis of his son's book business. See #285, 299.

Glashütte. A name that stands for watches of the highest quality, including very outstanding marine and pocket chronometers, tourbillon, and carousel watches of worldwide renown. Watchmaking in Glashütte goes

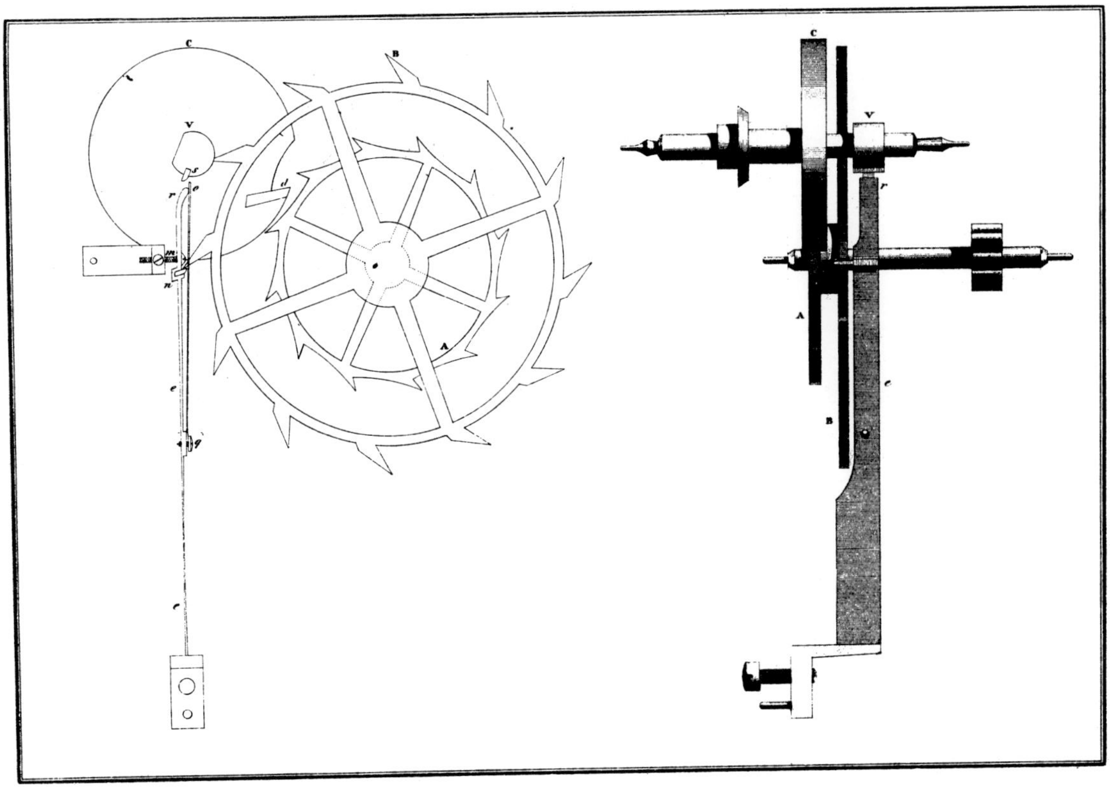

back to Ferdinand Adolf Lange and developed into numerous firms in a few years. The activities of the Glashütte firms up to 1948 are covered thoroughly in Herkner's monograph.[22] The most important firm, A. Lange & Sons, was nationalized by the DDR in 1948, a step that brought the existence of a century-old, highly significant enterprise to an end.[23 24]

F. A. Lange (1815-1875) began as an apprentice to Johann Friedrich Gutkaes in Dresden, then went to Paris in 1835 because of his particular interest in modern watchmaking, worked there for J. T. Winnerl until about 1840, and then returned to Dresden via Switzerland and England. On this trip he made contact with the best-known watchmakers.

Having returned to Dresden, he married the daughter of his former employer, Antonia Gutkaes, in 1842, and worked for her father's firm. In 1845, F. A. Lange gave up his secure position in Dresden to found a training school for watchmakers in Glashütte, the poorest village in the Erzgebirge, a task for which the Saxon Ministry of the Interior had offered him modest funding. Thanks to F. A. Lange's great personal dedication and talent for inspiring others, this became the basis of the important business in the title of Herkner's monograph. When F. A. Lange began to work on chronometers cannot be determined exactly; the literature disagrees by about ten years. According to Herkner, A. Lange & Sons first sent two marine chronometers to the German naval observatory in Hamburg for testing in 1886, and Richard Lange, F. A. Lange's son, made the first pocket chronometers in 1887. In 1892 a special chronometer escapement with a "short detent" was developed by Richard Lange. In 1897 a separate department for chronometers was set up, and shortly thereafter marine and pocket chronometers began to be delivered steadily to other countries. The production of tourbillon watches began in 1899. By 1904 a total of 210 pocket chronometers and 25 marine chronometers had been made, and by 1939 the total

89 Urban Jürgensen chronometer escapement with two escape wheels and spring.

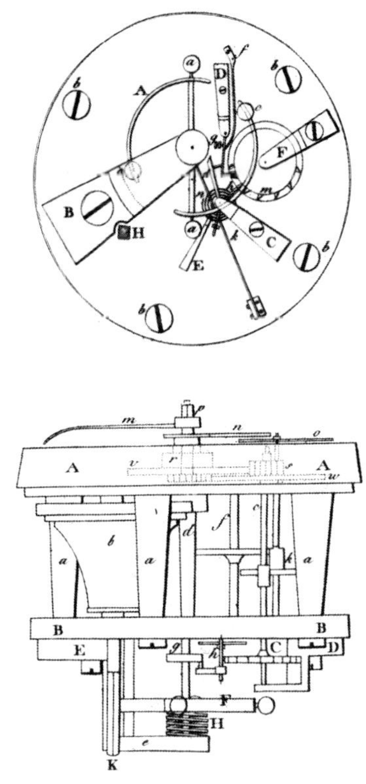

90 Working drawing of Urban Jürgensen's marine chronometers, after L. U. Jürgensen's book (2nd edition).

91 Marine chronometer, signed Kessels, Altona, Nr. 1287, eight-day running, with winding indicator; dial diameter ca. 112 cm; ca. 1847. *(after Dent)*

number of Glashütte chronometers had risen to 650; they had been sent to the Hamburg marine observatory and received 21 first, 20 second and 18 third places in Class I (see #341, 346, 354, 355).

Of particular importance to the successful development of watchmaking in Glashütte is the fact that F. A. Lange had gained several very noteworthy specialists as his colleagues, such as Moritz Grossmann (1826-1885), Studienrat Alfred Helwig (1886-1974), and Professor Ludwig Strasser (1853-1917).

Gutkaes, Christian (or Johann Friedrich, 1784-1845), court watchmaker in Dresden, maker of numerous interesting marine and pocket chronometers (and F. A. Lange's father-in-law); see #201, 202, 214.

Hardy, William (London, 1800-1830), a significant clock- and chronometer maker who won several important prizes and became famous for his special design to decrease the middle temperature error. See #134, 135.

Hohwü, Andreas (Amsterdam, 1840-1913), a renowned chronometer firm founded by Andreas Hohwü (1802-1885), son of a watchmaker in Gravenstein, Schleswig-Holstein, student of J. H. Kessel and Louis Breguet. Andreas Hohwü himself was an important maker of chronometers and regulators; later parts and movements made by Mercer and Johannsen were reworked there on a large scale. See #263.

Hughes, Henry & Son (1872-1941), a well-respected chronometer firm that extensively used parts and movements made by Dent, Gardner, Kullberg, Usher & Co., among others. See #387.

Jacob, Jean Aimé (St. Nicolas d'Aliermont, 1793-1871), an important French chronometer maker. See #245.

Johannsen & Co., Asmus (1856-1937), a very important British chronometer firm which extensively used movements by Kullberg and Mercer. See #281.

Jürgensen, Urban. Jürgensen is a name that has been closely linked with high- and highest-quality timepieces such as marine chronometers, pocket watches, and tourbillons since the second half of the 18th Century.

The originator of this fine reputation was Urban Jürgensen (1776-1830), son of the Copenhagen watchmaker Jörgen Jürgensen, who showed a particular inclination to his profession since his youth. He received a good training in his father's studio, then worked for J. F. Houriet in Le Locle, and later for Ferdinand Berthoud in Paris, finally learning British practices from Arnold in London. After the death of his father, Jörgen Jürgensen, in 1811 he helped his brother Frederik run the family business while simultaneously founding his own firm. Urban Jürgensen was a happy combination of the practical technician and scientist and the ambitious producer of high-quality instruments. He is said to have made a good number of outstanding chronometers for use on Danish merchant and naval ships, including some with his own particular escapement design (#89). They were essentially arrangements based on the experience he had gained in Paris and London. Unfortunately, it has not been possible to locate and photograph a single piece. The movement construction is shown in #90.

Urban Jürgensen originally recorded his experience and ideas in a Danish book published with the support of the King of Denmark; later it was improved and expanded with new ideas, and published in French.[25] This highly interesting work was reworked again by his son Louis Jürgensen[26] and finally published in German.[27] It is a remarkably clever, instructive book; its chapter on chronometers is still very readable today, as it presents good general insights into their functioning. But only the first French edition includes numerous details of Urban Jürgensen's experimental and scientific ideas, which are valuable in evaluating the insights of those times.

Urban Jürgensen's elder son, Louis-Urban Jürgensen (1806-1867), took over his father's business in Copenhagen and founded the Danish line; the younger son, Jules-Frederik (1808-1877), moved to Switzerland, founded his own firm in Le Locle in 1835, and became the ancestor of the Swiss line. His two sons, Jules-Urban II (1837-1894) and Jacques-Alfred Jürgensen (1842-1912), directed the business — generally jointly — until the latter died. In 1912 the Swiss firm passed out of the family, being directed by David Golay until 1925. He sold it to I. Heuer, who carried on production in Biel (Bienne) until 1932. In the great financial crisis of 1932 the business was closed, but in 1936 the American firm Aisenstein, Woronock & Sons, Inc. bought it, reopening the Biel factory in 1937. In 1942 it was taken over by Dave Woronock, who finally sold it all to Viktor Haff in 1960. According to Golay, the company operated as an international merchandising firm with its emphasis on high-quality products.

Several years ago a great exhibition was held in Le Locle, at which more than 100 watches made by the Jürgensen firm were displayed, having been borrowed from the USA. The catalog[28] published for this exhibition is an impressive document of the horological activities of this great dynasty of watchmakers. See #152, 153, 154, 267, 379.

Kelvin, Sir William Thomson (1824-1907), British physicist, since 1892 Lord Kelvin.

There are a number of businesses whose names consist of Kelvin and other family names, and which have been involved in the chronometer business. Essentially, they all appear to have used movements and parts made by Mercer and other chronometer makers, and thus they will only be named briefly here: Kelvin Ltd., Glasgow, 1907-1989; Kelvin & Baird, Newcastle-on-Tyne, 1913-1933; Kelvin, Bottomly & Hutton, London, 1905-1918; Kelvin & White, Glasgow, 1870-1914; Kelvin, White & Hutton, London, 1904-1937 and Boston, 1916-1933. See #305.

Kessels, Heinrich Johann (1781-1849). According to various contemporary sources, H. J. Kessels was a very significant personality, but there is relatively little to be found on him in the literature. Born in Maastricht, The Netherlands, on May 15, 1781, he studied first with a blacksmith, but soon went to Paris where he became apprenticed to A. L. Breguet. Then he worked in London for a time, and as of 1828 in Copenhagen, where he was supported particularly by King Frederik VI, who encouraged Kessels to settle in Altona, near Hamburg, as it was Danish territory then. Kessels later became a Knight of the Danish Order of the Danebrog, chronometer maker to the Royal Danish Navy, member of the Royal Society of Stockholm and the Mathematical Society of Hamburg. He died of cholera in Bristol in 1849. Although Kessels made a good number of excellent marine chronometers, noteworthy astronomical clocks, and regulators, very little is known of them. He is said to have supplied chronometers to the observatories of Nikolaiev, Dorpat, Moscow, and Pulkova, as well as Stockholm, Uppsala, Lund, Christiania, Krakau and other cities. Illustration 91 shows an eight-day marine chronometer begun by him and finished by his friend Dent. See also #211, 212, 213.

Kullberg, Viktor (1824-1890), was the most prominent chronometer maker of his time and produced chronometers under his own name as well as those of numerous other important chronometer firms. He was born in Visby, on the Swedish island of Gotland. He originally studied with a local chronometer maker, where he had to do all the work himself, but soon went to Louis-Urban Jürgensen in Copenhagen for further training. In 1851 the great international exhibition drew him to London, where he soon worked with the prominent British chronometer makers, specializing in chronometer escapements. In 1856, at the time of a great economic depression, he went into business on his own and soon attracted attention at public competitions and exhibitions with his crown-wound pocket watches and chronometers. Between 1860 and 1873 he won some fifteen honors, gold and silver medals, and first prizes from all the important institutions in Europe. In 1874 he became the premier Swedish and Norwegian chronometer maker and was honored with the Order of Vasa. Afterward he continued to gain the highest honors in Europe, America and Australia, and in 1882 the Astronomer Royal ranked one of his chronometers as the best chronometer that had ever been tested in Greenwich. From 1860 on, Kullberg used his own design for eliminating the secondary temperature error; these balances were generally regarded as the simplest and most reliable of all the types made until then. In general, his marine and pocket chronometers closely resembled the usual British types, though in arranging the chain and fusee he used the "reversed fusee" principle in which the radial force on the arbor pivots of the fusee are less, and thus friction and wear are reduced. Despite its considerable advantages, this style found little acceptance in Britain. It was used more often on the Continent, by Winnerl and several other French producers, Lange of Glashütte, etc. The high quality of his products was essentially a result of his outstanding workmanship. After his death, his two illegitimate sons and a nephew named Wennerstrom took over Kullberg's business. After the two Kullberg sons died, Wennerstrom directed the business alone. After his death the Swede Sanfrid Lundquist bought the firm and, with the widow of the younger Wennerstrom, kept it on the same distinguished level. See #261, 297, 301, 306.

Litherland, Peter, worked in Liverpool from 1790 to his death in 1805. He was a successful inventor in the realm of the pocket watch, and also produced chronometers.

Le Roy, with the same name as the famous Pierre Le Roy of the heroic epoch, but not related to him or his ancestors, there are two clockmaking firms in Paris, one presently called L. Leroy et Cie., the other known until 1960 as Leroy et Fils. The first firm produced noteworthy marine chronometers for many years and was the official horologer to the navy around 1880. Over the decades various changes took place in the ownership of the two firms, resulting in several name changes and, in connection with that, legal disagreements, which make an overview of the Leroy situation difficult. Ch. Allix and T. C. Sabrier recently published a chronological study[29] of the events connected with these two firm names over the course of about a century and a half, making information on the numerous name changes somewhat easier to come by. See #269, 308, 343.

Loseby, Edward Thomas, born in 1817, worked in London from about 1830 to 1853, later in Leicester until 1880. He was a very significant chronometer maker, known particularly for his designs for balances with additional mercury compensation. Beginning in 1842, he also built two small, bent thermometer tubes with mercury into the balance, in order to achieve an even and very extensive reduction of the middle temperature error (see #242). Originally his work pleased G. B. Airy, the Royal Astronomer at Greenwich, very much and gained him remuneration for the expenses of his work. Later, though, Loseby appears to have gone over Airy's head to the President of the Royal Society, Lord Northampton, for patronage, and thus incurred some displeasure on Airy's part. In his reports to the Admiralty, Airy did acknowledge the good qualities of Loseby's handling of the middle temperature error, but described them as a special use of the basic principle already formulated by Eiffe and thus not deserving special patronage for Loseby. Loseby was very disappointed by this negative judgment and the non-fulfillment of his hopes after the good test results achieved at Greenwich; he gave up chronometer making and moved back to Leicester, where he continued to make large clocks.

Details on Loseby's ideas, his chronometer with

auxiliary mercury compensation, and his very interesting auxiliary pendulum designs meant to make the period of the pendulum independent of the amplitude, were collected in a very readable brochure[30] several years after Loseby's death.

Margetts, George, was a master in London from 1799 to 1808. He is famous for complex astronomical-calendar pocket watches and chronometers, as shown in #99. Later he worked in partnership: Margetts & Hatton (1805-1811).

McCabe, James, (Belfast and London, 1760-1811). He was a noteworthy clock and chronometer maker, and his business was kept going by his sons and grandsons until 1883. See #194, 251.

Massey, Edward (London, 1772-1852), was a well-known clockmaker who also built marine chronometers.

Mercer. The name of an important chronometer business, founded in 1856 by Thomas Mercer (1822-1900) and as of this writing still successfully employed, in the construction of mechanical chronometers. The founder was eight years old when his father — a canal watchman in Lancashire — died unexpectedly and he was apprenticed to his grandfather, a local clockmaker. In 1843 he found a position at Russel's, a large clock-making firm in Liverpool, where he already produced anchor clocks and chronometer movements independently. Convinced that clockmaking in Lancashire was about to cease for lack of modernization, he decided to emigrate to America in 1854.

He already had his ticket in his pocket when some marine chronometers in the display of J. Fletcher in London impelled him to stop and go into the shop. This visit led directly to a job with this clockmaker, who then had a good reputation in London. In 1858 Thomas Mercer founded his own firm, and in 1874 he opened a workshop in St. Albans. Frank Mercer (1898-1970), Thomas Mercer's second son, took over the firm after his father's death. During World War I he received many contracts, but difficult times followed until 1939. Today Tony Mercer (born 1919), who has passed through all the levels of work up to that of master, runs the business as its president.

The monograph written by Tony Mercer is not only a good history of the firm over some 120 years, but also a very informative portrayal of the way chronometers were produced —especially in the latter half of the 19th Century. It is an unexpected experience for those who are not particularly informed to see from the long list of Mercer's customers how and when many of the old established chronometer makers obtained movements and parts from a few major manufacturers and combined them with balances and springs, likewise made by specialist firms such as Kullberg, Mercer and a few others, and then put their own names on them. See #280, 357, 358, 359.

Molyneux, Robert (London, 1800-1840), a good chronometer maker who patented several designs to deal with the middle temperature error and was highly respected for his especially superb fine mechanical work. See #186, 210, 222, 223.

Motel, Jean-François Henri (Paris Margny-les-Compiègnes, 1786-1859). For almost half a century, Motel had a great influence on the history of the chronometer in France. In spite of this, thorough evaluations of his achievements[31 32 33] have appeared only in the most recent years.

His outstanding and beautifully made chronometers exemplify, in fine mechanical terms, the best of the fine mechanical era. Around 1796, with special support from the government, he went to the already renowned School of Arts and Métiers in Compiègne. Louis Berthoud, who was paid by the state to train four talented apprentice watchmakers in chronometer building at the newly founded school in Argenteuil several years later, chose Motel as one of them in 1806. For some years thereafter, Motel lived in Louis Berthoud's house; he received an outstanding certificate from the Conservatoire des Arts & Métiers in 1813. In 1812 Motel received L. Berthoud's only book on clockmaking with a special written dedication from Berthoud as a mark of special recognition. When L. Berthoud died a few weeks after Motel's schooling ended, Berthoud's widow chose Motel to keep the business running. When Louis Berthoud died, a large number of marine chronometers were being made, and Motel finished them as well as making repairs. Motel's work was already valued so highly by the Ministry of the Navy that in 1823, at the age of 37, he was named the official naval clockmaker.

Since 1817 Motel owned his own business. From Breguet's business records it can be seen that Motel is named at times as having finished, or built the wheel trains of, Breguet's special clocks. It is not known, though, whether he worked as an employee of Breguet's or was already independent at that time. In any case, he was given the work of repairing marine chronometers as of 1819, and supplied complete chronometers to the navy as of 1821. It is only known that he lived on the Rue de l'Abbaye in Paris from 1830 to 1840, and on the Rue de Foin from 1850 on.

Just in the time between 1821 and 1832 Motel sold 85 chronometers to the navy for almost 200,000 francs, and during his life he built 200 chronometers. Despite this considerable volume of work, the outstanding quality of his work, and though a number of his beautiful chronometers can be found in the world's great collections, remarkably little is known about him. Only in the autumn of 1979 did J. C. Sabrier's extensive study make Motel's chronometric achievements known to a large circle of interested horologers. From these works, and thanks to generous support from private owners, it has been possible to illustrate Motel's works well. The special features of the individual chronometers have been noted in the captions. In general, the influence of Louis Berthoud can be seen clearly throughout Henri Motel's works. The first chronometers (up to No. 38) very much resemble those of L. Berthoud in overall construction, with the single difference that Motel preferred three-armed balances, while Berthoud generally used four-armed balances. In addition, Motel added an additional wheel on the side of the glass cover for front winding of the chronometer very early, while his oldest chronometers are still wound from the back.

Around 1826 Motel created his own caliber. It might be noted here that "caliber", when applied to watches, refers not only to the dimensions of the movement but also to its particular construction and the arrangement of wheels and bridges.

In his chronometers, Motel always used a chain and fusee and preferred a pivoted detent for the escapement, although he also made a few with spring detents. Motel's pivoted detent is practically identical to that developed by L. Berthoud (#86), a refined type that requires lubrication of the contact points between the passing spring and counterpivot. The caption to #86 indicates the significance of the individual parts; at first glance one may well reject this design for its complexity, but at closer study one recognizes the functional practicality of the details, which show that a high point of the fine mechanical period was attained here.

Motel usually used a conical balance spring for the oscillating system, but he also tried unusual combination shapes. The unusual weights for the balance ring, similar to those of L. Berthoud, are also noteworthy. In general, Motel's balances are designed for 18,000 beats per hour.

Motel often provided apparatus to keep his chronometers stable in transport. In many of his marine chronometers, unfortunately, the gimbal mountings have been removed and the movements built into simple wooden cases; this form is now called the "chronometer compteur." Motel also produced a number of especially well-made pocket chronometers.

Sabrier reports that in 1922 Motel's work records were still available, and that M. Querens[34] published a number of statistics on the cost of producing individual parts and making repairs. In 1827 Motel first displayed his products at an exhibition of French industrial products, receiving a silver medal and very laudatory comments from the jury. It is also noted that Henri Motel was basically more interested in producing outstanding movements than in scoring special triumphs. For these considerations, Motel made no concessions to the tastes of the times and was not influenced by the stylistic moves of his contemporaries. Motel probably made numerous attempts to achieve improvements but always stayed on his main course that, in his judgment, gave the best results. Thus as soon as he had created his new caliber, he never deviated from it. For this reason he also declined to accept special requests from private customers but concentrated on producing his basic type for the French Navy. See #170, 171, 172, 190, 191, 192, 206, 227, 240.

Pennington, Robert (London, 1780-1824), was a reputable chronometer maker who worked in partnership with William Howells. See #110, 138.

Robert, Henri (Paris, 1795-1874), was a student of Breguet and Perrelet and a very important French clock and chronometer maker who made numerous inventions. Named as horologer to the navy in 1847, he was also the author of several publications on clock and chronometer technology. See #208.

Seyffert, Johann Heinrich (Dresden, 1751-1818), a notable producer of chronometers and precision clocks, was the director of the mathematical-physical salon of Dresden as of 1801. See #125, 128.

Snellen, Willem, Dortrecht, end of the 18th Century. His few chronometer movements known to date suggest British training, perhaps even collaboration with Harrison. He may have been a son of the mayor of Dortrecht who bore the same name, was a master watchmaker and produced astronomical clocks. See #52, 53.

Ulrich, Johann Gottlieb (London, 1795-1895), a very interesting, extraordinarily inventive chronometer builder who suggested and carried out numerous improvements. He gained no particular success, probably because of the complexity of his ideas. See #176, 203, 204.

Vissière, Simon (Le Havre, 1822-1887), a very outstanding French chronometer producer, one of the first to use non-magnetic material for balance springs and balances themselves. See #241, 271.

Winnerl, Joseph Thaddeus, born in the Steiermark of Austria in 1799, died in 1886 in Paris, where he had settled in 1829. He was one of the most notable French watchmakers and chronometer makers. See #226, 243, 244, 250.

German Chronometer Makers

The production of marine chronometers in Germany has never yet been regarded as a problem. To be sure, the names of numerous masters are found on marine chronometers, especially those from the cities of northern Germany, but to this author's knowledge we have no overview of the numbers made by the individual masters, neither the number of chronometers delivered nor the years of their activity. Likewise there are scarcely any data on their schooling or training, or information on the use of these chronometers on major routes or important expeditions. The following list may indicate the extent of the subject and inspire studies of the indicated themes.

Names of German chronometer builders who, according to the annals of the Hydrographic Institute of Hamburg, submitted chronometers for testing between 1877 and 1939 and later:

Julius Albers, Bremerhaven, ca. 1876
Julius Assmann, Glashütte, 1911
Carl Bamberg, Berlin, 1956
Johannes Beyer, Bremen, 1956
E. Bröcking, Hamburg, ca. 1891
Wilhelm Bröcking, Hamburg, ca. 1876
Paul Erich Busse, Schnega, 1955
Chronometer Werke, Hamburg, 1909
(As of 1939 Chronometerwerke Gerhard D. Wempe)
Carl Cordes, Kiel, 1954
F. Denker, Hamburg, 1904
Hermann Diderich, Geestemünde, 1889
Wilhelm Gustav Ehrlich, Bremerhaven (Carl Ehrlich, father), 1876
J. Eisele, Kiel, 1894
Gebr. Eppner, Berlin; A. Eppner & Son, Berlin, 1876
L. Eschholz, Hannover, 1903
Friedrich Fey, Hamburg, 1953
Matthias Frantzen, Münster, end of the 18th Century
Walter Frenzel, Berlin, after 1950
Moritz Gerlin, Rostock, 1880
C. Gerstenberger, Glashütte, 1928
R. Griesbach, Glashütte, 1910
Ernst Hajer, Jever, no date
Franz Happe, Kiel, 1952

L. Jensen, Glashütte, 1898
W. Iversen, Hamburg, 1876
Adolf Kittel, Altona, 1876
Theodor Knoblich, Hamburg, 1876
Knoblich & Denker, Hamburg, 1889 (later Th. Knoblich Successors,
A. Meier Proprietor, Hamburg, and Knoblich Successors Sternberg, Hamburg)
E. Koopmann, Bremen, 1909
P. Krahmer, Wurzen, 1886
Kreuzinger & Leutert, Hamburg, after 1945
Eugen Kulms, Münster, Krille, Altona, Glashütte, 1905
Louis Kurtz, Münster, 1905
Ernst Kutter, Stuttgart, 1876
A. Lange & Sons, Glashütte, 1887
Friedrich Leutert, 1955
Franz Lidecke, Geestemünde (later Wesermünde), 1899
F. L. Löbner, Berlin, 1902
A. Mager, Brake an der Weser, 1880
J. H. Martens, Freiburg, 1883
Mierendorf, 1881 Walter Möller, 1954
Möller & Friedrichs, no date
L. Nieberg, Hamburg, 1876
Theodor von Pavlicek, Hamburg, 1922
Erich (or Ehrich) Petersen, Hamburg-St. Pauli, 1884
Uwe Peschlow, Lübeck, no date
Matthias Petersen, Altona, 1876
Walter Prell, Gera, Glashütte, 1936
Johannes Raabe, Glashütte, 1917
Max Richter, Berlin, no date
U. F. P. Sackmann, Altona, 1876
F. Schlesicky, Frankfurt am Main, 1895
Theodor Schmidt, Nordenham, 1914
J. Schnoor, Kiel, 1876
A. Schuchmann, 1904
F. Schuchmann, Wilhelmshaven, 1903
F. & A. Schuchmann, 1903
C. Schweizer, Munich, no date
Strasser & Rohde, Glashütte, 1909
Stübner, Glashütte, 1909
J. D. Thies, Hamburg, 1881
Tiede, Berlin, 1879
F. N. Tietz, Kiel, Karl Tietz, Proprietor, 1930
Union, Glashütte, 1910
Fr. Vetterlein, Glashütte, 1916
G. Ph. Völling, Rostock, 1876
Conrad Wiegand, Peine, 1907

The few of these many chronometers that have been accessible to date show noteworthy workmanship, simple, clear, knowledgable construction, and have proved themselves well according to test results. Yet aside from the valuable (but publicly unknown) reports in the *Annals of Hydrology and Maritime Meteorology* and in other German technical and horological literature, very little attention has been devoted to them, so that to date no critical evaluation or corresponding tradition has been able to develop.

With few exceptions, German scientists have not been generally drawn to study problems of time measurement, so basic questions have not been solved, such as has been the practice in France and Britain. In the latter half of the 19th Century, only Glashütte, Saxony, and the clock-making school located there had achieved the truly highest international fame over a long period of time, and in the purely scientific realm of time research, Professor Schuller of Göttingen with his experiments on pendulum influences, and Professor Albert Einstein with his studies of time paradoxes in relatively opposed systems, have achieved great independent deeds.

Chronometer Makers in the USA

In the USA numerous entrepreneurs turned to the production and sale of marine chronometers soon after the middle of the 19th Century. They included T. S. & J. D. Negus, John Bliss & Co., D. Eggert & Sons in New York, as well as Wm. Bond & Son in Boston, of whom very little is known today.

M. E. Whitney,[35] the first to describe American chronometers in depth, notes that as a chronometer specialist at the U. S. Naval Observatory almost forty years ago he was sent chronometers by these masters —sometimes thoroughly worn out — with requests for repairs if possible. Whitney made the suggestion that they be preserved as documents, which was unfortunately declined by the person then in charge.

1,2,3. Compare the footnotes in Chapter 3.1 (Harrison).
4. Anonymous, *Outline of Principal Inventions,* Journal of Philosophy, Chemistry and Arts, London, 1806.
5. Abraham Rees, *Cyclopaedia or Universal Dictionary of Arts, Sciences and Literature,* London, 1820. Reprinted as: Rees' Clocks, Watches and Chronometers, Newton Abbot, 1970.
6. G. B. Airy, By order of the Lords Commissioners of the Admiralty. *Account of Improvements in Chronometers,* made by Mr. John Sweetman Eiffe, with an Appendix containing Mr. Robert Molyneux's specification . . . , London, John Murray, 1842.
7. Ed. J. Dent, Brit, Patent No. 9302 of 1842, *On the Errors of Chronometers, and Explanation of a New Chronometer- Compensation's Balance.*
8. Ch. Ed. Guillaume, *La Compensation des Horologes et des Montres,* Paris, 1923.
9. Rottok, *Chronometerprüfungen bei der französischen, englischen und amerikanischen Marine,* Annalen der Hydrographie, Berlin, 1917.
10. G. Glaser, *Die Berechnung, Dimensionierung und Anpassung von Triebfedern für Uhren,* Die Uhr 2, 3, 4 1963.
11. Ed. Phillips, *De la Compensation des Temperatures dans les Chronometres,* Paris, 1880. It is noteworthy that Phillips offers no quantitative data for the influence of temperature on the elasticity coefficient in this work, although he himself called the neglect of this a weakness in other works.
12. J. C. Sabrier, *Henri Motel et la Chronometrie Française de la Premiere Moitié du XIX,* Ancaha no. 25, Eté 1979.
13. Tony Mercer, *Mercer Chronometers; Radical Tom Mercer and the House he founded,* Ashford, Kent, 1978.
14. E. M. Barraud, *Barraud, the Story of a Family,* London, 1967.
15. Cedrik Jagger, a) *Paul Philipp Barraud,* 1750-1829, A.H., London 1968; b) *Paul Philipp Barraud, — The Supplement,* A.H., London, 1979.
16. L. Berthoud, *Entretiens sur l'horlogerie à l'usage de la marine,* Paris, 1812.
17. Kurt H. Egli, *Inspiration aus alten Uhren für Neukonstruktionen,* CIC-Konferenzberichte. Vol. 3, 1979.
18. George Daniels, *The Art of Breguet,* London & New York, 1975.
19. Paul M. Chamberlain, *It's about Time,* New York, 1941.
20. Described in the Horological Journal, London, in 1859-60-64.
21. Vaudrey Mercer, *Edward John Dent and his Successors,* A.H., London, 1977.
22. Kurt Herkner, *Glashütte und seine Uhren,* Dormagen, 1978.
23. O. Thielemann reported in *Uhren und Schmuck,* No. 17, 1980, that chronometers are being produced at the VEB Uhrenfabrikin Glashütte at about fourfold capacity, and mentioned, among other things, the collaboration of Soviet trainees.
24. See Martin Huber, *Die Uhren von A. Lange & Söhne, Glashütte, Sachsen,* Munich, 1979.
25. Urban Jürgensen, Horologer, *Prinzipes géneraux de l'exactemésure du temps pour les horloges,* Copenhagen, 1805.
26. Louis-Urban Jürgensen, *Memoires sur l'Horlogerie exacte,* Paris, 1842.
27. Urban Jürgensen, *Allgemeine Grundsätze der genauen Zeitmessung durch Uhren,* Leipzig, 1840.
28. R. Delapierre, *Les Jürgensen, Une Dynastie de Grands Horlogers,* Le Locle, 1974.
29. Ch. Allis & J. C. Sabrier, *Le Roy et Leroy,* Horlogerie Ancienne No. 27, 1980.
30. J. J. Farmer, *An Account of the Improvements in Chronometers, Watches and Clocks by E. T. Loseby,* Coventry, 1852/London 1914.
31. J. C. Sabrier, *Henri Motel et la Chronometrie Française de la première moitié du XIXe,* ANCAHA No. 25, Eté, 1979.
32. Querens, *Henri Motel, Horloger de la Marine,* L'Horloger 1922, pp/118ff.
33. A. Chapiro, *Was Henri Motel a Pupil of Breguet?,* AH, London, 1973.
34. M. Querens, L'Horloger, 1922.
35. Marvin E. Whitney, *The Ships Chronometers,* 1985.

92a Pocket chronometer in gilded silver case (hallmark 1776-77), signed "John Arnold invt. et fecit No. 28." White enamel dial, 63 mm diameter.
Clockmakers Company Museum, Guildhall, London

92b Back of movement of pocket chronometer no. 28 by J. Arnold, chronometer escapement with detent and compensated balance (S-type), the oldest extant watch by J. Arnold with detent escapement.

93a Very early marine chronometer by J. Arnold, London, no. 12, in octagonal mahogany case, 1782; 160 mm case width. One-day going, steel balance spring, chronometer escapement with detent, steel compensated balance, white enameled dial.
British Museum, London, ex Ilbert collection

93b Back of movement of marine chronometer no. 12 by J. Arnold, balance type-modified double T design.

94a Marine clock, signed "John Arnold & Son", No. 69, London, ca. 1875, in octagonal wooden case.
Asprey, London

94b Back of movement of marine chronometer "J. Arnold & Son" no. 69, chronometer spring detent escapement and Arnold's Z balance.

95a Silver pocket chronometer. signed "John Arnold No. 59, London", 1784, 75 mm diameter, thoroughly described in *Alte Uhren* No. 4, 1978.
University Observatory, Vienna

95b Movement of pocket chronometer "J. Arnold no. 59", movement modernized in Amsterdam in 1874.

96a Pocket chronometer in 22-karat gold case with outer case, signed Thomas Wright, in the Poultry, London 1784, 56 mm diameter, white enameled dial, pierced gold hands.
Bobinet, London

96b Rear view of movement of pocket chronometer (96a) with large cap jewel in cock, regulator, movement no. 2228, with full plates, gilded; jewel bearings for the fusee, Earnshaw escapement with combined escapement and discharge spring, patent number 1354 of 1783 (inventor Thomas Earnshaw); 3-armed compensated balance, blued flat steel spring (thoroughly described in Ant. Hor. IX, 1976, pp. 785 ff.)

96c Rear of the top plate of pocket chronometer (96a), disassembled.

96d Enlarged view of chronometer spring.

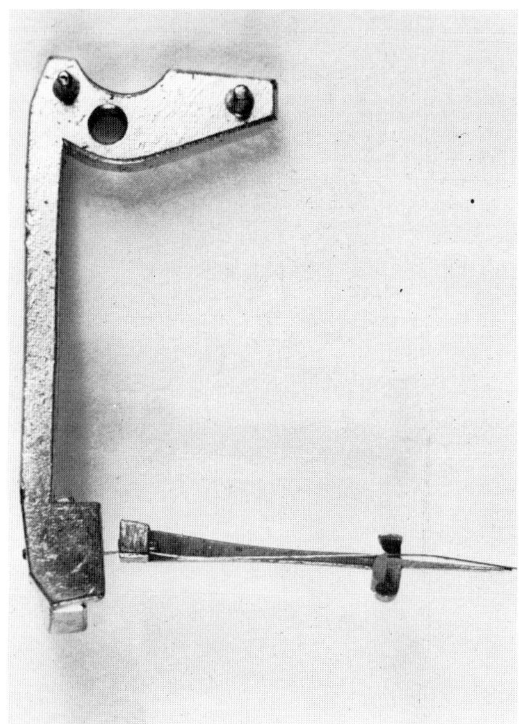

97a Pocket chronometer in silver case and outer case, signed "L. Kendall London, B + y" (HM 1786-87), 55 mm diameter, detent escapement, Harrison's remontoir, brass balance with spiral bimetallic compensation.
Clockmakers Company Museum, Guildhall, London

97b Rear of movement of Kendall's pocket chronometer.

98a Pocket chronometer in silver case with outer case, signed "Thos. Reid, Edinburgh, Chron. 26", 1788, 59 mm diameter, white enamel dial with seconds in the center and two separate dial circles for hours and minutes, gold hands.
Bobinet, London

98b Rear view of movement of Reid's pocket chronometer, gilded movement with full plates, escapement with Arnold's spring detent, escape wheel of polished steel, three-armed balance of brass with two added compensating segments.

99a Unusual bracket clock with chronometer movement, signed Margetts, London, No. 80, ca. 1790; white enameled dial, 114 mm diameter, Chinese rosewood case richly carved with Chinese landscapes, cliffs and figures.
British Museum, London, ex Ilbert collection

99b Movement of chronometer signed "Margetts, London no. 80", Earnshaw escapement, movement with steel pillars, originally equipped with a remontoir.

100b Movement of the chronometer by Bullock, one-day going, lever escapement after Mudge, very light 2-armed balance with unusual temperature compensation, regulator acting on balance spring.

a Dial of a marine chronometer by James Bullock, ndon, ca. 1790, in mahogany case, 190 mm length ngles.
tish Museum, London, ex Ilbert collection

101a Marine chronometer in octagonal mahogany case, signed E. Norton, London, ca. 1790; enameled dial 16.5 cm diameter with separated numeral circles for hours, minutes and seconds, with chain and fusee, Earnshaw escapement, with flat hairspring, original remontoir is missing.
British Museum, London, ex Ilbert collection

101b Back of movement of Norton's marine chronometer.

101c Side view of movement of Norton's marine chronometer.

101d Details of the Earnshaw escapement, escape wheel, spring detent and balance arbor.

▽ 101 d

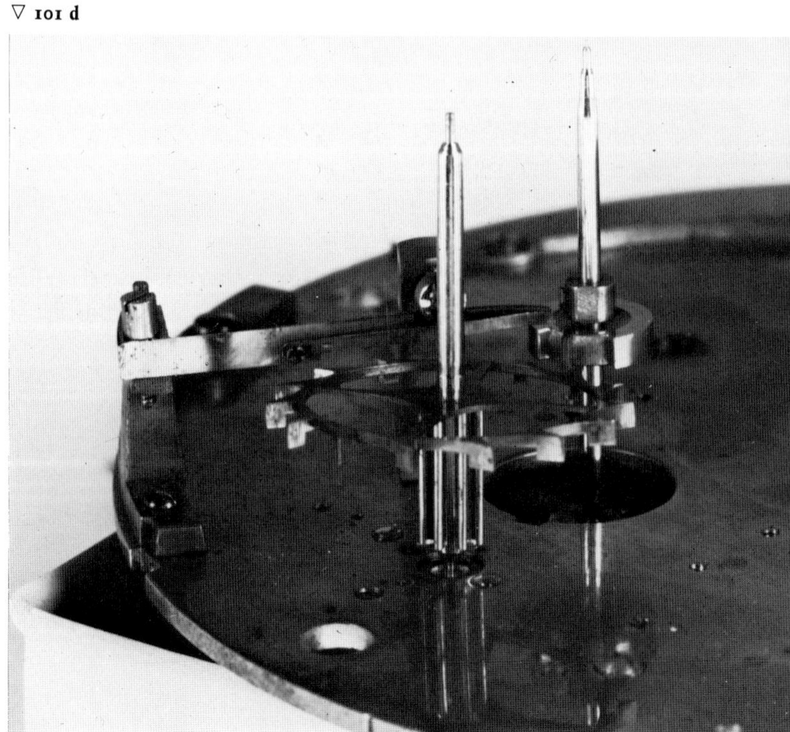

102a Pocket chronometer with silver case and outer case, signed "Thos. Earnshaw, London invt. et fecit", no. 58 (HM 1791/92), 48mm diameter.
Clockmakers Company Museum, Guildhall, London

102b Back of movement of pocket chronometer no. 58 by Earnshaw, movement number 1514m chronometer spring detent escapement, two-armed bimetallic balance.

101e Under-the-dial view.

103a Marine clock, signed "Josiah Emery (Charing Cross) London, No. 1"; unusual brass case with additional wooden carrying case, enameled face with three separate dials, gold hands. (Of the four identical specimens submitted to Greenwich for testing in 1792, this is the only one that is still extant.)
Clockmakers Company Museum, Guildhall, London

103b Side view of the movement of Emery's chronometer, unusual constant force escapement with two cylindrical springs that are tensed in sections by the escape wheel and release directly on the balance arbor. The balance has four bimetallic arms, roller bearings, and diamond cap jewels on its ends.

104a Gold pocket chronometer no. 2390, montre marine no. 16, signed Louis Berthoud, 1792; gold case with cancelled catch.
Th. Beyer Museum of Timekeeping, Zürich

104b Back of the movement of L. Berthoud's pocket chronometer, with "anti-friction roller" bearings for the balance arbor on the back plate.

104c Under-the-dial view of L. Berthoud's pocket chronometer.

104d Large-scale view of the escapement bimetallic compensated balance, riveted, not visible in the picture.

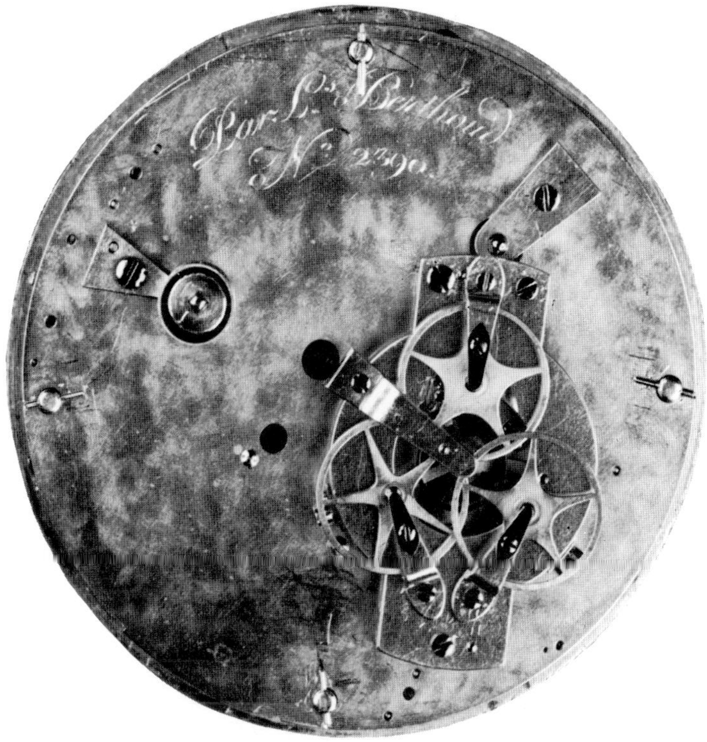

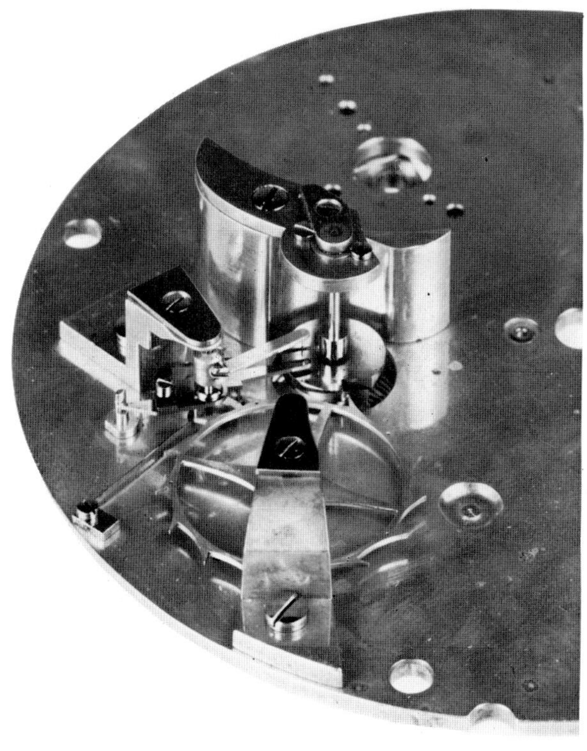

105a Marine chronometer (MM), signed Ferdinand Berthoud, no. 52 (ca. 1793), Paris, movement very much refined from Berthoud's earlier work; unusual gimbal mounting, vertical dial. Merchant marine version.
Arts et Métiers, Paris

105b Back of the movement of MM no. 52, very simplified construction, chronometer escapement, balance arbor in ruby bearings with similar cap jewels; two compensating weights on bimetallic strips built into the balance; direct spring drive.

106a Marine chronometer, signed John Arnold & Son, London, no. 68, ca. 1793; in carrying case with bronze handles.

106b Back of the movement of marine chronometer no. 68 by Arnold & Son, with typical Arnold Z balance.

106a

105b

106b

107a Marine chronometer, signed John Arnold & Son, London, no. 14/104, 1790, in octagonal mahogany case, 168 mm case width, silvered dial.
British Museum, London, ex Ilbert collection

107b Back of the movement of marine chronometer no. 14/104 by Arnold & Son, with typical Arnold Z balance, cylindrical balance spring made of gold, eight-day going.

108 Rear view of a marine chronometer movement by John Arnold & Son, no. 26, ca. 1791, octagonal mahogany case 160 mm wide, 30-hour movement, bimetallic balance in final Z construction, chronometer spring detent escapement, helical balance spring.
British Museum, London, ex Ilbert collection

109a Marine chronometer, signed "John Arnold & Son, London No. 78/168, invt. & fecit", ca. 1795; mounted directly in octagonal hardwood case, silvered dial.
Clockmakers Company Museum, Guildhall, London

109b Movement of marine chronometer no. 78-168 by Arnold & Son, eight-day movement, chain and fusee, spring detent escapement, compensated balance (Z type), helical steel spring with terminal curves under the balance to unburden the lower bearing.

110a Marine clock from the workshop founded (ca. 1796) by Thomas Mudge's son before his death, signed on the plate "Pennington, Pendleton and Others for the Son of the Inventor, No. 11"; with silver applique on the dial, one dial each for seconds and for minutes and hours together.
National Maritime Museum, Greenwich, London

110b Rear view of the marine chronometer, now equipped with Earnshaw escapement and balance, originally with Mudge's constant force escapement.

107a

107b

108

109a

109b

111a Marine chronometer movement by Thomas Earnshaw, London, no. 245, in mahogany case 190 mm wide, ca. 1795; one-day movement, silvered dial.
British Museum, London

111b Movement of Earnshaw's marine chronometer; helical balance spring, bimetallic balance with typical Earnshaw weights and adjusting screws.

112a Marine chronometer by Thomas Earnshaw, London, no. 509, ca. 1793-95, in gimbal ring, removed from its case.
British Museum, London, ex Ilbert collection

112b Movement of the Earnshaw marine chronometer, one-day going, signed "Thos. Earnshaw, Inv. et fecit, No. 509" and also "No. 2853."

113 Marine chronometer, signed Ferdinand Berthoud, no. 65, Paris 1796; case diameter 61.5 mm, movement height 42 mm; with large full balance and built-in compensating strip.
Musee International d'Horlogerie, La Chaux-de-Fonds

114a Small marine chronometer, signed Louis Berthoud no. 38, Paris 1799; silver case, specially prepared for Captain Boudin's Australian expedition with the ship Naturalista.
Antiquorum, Geneva

114c Inside of plate from L. Berthoud's marine chronometer no. 38, with escapement, movement diameter 70 mm.

114b Back plate of L. Berthoud's marine chronometer no. 38, with three-armed balance, chronometer detent with locking jewel, platinum screws in the balance.

114d Enlarged view of the balance with arbor and conical spring of L. Berthoud's marine chronometer no. 38.

115a

115b

115c △

115d △

115e ▽

115a Marine chronometer, signed Breguet no. 428, ca. 1800; in oak case 153 mm long, noteworthy achievement of the master, with numerous mechanical features. The specialties of this interesting developmental work were studied by A. G. Randall in the April 1979 issue of *Alte Uhren*.
British Museum, London, ex Ilbert collection

115b Enlarged view of the dial of Breguet's marine chronometer no. 428: dial circle for hours, minutes and seconds, plus three small rings to register the number of weak and hard shocks plus winding indication; winding from above, through the dial.

115c The movement of Breguet's marine chronometer no. 428 after removal of the dial.

115d Side view of the movement of Breguet's marine chronometer no. 428, 30-hour running with constant-force escapement, helical balance spring without terminal curves.

115e Movement of Breguet's marine chronometer no. 428 after removal of the top plate; the two parallel-working barrels that Breguet liked to use are easily seen.

115f Movement of Breguet's marine chronometer no. 428, front plate removed, showing the common winding apparatus for the two springs.

115g Bottom plate of Breguet's marine chronometer no. 428, with two-armed balance.

115h Vertical view of the bottom plate of Breguet's marine chronometer no. 428.

115i View of the plate of Breguet's marine chronometer no. 428, with the constant-force escapement.

115k Individual parts of the constant-force escapement of Breguet's marine chronometer no. 428.

115l Schematic drawing of the constant-force escapement of Breguet's marine chronometer no. 428; this escapement has been described as the "hammer-and-nail" system.

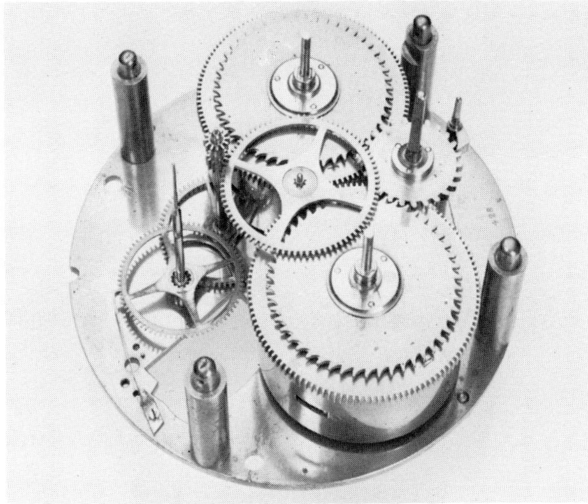

115f △

115g △

115h ▽

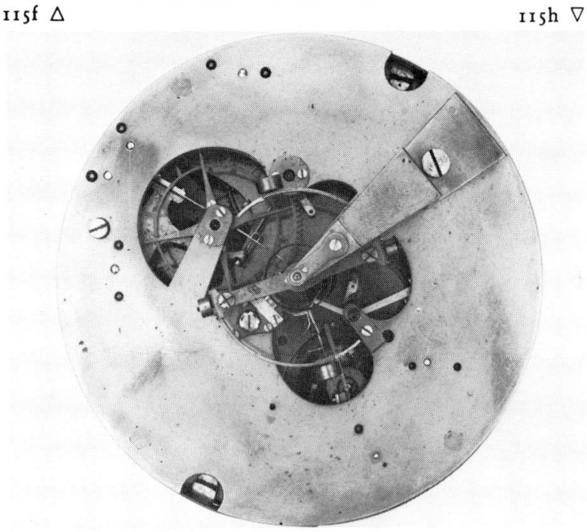

115i ▽

◁ 115l

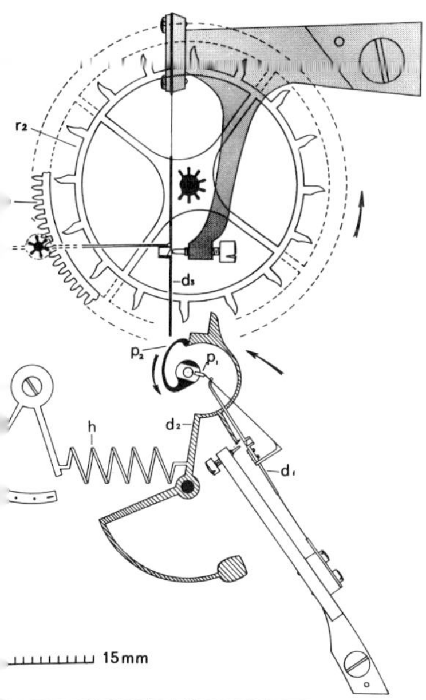

CONSTANT FORCE ESCAPEMENT.
A. L. BREGUET.
No. 428 circa 1798.

115k ▽

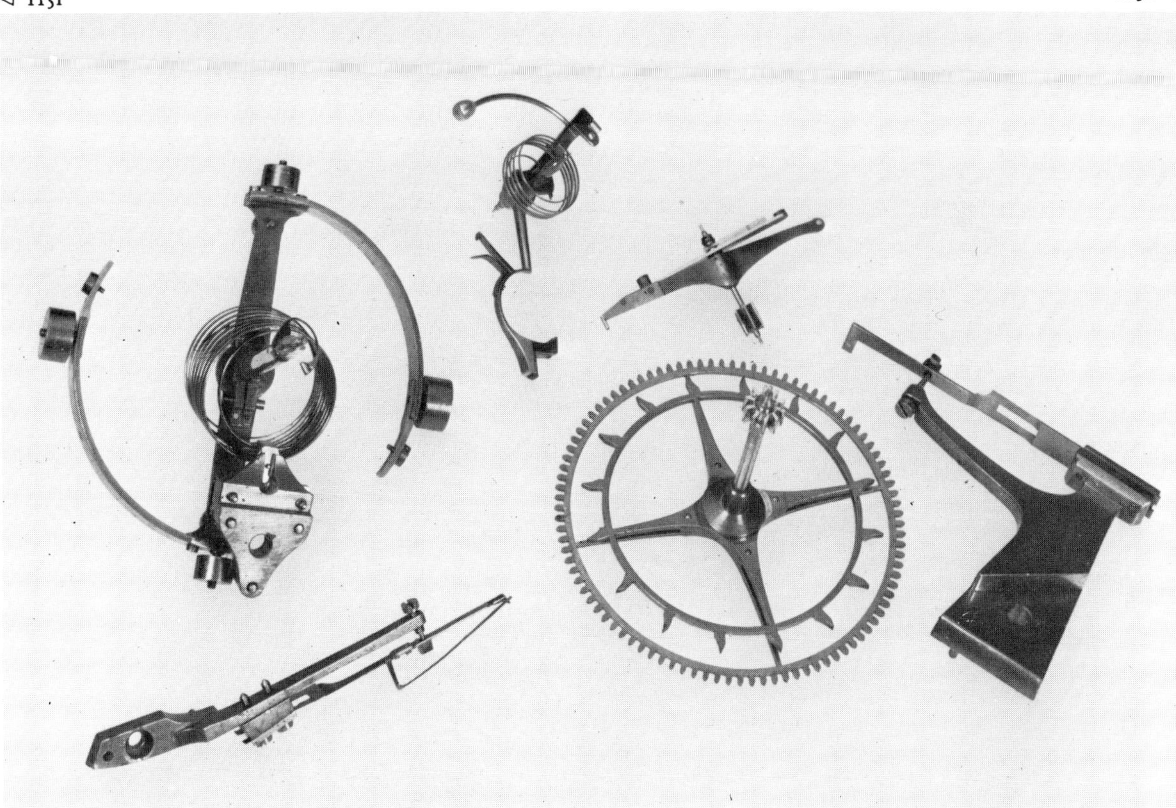

116a Dial of a pocket chronometer, signed "Chas. Haley, invt. et fecit London P/DN", end of the 18th Century; white enameled dial, 53 mm diameter.
Clockmakers Company Museum, Guildhall, London

116b Rear view of the movement of Haley's pocket chronometer, diameter 53 mm, with chronometer spring, helical balance spring, two-armed balance and constant-force escapement.

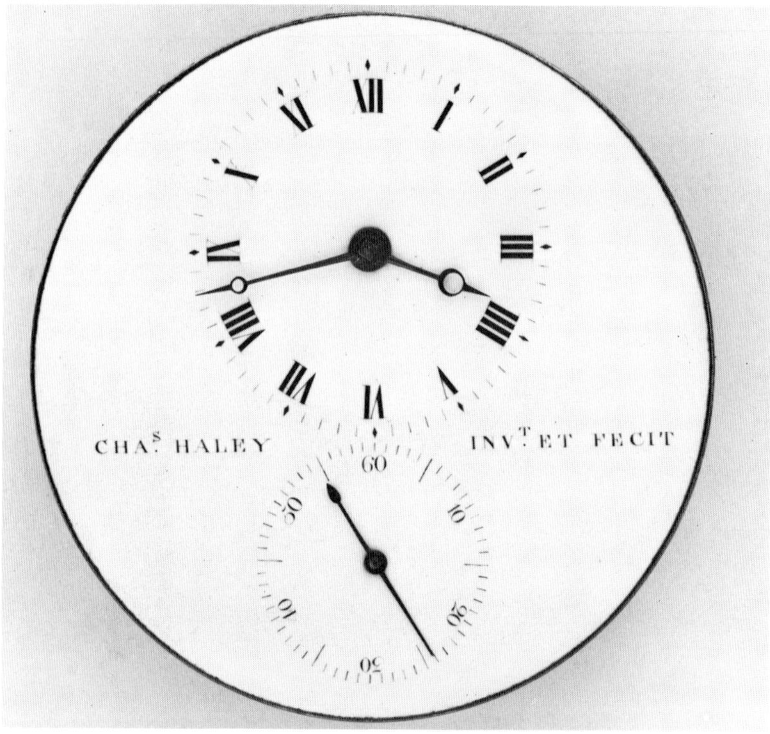

117a Pocket chronometer in silver case, signed "Barraud 32/5816, London"; 56 mm diameter, enameled dial (case and movement seem to have been united later); presumably the movement was built about 1805-1810 and delivered in a gold case.
Clockmakers Company Museum, Guildhall, London

117b Rear view of the movement of Barraud's pocket chronometer, the movement showing all of Earnshaw's typical features, such as "sugar-tongs" compensation, spring detent escapement and flat hairspring; the compensating element is fastened to a pusher so as to influence the rate in the manner of a regulator.

118a Pocket chronometer (deck watch), signed Thomas Earnshaw, London, no. 580, ca. 1800; 30-hour movement.
Privately owned

118b Rear view of the movement of Earnshaw's pocket chronometer no. 580 and added no. 3023, full balance, flat spiral spring, "sugar-tongs" compensation.

119a Pocket chronometer in silver case (HM 1793), enameled dial signed "JRA" (John Roger Arnold) no. 1677.
Privately owned

119b Rear view of the movement of the pocket chronometer signed by John R. Arnold, London, inv. et fec., with two-armed compensated balance (Z-balance), chronometer escapement after Arnold, chain and fusee.

120a Marine clock, signed "Horl. Longit — par Ferdinand Berthoud", no. XXXII, dial diameter 10 cm, with three separate dials for hours, minutes and seconds, now mounted in a glass cylinder for easy viewing of the movement; ca. 1800.
Arts et Métiers, Paris

120b Side view of the marine clock of F. Berthoud, no. XXXII.

121 Pocket chronometer in silver case, by Armand, ca. 1800.
A- Back of the movement, bimetallic compensated balance, ball-shaped hairspring.
B- Overall view, opened.
Kobstadtmuseet Åarhus

122 Side view of the movement of an unusually built marine chronometer with barrel, chain and fusee, two-armed compensated balance, after Arnold, and helical balance spring; signed J. R. Arnold, London 295 1800/1805.
Privately owned

123a Movement of a marine chronometer by Brockbanks, London, no. 436, ca. 1800. in octagonal mahogany case, 175 mm wide; removable section of movement, brass handles on the sides.
British Museum, London, ex Ilbert collection

123b Rear view of the movement of Brockbank's marine chronometer no. 436; movement with chain and fusee, one-day going, top plate and escapement added later, bimetallic balance, two-armed helical steel spring.

124a Movement of a marine chronometer, signed "Arnold & Son, London, no. 106/196:; square mahogany case, 24 cm wide, ca. 1800; 56-hour running, silvered dial, movement and case with later additions; gimbal mounting.
British Museum, London, ex Ilbert collection

124b Rear view of the movement of marine chronometer by Arnold & Son, no. 106/196; unusual compensated balance, chronometer spring detent escapement, helical balance spring with terminal curves.

125a Marine chronometer signed "Heinric Seyffert, Dresden", 1804, in cylindrical case, 148 mr diameter, with octagonal carrying case.
Privately owned, Lucerne

125b Dial of the Seyffert chronometer, with separate rings for hours, minutes and seconds, winding from above.

125c Rear view of the movement of the Seyffert chronometer, with fine adjustment for the balance and built-in thermometer.

125d Movement of the Seyffert chronometer, large undivided balance, two flat spiral hairsprings.

125e Fine adjustment of the Seyffert marine chronometer, fusee and chain.

126a Pocket chronometer in silver case (deck watch), signed Brockbanks London no. 422 (hallmark 1801), movement no. 422, 63.5 mm diameter.
Privately owned

127a Pocket chronometer in silver case, signed "Brockbanks London" no. 519, 64 mm diameter, white enamel dial, gold hands, described in Ant. Hor. Journal, Sept. 1963.
English, private collection

126b Rear view of the movement of Brockbank's pocket chronometer no. 422; Peto cross-detent escapement with added protective device against overswinging of the balance; see #88 for the special arrangement of the discharge spring.

127b Rear of the movement of Brockbank's pocket chronometer no. 519, plates fully gilded, jewel bearings for the fusee, three-armed compensated balance with buffer blocks to prevent overswinging, Peto cross detent escapement.

128a Pocket chronometer by Johann Heinrich Seyffert, no. 5, Dresden 1803; silver case, in wooden box, enameled dial.
Th. Beyer Museum of Timekeeping, Zürich

128b Back of the movement of J. H. Seyffert's pocket chronometer no. 5, fire-gilded movement with chronometer escapement and balance after Arnold, helical balance spring.

129a Pocket chronometer, signed George Jamison, Charing Cross, London, ca. 1802; the gold case (hallmark 1802) by Thomas Howell of Bath; in wooden box with key; enameled dial, with Celsius thermometer scale near the 12.
Th. Beyer Museum of Timekeeping, Zürich

129b Rear of the movement of G. Jamison's pocket chronometer; fire-gilded movement with chain and fusee, chronometer escapement, helical balance spring, large sapphire bearing on winding.

130a Marine chronometer in wooden case, without gimbal mounting, signed Jn. U. Etter (Bonn), ca. 1800-05; enameled dial, 53 mm diameter, minute in the center, separate minute and hour indication, small second near "40", 30-hour running.
The Time Museum, Rockford, Illinois

130b Movement of Etter's marine chronometer; lever escapement after Grant with exactly equal movement and resting and two opposite crown-geared escape wheels; Arnold's "SS" balance, helical balance spring, chain and fusee.

131a Marine chronometer, signed James Hatton, London, no. 286, ca. 1805; eight-day movement, mahogany case, dial diameter 12.5 cm.
Asprey, London

131b Rear view of the movement of James Hatton's marine chronometer no. 286, Earnshaw spring detent escapement, segmented balance weights, barrel, chain and fusee.

132a Dial of a small marine chronometer, signed Louis Berthoud, Paris, no. 49, ca. 1805; mahogany case, 122 mm side edges, movement in machined silver case, separate dial rings for hours, minutes and seconds.
British Museum, London, ex Ilbert collection

132b Movement of marine chronometer no. 49 by L. Berthoud; detent, conical balance spring, 24-hour movement.

133a Marine chronometer, signed J. R. Arnold, London; test version, ca. 1803, completed with dial and numbered 425, ca. 1815 and delivered to the British Admiralty.
Wood, Edinburgh

133b Movement of marine chronometer no. 425 by J. R. Arnold; direct drive from barrel, spring signed "Robt. Clark, 1803"; Arnold's escapement and balance.

134a Marine chronometer, signed "Wm. Hardy invt. et fect., London No. 2", early 19th Century; wooden case with gimbal mounting.
Clockmakers Company Museum, Guildhall, London

134b Movement of Hardy's marine chronometer no. 2, spring detent escapement, Hardy's compensated balance, direct drive from the barrel.

135a Dial of a small marine chronometer, signed "William Hardy, London, no. 53", mahogany case 75 mm wide, ca. 1805; 24-hour going, silvered metal dial.
British Museum, London

135b Back of the movement of W. Hardy's marine chronometer no. 53; helical steel balance spring, Hardy's patented balance with one crossarm and two weight arms ending in radial surfaces bending inward or outward.

4a △

134b ▽

135a △

135b ▽

136a Pocket chronometer, signed Thomas Earnshaw, London, no. 777, ca. 1807; gold double case.
Th. Beyer museum of Timekeeping, Zürich

136b Rear view of the movement of Earnshaw's pocket chronometer no. 777, movement with chain and fusee, chronometer escapement, three-armed flat, highly polished steel balance with flat hairspring, regulator device with bimetallic "sugar-tongs" temperature compensation.

137a Pocket chronometer, signed "Abraham Louis Breguet, Paris, 1809", no. 224; this watch was used by Alexander von Humboldt on his travels through South America.
Th. Beyer Museum of Timekeeping, Zürich

137b Movement of pocket chronometer no. 224 by Breguet, with chain and reversed fusee, chronometer spring detent escapement after Arnold, three-armed bimetallic compensated balance, helical balance spring, "parachute" shock repeating.

138a Pocket chronometer in 18-karat gold case and outer case (hallmark 1808), signed "Pennington London 117/511", 65 mm diameter, white enameled dial, gold hands.
Privately owned in Britain

138b Rear view of the movement of Pennington's pocket chronometer no. 157/511, gilded movement with full plates, Arnold's spring detent escapement, Pennington's balance, blued helical hairspring, "silent" 5-minute repeating mechanism after Stogden.

138c Under-the-dial view of Pennington's pocket chronometer shows the repetition mechanism.

139a Pocket chronometer (deck watch) in silver outer case, signed "Brockbanks London, no. 4961", ca. 1810.
National Maritime Museum, Greenwich

139b Movement of Brockbank's pocket chronometer with outer case and dust cover, 40-hour going, three-part balance with segmented weights, flat hairspring, plus fine adjustment with regulator.

140a Marine chronometer by Brockbanks, London, no. 632, ca. 1810, in rosewood case 175 mm wide, white enameled dial.
British Museum, London, ex Ilbert collection

140b Movement of Brockbank's marine chronometer, with Brockbank's three-armed balance, two-day movement.

141a Marine chronometer, signed Breguet et Fils no. 2068, third series, 1816; hour and minute hands from the, small second dial, in mahogany case with gimbal mounting, silver-plated.
Ineichen, Zürich

141b Rear view of the movement, Robin escapement working on both sides, three-armed compensated balance, blued helical steel spring with terminal curves, chain and fusee.

142 Marine chronometer movement by A. L. Breguet, no. 3529; two parallel-working spring barrels without fusee, two-armed compensated balance, small silvered dial, 8.8 cm diameter, ca. 1810.
Bobinet, London

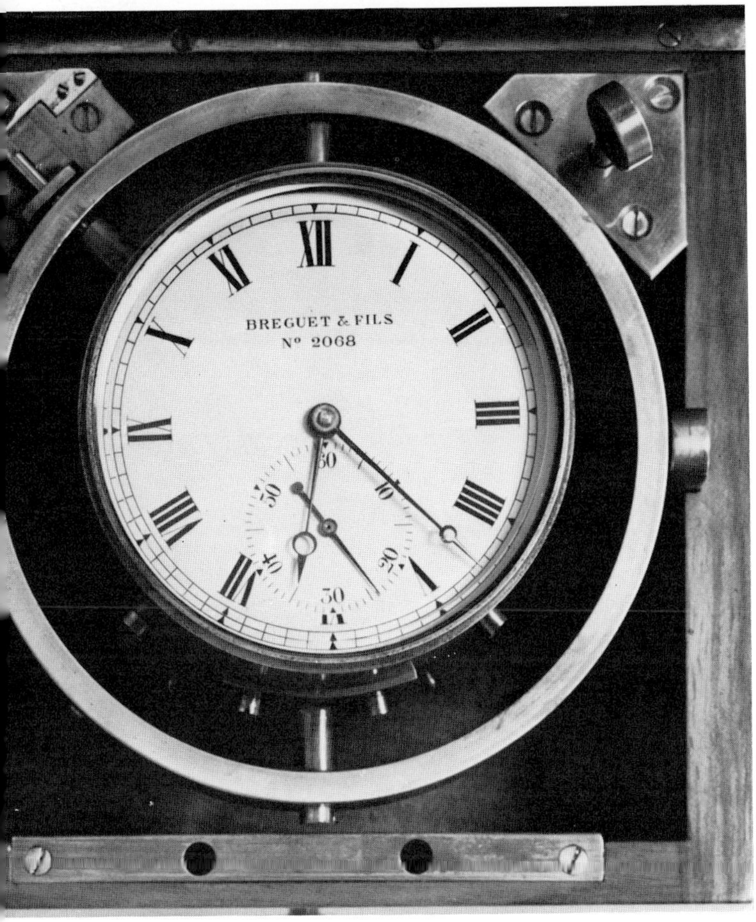

143a Marine chronometer, signed Frodsham & Baker no. 2503, ca. 1810; dial diameter 8.5 cm, with on-off dial, in mahogany case with gimbal mounting.
Asprey, London

143b Movement of the chronometer, signed Baker & Frodsham; compensated balance with unusual combination of weights, helical balance spring.

141a △ 143a ▽ 143b ▽

144a Dial of a marine chronometer, signed "Louis Berthoud no. 127", ca. 1810; separate dials for hours, minutes and seconds; mahogany case.
British Museum, London, ex Ilbert collection

144b Back of the movement of marine chronometer no. 127 by L. Berthoud, one-day movement, typical Berthoud bridge over the balance.

145 Back of the movement of a marine chronometer signed "J. R. Arnold, London, invt et fecit, no. 379"; Arnold's Z balance, bimetallic compensation, helical blued steel hairspring between balance and plate.
National Maritime Museum, Greenwich

146 Pocket chronometer, signed "Thos. Earnshaw invt. et fecit, no. 630, London"; originally fitted with lever escapement, later rebuilt with "English" lever; Earnshaw's compensated balance, helical steel hairspring, silver outer case (HM 1814-15), movement diameter 4.5 cm.
National Maritime Museum, Greenwich

148a Pocket chronometer, signed "Brockbanks, London no. 700", in gold double case (hallmark 1812-13), four-colored, with chased floral ornamentation, gold dial with inlaid second dial, the hour ring is also enameled in blue, diameter 61 mm.
Clockmakers Company Museum, Guildhall, London

148b Rear view of the movement of Brockbank's pocket chronometer no. 700, three-armed bimetallic compensated balance, spring detent escapement, helical balance spring and striking on three gong springs.

147 Movement of a pocket chronometer, signed "Brockbanks London, no. 778", with chain and fusee, three-armed compensated balance with helical weights, in silver case (hallmark 1812-13), with additional engraving "No. 2285, reparé par Breguet."
Privately owned in Britain

149a Movement of a marine chronomete
by A. L. Breguet, ca. 1815, "echappemen
naturel", three-armed compensate
balance, chain and fusee.
Privately owned

149b Enlarged view of the three-armed compensated balance of Breguet's marine chronometer.

150a Dial of an unusually small marine chronometer, signed "Charles Young, London, no. 503", ca. 1817, mahogany case 95 mm wide; gilded dial with silvered second dial on it.
British Museum, London, ex Ilbert collection

150b Movement of marine chronometer no. 503 by Charles Young, with three-armed balance, six segmented weights and Breguet hairspring, one-day movement.

151 Rear of the movement of a pocket chronometer, signed "Reid & Son, London, no. 7812", ca. 1810; unusual combination of a spring escapement after Earnshaw, with Arnold's "Z" balance.

152a Marine chronometer, wooden case with silver pocket chronometer with gimbal mounting built in, signed "Urban Jürgensen" (Copenhagen), ca. 1815; silver case, enameled face with three separate dials for hours, minutes and seconds.
Privately owned

152b Side view of the movement of Urban Jürgensen's pocket chronometer, two-armed compensated balance, helical balance spring.

152c Rear view of the movement of U. Jürgensen's pocket chronometer with two-armed compensated balance and ten adjusting screws on each half of the balance.

153a

153a-b Marine chronometer by Urban Jürgensen, in mahogany case, ca. 1820; signature on mounting "Urban Jürgensen no. 44, Kjöbenhavn, no. 44 fecit"; contains a pocket chronometer, enameled dial with hour and minute hands from the center and small second dial.
Privately owned

153c Enlarged view of the pocket chronometer movement from the marine chronometer signed "Urban Jürgensen XLIV", gilded plates, three-armed compensated balance.

3b △

154 Marine chronometer by Urban Jürgensen, one of the few pieces by this master that has existed until recent times, but unfortunately was stolen a few years ago; hour and minute hands from the center, small but clearly marked second dial.
"De Gamle By" Museum, Aarhus

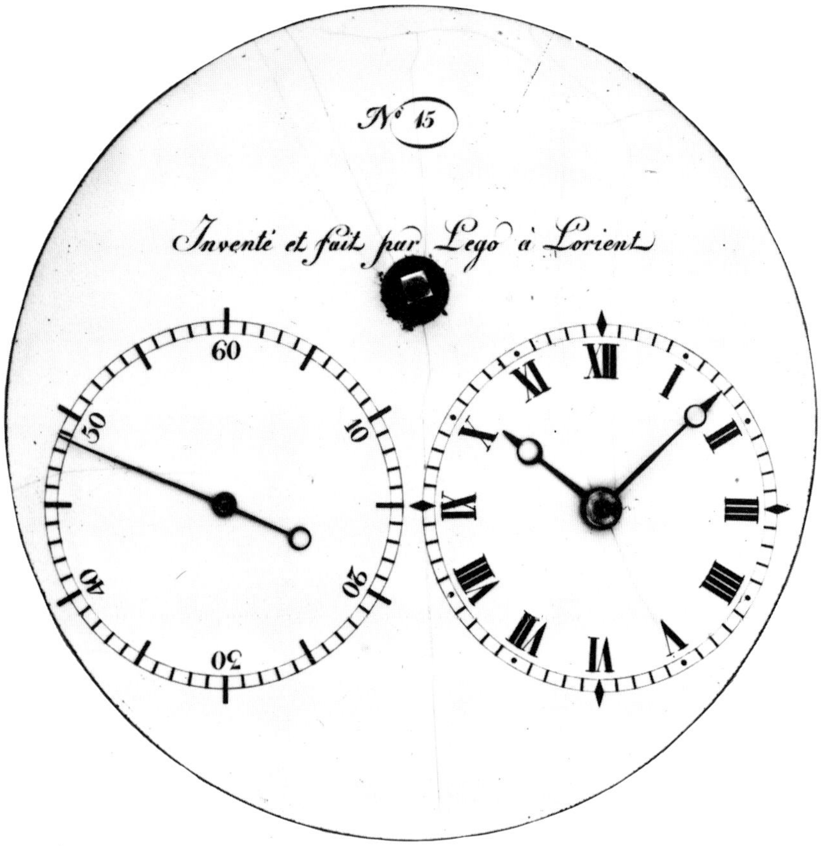

155a △

▽ 155b

156a

15

155a Dial of a marine chronometer in a mahogany case, signed "Inventé et fait par Lego à 1, Orient, No. 15", ca. 1818; case 125 mm wide, white enameled dial.
British Museum, London, ex Ilbert collection

155b Rear view of the movement of marine chronometer no. 15 by Lego, with unusual structure of the movement built up on the front plate, direct barrel drive without chain, four-armed compensated balance.

◁ 157a

157b △

158a △

156a Marine chronometer in mahogany case with elliptical gimbal ring, signed "A. L. Breguet, Paris", ca. 1818; with large enameled dial, one circle for hours and minutes and a separate one for seconds, two front winding holes.
Privately owned

156b Movement of the marine chronometer by A. L. Breguet, two barrels for direct drive without chain and fusee, two-armed compensated balance with helical balance spring, movement built up on front plate with bridges.

157a Silver pocket chronometer, signed "Barwise, London", no. 7604 (hallmark 1816).
London art trade

157b Movement of the Barwise pocket chronometer, with two-armed compensated balance.

158a Marine chronometer, signed "E. Emanuel, Portsmouth", dial diameter 127 mm, in mahogany case with gimbal mounting, ca. 1820.
Rijksmuseum "Nederlands Scheepvaart Museum", Amsterdam.

158b Movement of the marine chronometer by Emanuel, eight-day movement, compensated balance, helical balance spring with only six windings, chain and fusee.

158b ▽

159a △

159b △

160a ▽

160b

161a △

161b △

159a Movement of Barraud's marine chronometer no. 986, at left the bridge mounting for the large barrel, eight-day movement, plus an auxiliary plate for the balance and escapement, two-armed compensated balance, with auxiliary compensation after Molyneux, rebuilt after 1840.

159b Marine chronometer, signed "Barraud's No. 986", mahogany case with gimbal mounting, movement turned for removal, unusual permanent winding apparatus, probably once owned by the East India Company, ca. 1825
British art trade

160a Marine chronometer in mahogany case, signed "Barraud, London, No. 1675", ca. 1835, with up-down indication in a segment of the dial.
Musée International d'Horlogerie, La Chaux-de-Fonds

160b Movement of marine chronometer no. 1675 by Barraud, large movement with chain and fusee for eight-day going, separate plate for escapement and balance.

161a Movement of Barraud's marine chronometer no. 2/570, movement with eight-day going, spring detent escapement, plate for the escapement set on the bottom plate, Molyneux-type balance to eliminate the middle temperature error.

161b Marine chronometer with gimbal mounting, in mahogany case, ca. 144 mm long; signed "Barraud's London No. 2/570:; ca. 1824, 24-hour dial.
British Museum, London, ex Ilbert collection

162 Movement of a marine chronometer, signed "Barraud, London No. 731", ca. 1820, eight-day running, movement diameter 12.7 cm, permanently built-in winding with two wings folding out, special plate for escapement and balance.
National Maritime Museum, Greenwich

163a Dial of a marine chronometer in square mahogany case 165 mm wide, signed "Barrauds, London, No. 982", ca. 1824; silvered dial with winding indication.
British Museum, London, ex Ilbert collection

163b Movement of Barraud's marine chronometer no. 982, with chain and fusee, eight-day movement, plate for escapement and balance, very modernized.

164a

164b

Opposite page, below
164a Marine chronometer, signed "Berthoud Frères No. 215" (Sons of Louis Berthoud), Paris, ca. 1820; mahogany case, lever set in case with rhombic brass plate, brass inner case, gimbal mounting, movement in its own closing brass case for use as deck watch, original key, silvered dial with three steel hands.
Th. Beyer Museum of Timekeeping, Zürich

164b Marine chronometer case, gimbal mount, movement removed.

164c Movement and upper plate of marine chronometer no. 215 by Berthoud Frères, dial removed, all bearings with steel caps, fine workmanship, chain and fusee, wheels and pinions with large numbers of teeth.

164d Movement of the Berthoud Frères marine chronometer no. 215, bottom plate removed, barrel, fusee and wheels.

164e Movement of the Berthoud Frères marine chronometer no. 215; three-armed compensated balance, helical steel balance spring without terminal curves, diamond cap jewel, four massive steel pillars with steel nuts.

164f Enlarged view of the escapement of Berthoud Frères marine chronometer no. 215, chronometer pivoted detent escapement, ruby bearings.

164c △

164e △

164d ▽

164f ▽

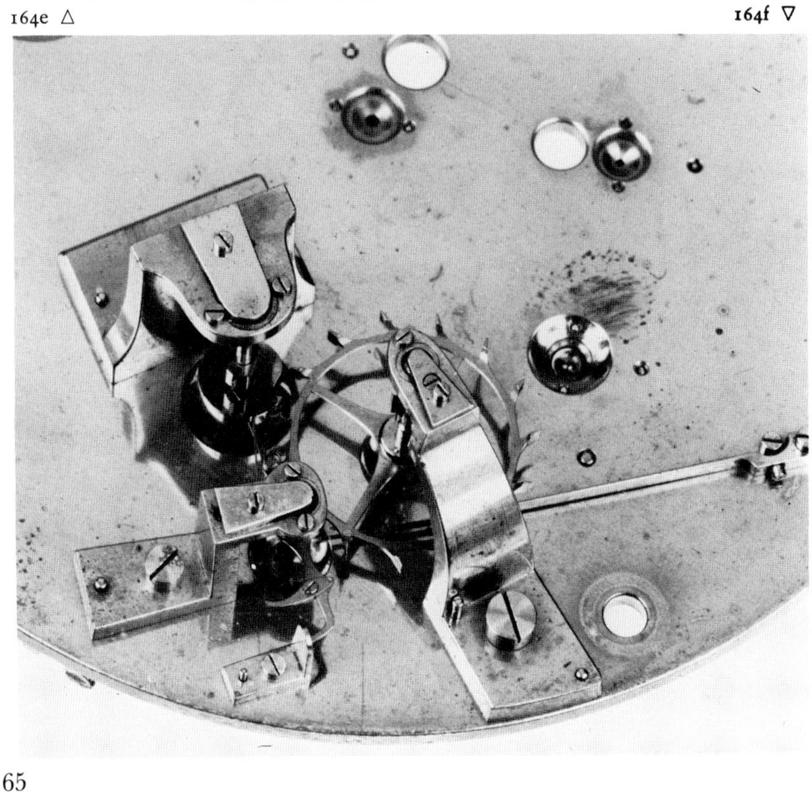

165a Marine chronometer in mahogany case, signed "A. L. Breguet, No. 5028, Paris", ca. 1820, with unusual large gimbal mounting, enameled face with separate dials, one for hours and minutes, the other for seconds.
Privately owned

165b Movement of marine chronometer no. 4028 by Breguet, with two-part compensated balance, helical balance spring and chronometer spring detent escapement, 40-hour going.

166a Marine chronometer in mahogany case with large gimbal mount, unsigned, but probably from the circles of the great French masters around 1820; separate dials for hours, minutes and seconds on enameled face; thorough description of this interesting chronometer in *A Tour-de-Force in Constant Force,* Ant. Hor., December 1974.
Privately owned

166b Dial of the marine chronometer with gimbal ring.

166c Movement, showing constant-force escapement after removal of balance and its bridge.

166d Movement of the marine chronometer, with balance.

166e Enlarged view of the three-piece compensated balance of the marine chronometer; hemispheric balance spring, one second beat.

6b △ 166c ▽ 166d △ 166e

167a Traveling chronometer in mahogany case, signed "Honoré Pons" (St. Nicolas d'Aliermont), ca. 1820; silver dial for vertical use. *Arts et Métiers, Paris*

167b Side view of the movement of H. Pons' traveling chronometer, bimetallic balance and helical steel balance spring.

167c Rear view of the movement of H. Pons' traveling chronometer, showing arrangement of the escape wheel and unusual lock for it, plus rear winding.

168a Dial of a small marine chronometer, signed "Gibbon's Patent Remontoire", ca. 1820; mahogany case 165 mm wide.
British Museum, London, ex Ilbert collection

168b Movement of the marine chronometer by Gibbon, compensated two-armed balance after Earnshaw.

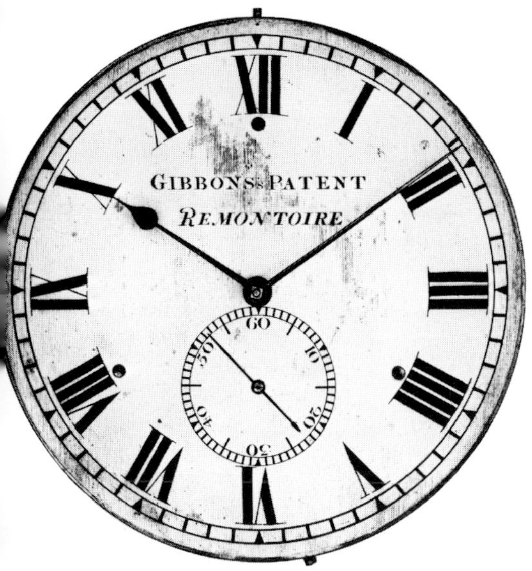

169a Pocket chronometer in silver case, signed "Dent No. 6003, London", ca. 1840, dial diameter 52 mm.

169b Rear view of Dent's pocket chronometer no. 6003, two-armed compensated balance, one-day going.

170a Marine chronometer in wooden case without gimbal mounting, signed "H. Motel, Paris, No. 25", ca. 1820.
Time Museum, Rockford, Illinois

170b Movement of the marine chronometer, plate signed "Construite par Hri Motel" and no. 25, unusual bridge for the compensated bimetallic balance, sliding adjusting weights, semicircular scale below, twice 0 +/-30 degrees.

171a Marine chronometer in mahogany case with gimbal mounting, signed "Hri Motel, No. 38, Prais", pre-1823; front winding by a lever in the brass rim around the dial.
Privately owned

171b Enlarged view of the dial of H. Motel's marine chronometer no. 38.

171c Rear of the movement of H. Motel's marine chronometer no. 38, four-armed compensated balance, conical hairspring, chronometer rocker after L. Berthoud.

171d Side view of the marine chronometer movement of H. Motel's no. 38, seen diagonally from above toward the balance bridge, showing conical hairspring.

172a Marine chronometer in mahogany case with gimbal mounting, 160 mm long, silvered dial, sliding lid on case, signed Motel no. 44, Paris, ca. 1830.
British Museum, London, ex Ilbert collection

172b Movement of marine chronometer no. 44 by Motel, 24-hour going, chronometer escapement with detent.

173 Movement of a pocket chronometer, signed "John Riviere, Bath, N0. 1246" with added "Cummins's Imp. Chronometer", ca. 1825; escapement with combination of Arnold's and Earnshaw's ideas, two-armed compensated balance with added Molyneux-type apparatus to eliminate the middle temperature error, scale under balance for amplitude control.
Privately owned

174 Movement of a pocket chronometer, signed "Barrauds, London No. 858"; ca. 1850, with two-armed balance and added Molyneux-type compensation to eliminate the middle temperature error, chronometer spring detent escapement.
Privately owned

175 Side view of the first chronometer built by William C. Bond (born 1789) in the USA, finished in Boston in 1812. He used the weight drive because he could get mainsprings from England, which he didn't want to do because of both patriotism and the Napoleonic Wars in Europe. Otherwise the movement much resembles British examples. Bond used the chronometer spring detent escapement after Earnshaw, with the balance between the plates.
Smithsonian Institute, Washington D.C.

176 Marine chronometer by John Gottlieb Ulrich (1795-1875), ca. 1828, with Ulrich's constant-force escapement. See #204b.
The Mariners Museum, Newport News, Virginia.

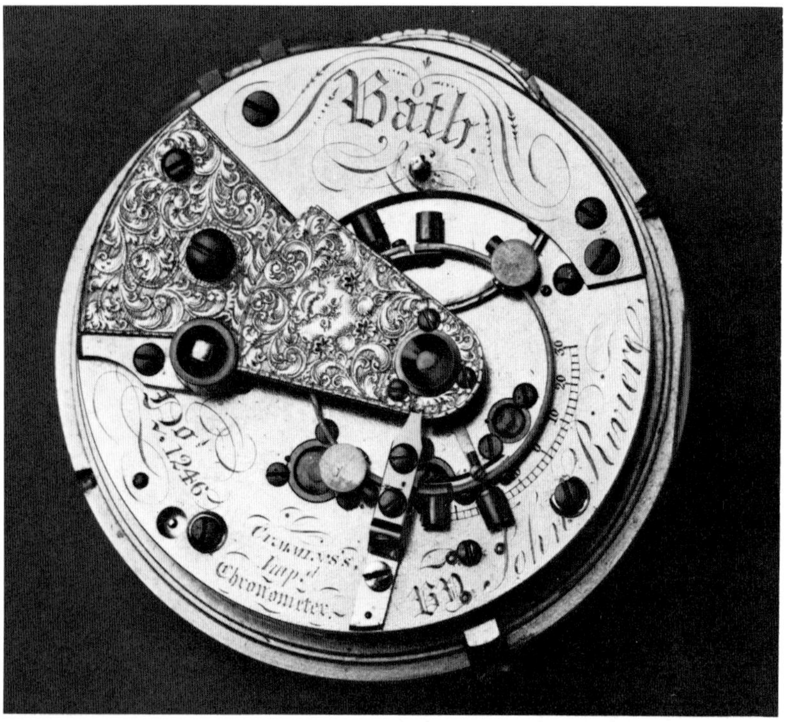

177a Marine chronometer in mahogany case, signed Arnold No. 435, with gimbal mounting, ca. 1815.
British art trade

177b Movement of Arnold's marine chronometer no. 435; compensated two-armed balance, helical hairspring with terminal curves, one-day running.

178a Marine chronometer in lacquered brass case, signed "John Roger Arnold, No. 565", London, ca. 1825; dial diameter 78 mm.
Privately owned in Switzerland

178b Back of the marine chronometer no. 565 by Arnold, two-day going, Arnold balance spring, new balance (type U and O by Arnold), helical gold balance spring with terminal, balance and escape wheel plus second-wheel arbor mounted in ruby bearings with cap jewels.

77a △ 178a ▽ 177b △ 178b ▽

179a Pocket chronometer, signed "Breguet et Fils", with mahogany case and gimbal mounting, Paris, ca. 1825; enameled dial with two intersecting semicircles, one for minutes and hours, the other for seconds, 6.2 cm diameter.
German Museum, Munich

179b Enlarged view of the movement of the pocket chronometer with two separately wound but parallel-functioning main spring barrels, each with a Breguet lock.

180a Pocket chronometer in silver case, signed Duchemin à Paris, No. 342, with enameled dial, ca. 1825.
Privately owned

180b Movement of Duchemin pocket chronometer no. 342, ca. 70 mm diameter, with two separately wound but parallel functioning mainsprings, obvious influence of Breguet; Duchemin is not cited by Tardy.

181a Marine chronometer in round metal case, signed "Breguet et Fils, No. 4514", ca. 1826; silvered dial, blued steel hands, unusual, partly intersecting circles, 70 mm diameter.
Privately owned

181b Movement of marine chronometer no. 4514 by Breguet et Fils, with two separately wound but parallel functioning mainspring barrels, compensated balance, helical hairspring.

182a Marine chronometer in mahogany case with unusual gimbal suspension, signed "Breguet et Fils No. 4310" (third series, begun in 1806 as no. 431 but finished only in 1826); enameled dial with three separate circles for hours, minutes and seconds, front winding.
Privately owned

182b Movement of the Breguet marine chronometer. two-day going, two parallel barrels, special lever escapement with exactly equal-lift; a very interesting experimental piece by Breguet.

182c The view of the "rocking bearing" drive via two toothed wheels with number of teeth differing by one ("Ferguson's paradox").

182a

182b

182c ▷

182d Enlarged view of the two-armed compensated balance of Breguet's marine chronometer, balance pivot runs in a "rocking bearing" of sapphire for better distribution of lubricants.

182e Side view of the balance system of Breguet's marine chronometer, helical hairspring and special pinion for the two drive wheels of the "rocking bearing."

182f Enlarged view of the lever escapement removed from Breguet's marine chronometer.

182g Enlarged view of the balance system of Breguet's marine chronometer.

2d △

182e ▽

183a Marine chronometer in mahogany case with gimbal mounting, signed "Brockbanks & Atkins, No. 1007", London, ca. 1825-30; silvered dial with winding indicator.
Danish Commercial and Maritime Museum, Kronborg Castle

183b Rear of the movement of Brockbanks & Atkins marine chronometer no. 1007, 48-hour going, Earnshaw-type balance, blued steel helical hairspring, chronometer spring detent escapement, chain and fusee.

184a Enlarged view of a chronometer in protective case, removed from its box, signed "Brockbank & Atkins, London No. 1191", ca. 1825, dial diameter 11.3 cm.
National Museum of History & Technology, Smithsonian Institute, Washington D.C.

184b Back of the movement of Brockbank & Atkins marine chronometer, eight-day movement with chain and fusee, chronometer spring detent escapement, two-armed bimetallic compensated balance, separate plate for escapement and balance.

185a Marine chronometer in mahogany case with oval gimbal mounting, signed "Breguet et Fils No. 3151", ca. 1825.
Asprey & Co., London

185b Movement of the Breguet & Fils marine chronometer, separately wound, parallel functioning spring barrels, two-armed compensated balance, helical hairspring, chronometer spring detent escapement.

186a Marine chronometer in wooden case with gimbal mounting, signed "Robt. Molyneux, London", ca. 1827-28, with 24-hour dial, separate second plate, overall diameter 10.5 cm.
National Maritime Museum, Greenwich

186b Movement of the Molyneux marine chronometer, Earnshaw balance, chronometer spring detent escapement, movement with chain and fusee.

187a Marine chronometer in walnut case, 235 mm wide, sliding lid, unusual gimbal mounting, signed "Breguet et Fils, No. 4591", ca. 1828.
Privately owned, Lucerne

187b Movement of Breguet et Fils marine chronometer no. 4591; two-day movement without chain and fusee, two-armed compensated balance with two regulating and two fine adjusting screws, chronometer spring detent escapement, white helical hairspring without regulator or terminal curves, spring attachment block with three setting screws for adjustment of the hairspring's height and circumference without bending it, angled lifting jewel.

187a △

188a ▽

187b △

188b

189a Marine chronometer in mahogany case with observation window in lid, signed "Breguet et Fils, No. 3073", ca. 1830, *Privately owned*

189b Rear view of the movement of the Breguet et Fils marine chronometer no. 3073, with two separately wound, parallel functioning spring barrels.

188a Marine chronometer in mahogany case, 255 mm long, signed "Breguet et Fils, Paris, No. 2741", 1813-1822; silvered dial, minute and hour hands at left, small second above in the center, independent second dial as stopwatch with minute counter at right, up-and-down indication for both spring barrels at bottom.
British Museum, London, ex Ilbert collection

188b Movement of Breguet et Fils marine chronometer no. 2741, eight-day movement, chain and fusee, lever escapement, two-armed compensated balance, helical hairspring with terminal curves, wheels and independent second driven by a geared barrel (foreground).

189c Enlarged view of the going system of Breguet et Fils marine chronometer, two-armed compensated balance, helical hairspring with terminal curves, balance arbor with "parachute" shock absorber.

190a Dial of a marine chronometer in mahogany case without gimbal mounting, signed "Hri Motel Hger de la Marine Rle No. 103", Paris, ca. 1830; with lever in brass ring around the dial, winding under it.
Privately owned

190b Rear view of the movement of Motel's marine chronometer no. 103, with conical balance spring, chain and fusee. Pivoted detent after L. Berthoud and three-armed compensated balance with 18,000 beats per hour.

190c Movement of Motel's marine chronometer no. 103.

190d Rear view of the movement of Motel's marine chronometer no. 103, with bridge and balance removed to show escape wheel and detent.

191a Marine chronometer, signed "Hri Motel Hger de la Marine Rle No. 106", Paris, ca. 1830, with lever in brass ring around dial, winding under it. *Privately owned*

191b Movement of Motel's marine chronometer no. 106, two-armed balance with unusual round adjusting weights, conical hairspring, escapement and balance combined in a single system.

191c Rear of the movement of Motel's no. 106, brid wheel and unusual chronometer spring detent escapement in Arnold's manner, with pulling tension.

191d Enlarged view of the balance, escape wheel and chronometer spring detent of Motel's chronometer.

192a Enlarged view of the dial of a marine chronometer in mahogany case with gimbal mounting, signed "Hri Motel No. 119", Paris, ca. 1830; silvered dial with three separate circles for hours, minutes and seconds; diameter 9 cm, side winding from the front.
Privately owned

192b Movement of Motel's marine chronometer no. 119, four-armed balance with segment-shaped regulating weights, movement with chain and fusee.

192c Enlarged view of the front winding of Motel's marine chronometer no. 119.

193a Pocket chronometer in silver case, signed "M. Julien Huguenin", Paris, ca. 1830; dial diameter ca. 59 mm, probably the masterpiece of a Neuenburger living in Paris.

193b Movement of Huguenin's pocket chronometer, 32-hour going, chronometer pivoted detent escapement after L. Berthoud, geared barrel, winding through the lid, 3-armed compensated balance with movable weights and three platinum regulating screws, blued helical hairspring without regulator; the whole mechanism is mounted in rubies, balance and detent with cap jewels.

194a Marine chronometer in mahogany case with gimbal mounting, dial diameter 8.5 cm, with convex glass lid, signed "James McCabe, No. 172, London", ca. 1830.
Asprey, London

194b Movement of McCabe's marine chronometer No. 172, with chronometer spring detent escapement, blued steel balance spring, balance arbor mounted with diamond end-stone.

195a Chronometer in wooden case, signed "John Antes, London", ca. 1825-30; with enameled dial and three circles for hours and minutes, seconds, and date.
Christie's, London

195b Movement of Antes' chronometer, chain and fusee, bimetallic strip for temperature compensation below.

196a Marine chronometer in mahogany case with gimbal mounting, signed "Morris Tobias, London, No. 739", ca. 1830-40; dial diameter ca. 95 mm.
National Scheepvaartmuseum, Antwerp

196b Movement of Tobias' marine chronometer no. 739, two-day going, movement with chain and fusee, movement diameter 79 mm, two-armed compensated balance and helical hairspring.

197a Movement of a marine chronometer in a leather case, 16 cm high, 15 cm wide, signed "Joseph Banister, Colchester, Patent No. 1", post-1836, winding indicator.
British Museum, London, ex Ilbert collection

197b Rear view of Banister's marine chronometer, with 48-hour going, helical hairspring, bimetallic compensated balance with large and small adjusting screws, unusual duplex escapement.

198 Movement of a marine chronometer, signed "Westwood, London, No. 6056", ca. 1830, originally had Earnshaw's spring detent escapement, was rebuilt by Molyneux in 1840 and fitted with a new balance and hairspring in 1879. Unusual movement structure. *National Maritime Museum, Greenwich*

199a Marine chronometer in mahogany case with gimbal mounting, signed "Widenham, London, No. 944", eight-day movement.

199b Dial of Widenham's marine chronometer no. 944, with winding indicator.

199c Movement of Widenham's marine chronometer, with separate plate for the balance assembly.

201a Silver pocket chronometer, signed "Fr. Gutkaes in Dresden", no. 21, ca. 1830, with three separate dial circles for hours, minutes and seconds.
Privately owned

201b Rear view of the movement of Gutkaes' pocket chronometer, fire-gilded movement with chain and fusee, chronometer spring detent escapement, three-armed compensated balance, balance arbor mounted in diamond cap jewel, helical hairspring.

199d Enlarged view of the plate for the swinging system of Widenham's marine chronometer.

200a Marine chronometer in mahogany case with gimbal mounting, signed "Widenham, London, No. 1181", ca. 1830; dial with recessed second ring and winding indicator, eight-day going. *German art trade*

200b Movement of Widenham's marine chronometer no. 1181; movement with chain and fusee, additional plate for escapement and balance.

202a Marine chronometer, signed "Fr. Gutkaes No. 48, Dresden", ca. 1830; enameled dial, ca. 6 cm diameter, with three dial circles for seconds, minutes and hours. *German Museum, Munich*

202b Back of the movement of Gutkaes' marine chronometer, 40-hour movement with chain and fusee, winding from rear, bimetallic balance with oval weights.

202c Diagonal rear view of the movement of Gutkaes' marine chronometer, showing balance with helical hairspring and escape wheel position over the plate, chronometer spring detent escapement in Earnshaw's manner.

202d Side view of the movement of Gutkaes' marine chronometer, third, fourth and escape wheels mounted in bearings, balance pivot in diamond cap jewel, lifting stone is hemispherical, discharging stone triangular, both rubies.

203a Chronometer, signed "Ulrich & Co., London", no. 2, ca. 1840, also marked on dial "Edwards & Hunter, Cornhill"; two separate silvered raised dials, one for hours, minutes and winding indication, the other for seconds, brass main dial.
British Museum, London, ex Ilbert collection

203b Side view of the movement of Ulrich's chronometer; chain and fusee, 30-hour going, inverted beam balance and very large helical hairspring.

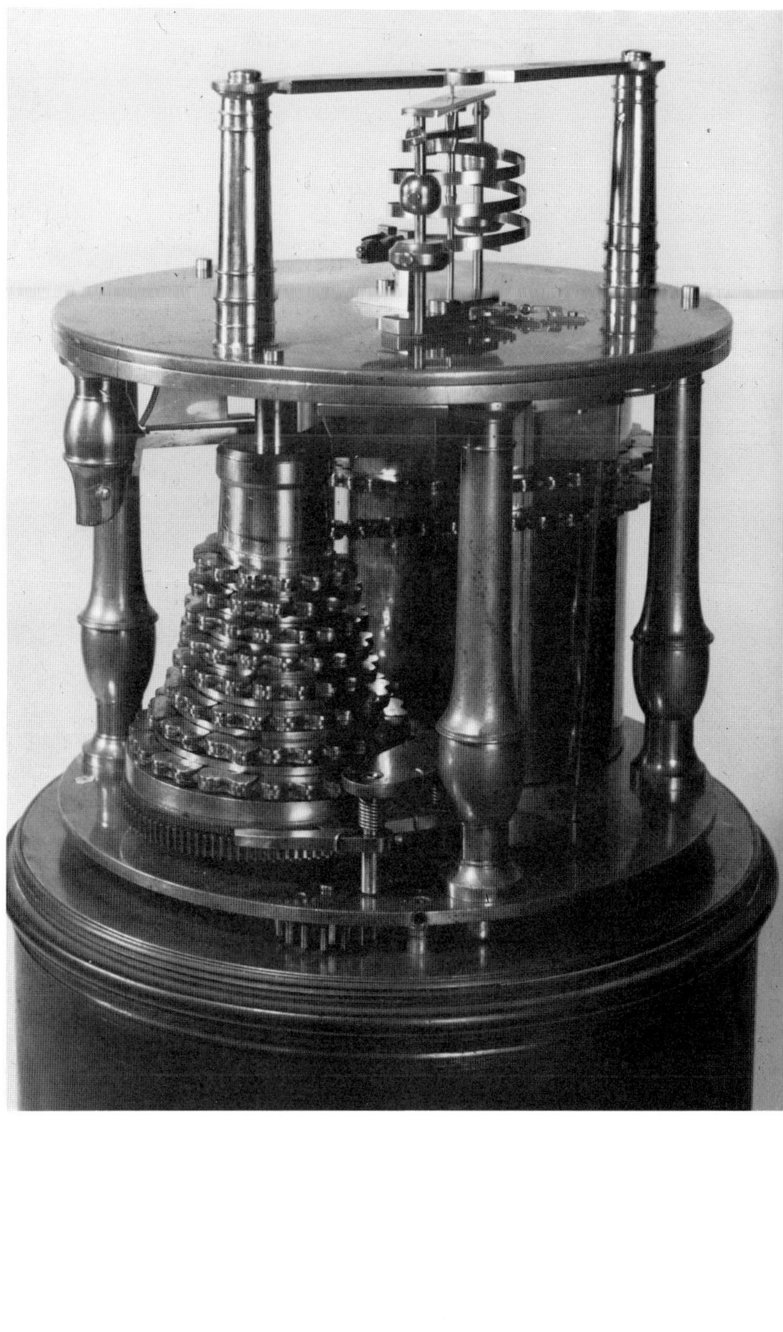

204a Marine chronometer in mahogany case, 35 cm long, large gimbal mount, signed "Josh. Croucher, London", ca. 1830; dial diameter 16.5 cm, with raised dial for hour and minutes, winding indicator, and second dial; winding through the dial.
Clockmakers Company Museum, Guildhall, London

204b Side view of movement of Croucher's marine chronometer, with chain and fusee plus Ulrich's patented constant-force escapement, a very complex system with seven springs.

204c Rear view of movement of Croucher's marine chronometer, with Ulrich's constant-force escapement, special compensated balance of brass and platinum.

205a Marine chronometer in mahogany case, signed "Charles Frodsham, élève de Parkinson & Frodsham, No. 1500, London & Liverpool", ca. 1835 (?), dial with hour and minute from the center, separate second dial and up-and-down indicator.
Privately owned

206a Dial of a marine chronometer (without case), signed "Hri Motel, No. 205, Paris", ca. 1835; enameled dial with hour and minute from the center and separate second dial.
The Time Museum, Rockford, Illinois

206b Rear of the movement of Motel's marine chronometer, with chain and fusee, separate plate for escapement and balance.

205b Side view of the movement of Frodsham's marine chronometer, two-day movement with chain and fusee, compensated balance, with helical steel hairspring.

207a Marine chronometer in wooden case with large gimbal mount, signed "Perrelet et Fils, Paris"; Louis-Frédéric Perrelet (1781-1854), grandson of Ab. L. Perrelet, ca. 1835; dial with three separate circles for hours, minutes and seconds.
Arts et Métiers, Paris

207b Rear view of the movement of Perrelet's marine chronometer, unusual construction, using several cocks or bridges instead of a bottom plate, three-armed compensated balance, chronometer pivoted detent escapement, going barrel.
Opposite page, below

207a △

208a ▽

208b ▽

208c ▽

△ 2

208a Marine chronometer in wooden case with gimbal mounting, signed "Henri Robert à Paris", ca. 1835; dial diameter 72 mm, with three dials for seconds, hours and minutes, plus up-and-down indication.
Musée International d'Horlogerie, La Chaux de-Fonds

208b Dial of Robert's marine chronometer.

208c Rear of the movement of Robert's marine chronometer; direct drive from barrel, two-armed compensated balance, chronometer pivoted detent escapement.

209a Marine chronometer, signed "Eugenio Couillaut, No. 3, Madrid," ca. 1835; silvered dial, ebonized case 180 mm long.
British Museum, London, ex Ilbert collection

209b Back of the movement of Couillaut's marine chronometer, two-day going, two-armed compensated balance with helical hairspring, unusual escapement with constant force, geared barrel.

210a Marine chronometer in mahogany case with gimbal mounting, signed "Molyneux & Sons, London, No. 1558", ca. 1832; dial with main ring for hours, minutes, separate second ring with winding indicator.
Wood, Edinburgh

210b Rear view of movement of Molyneux' marine chronometer; two-armed compensated balance after Earnshaw, before Molyneux' improvements to decrease the middle temperature error.

a △

210a ▽

209b △

210b

211a Silver pocket chronometer with wooden case, signed "Kessels, Altona, No. 1303", ca. 1830, enameled dial.
Th. Beyer Museum of Timekeeping, Zürich

212a Marine chronometer, signed "Kessels, Altona, No. 1319", ca. 1840; in mahogany case 10 cm long.
Historical Clock Museum, Wuppertal

211b Back of the movement of Kessels' pocket chronometer, fire-gilded movement, chain and fusee, two-armed compensated balance with numerous coarse and fine adjustment screws, chronometer spring detent escapement.

212b Movement of Kessels' marine chronometer, bimetallic balance, helical hairspring and chronometer spring detent escapement.

213a Observation watch in mahogany case, signed "Kessels, Altona, No. 1413", ca. 1842.
City Historical Museum, Düsseldorf

213b Rear view of the movement of Kessels' observation watch, movement with chain and fusee, two-armed bimetallic compensated balance with regulator and scale for adjustment, duplex escapement.

214a Silver pocket chronometer, signed Fr. Gutkaes, Dresden, ca. 1840.
Privately owned in Germany

214b Rear view of the movement of Gutkaes' pocket chronometer, winding from the rear, chain and fusee, bimetallic compensated balance, chronometer spring detent escapement.

215 Movement of a pocket chronometer, signed "W. Bröcking, Hamburg", ca. 1840; two-armed bimetallic balance, chain and fusee, chronometer spring detent escapement.
Privately owned

216 Movement of a marine chronometer; signed "Arnold & Dent, London", no. 1042, two-day going, bimetallic Hardy-type balance with raised vertical screw weights, chronometer spring detent escapement, movement diameter 9 cm.
National Maritime Museum, Greenwich

217a Marine chronometer in wooden case, signed "James F. Cole, Invt. London AD 1840"; dial diameter 80 mm, three separate dials for seconds, minutes and hours, three-day going.
Clockmakers Company Museum, Guildhall, London

217b Back of the movement of a marine chronometer by Cole, direct drive from the barrel, Cole's unusual "double rotary detached escapement"; flat hairspring with four over-coils, after Breguet.

217c Drawing of Cole's escapement, which prevents "galloping."

218 Movement of a marine chronometer, signed "E. J. Dent, No. 1771", with glass balance and hairspring, ca. 1840.
Photo by V. Mercer

219 Marine chronometer movement, signed "French No. 5722", dial diameter 9 cm, ca. 1840; two-day movement, two-armed bimetallic compensated balance, Earnshaw spring detent escapement, helical hairspring.
Asprey, London

220 Rear of the movement of a marine chronometer, signed "Barraud, London", ca. 1840 (?); eight-day movement with chain and fusee, two-armed bimetallic balance, helical hairspring.
Photo by T. Mercer

221a Marine chronometer in mahogany case, 178 mm long, signed "Barwise, London, No. 4307", ca. 1840.
British Museum, London, ex Ilbert collection

221b Movement of Barwise's marine chronometer. two-armed compensated balance, chronometer spring detent escapement.

222a Silver pocket chronometer (HM 1840), signed "Robert Molyneux, London, No. 2280" ; dial diameter 62 mm.
British private collection

222b Rear view of the movement of Molyneux' pocket chronometer, gilded movement with additional plate for escapement and balance, chronometer spring detent escapement, helical hairspring, 56-hour going.

223b Rear of the chronometric thermometer by Molyneux.

223a Molyneux's chronometric thermometer no. 3, a unique instrument developed at the astronomer Airy's suggestion for the chronometer room of the Greenwich observatory; essentially a normal chronometer movement in which the effects of the bimetallic components in the balance are reversed so that the influence of temperature variations in the observation interval — such as one week — are summarized. Extreme values of surrounding temperatures can be measured simply with max./min. thermometers, but this instrument showed the temperature time integral easily. Molyneux's instrument was used at Greenwich until well into the 20th Century.
Royal Greenwich Observatory, Herstmonceux

224a Marine chronometer, signed "Brockbank & Atkins, London, No. 1217", ca. 1835; mahogany case, normal gimbal mounting, gold hands, up-and-down indication.
British art trade

224b Movement of the Brockbank & Atkins marine chronometer, eight-day movement, barrel with chain and fusee, escapement and balance on a separate small plate.

225a Marine chronometer in wooden case, signed "E. Barthet, Maresille, No. 10", ca. 1840; dial diameter 80 mm, central minute, additional dials for hours, seconds and Réaumur degrees. The chronometer was found on an abandoned ship in 1846.
Clockmakers Company Museum, Guildhall, London

225b Back of the movement of E. Barthet's marine chronometer, movement with chain and "reversed fusee", splendid workmanship, two-armed compensated balance, barrel-shaped hairspring, chronometer spring detent escapement.

226a Marine chronometer in wooden case with gimbal mounting, signed "Winnerl No. 441", small second dial by the XII.
Wood, Edinburgh

226b Back of the movement of Winnerl's marine chronometer, two-armed compensated balance, chain and "reversed fusee".

225a △ 226a ▽ 225b △ 226b

227a Marine chronometer mounted as "compteur marine", signed Hri Motel No. 257", Paris, ca. 1845.
Th. Beyer Museum of Timekeeping, Zürich

227b Movement of Motel's marine chronometer, refined, richly structured construction, barrel with chain and fusee, separate balance assembly, four-armed compensated balance with conical hairspring and special adjustable attachment of the spring's terminal curves.

227c Enlarged view of the balance assembly shows the escape wheel with pierced teeth for better oil retention and unusual detent.

228a Marine chronometer in mahogany case, signed "Victor Gannery", no. 26, ca. 1850, large gimbal mount, dial for hours and minutes, two additional scales for seconds and winding indication.
Privately owned

228b Marine chronometer by Gannery, showing the movement removed with its own protective casing and spring mounting.

228c Movement of Gannery's chronometer, with two-armed compensated balance, reversed fusee.

229 Marine chronometer in mahogany case with gimbal mounting, signed "Victor Gannery, No. 13", ca. 1840, 49-hour movement.
Arts et Métiers, Paris

228b △

228a △

228c ▽

229

230a Marine chronometer in mahogany case, signed "George Muston No. 656, London-Bristol", 1842; with second dial near the XII and winding indicator near the VI, plus enclosed Greenwich certificate of January 8, 1842.
Bobinet, London

230b Movement of G. Muston's marine chronometer, eight-day movement with chain and fusee, barrel extends out over bottom plate and is held by special bridge; two-armed bimetallic balance with movable adjusting screws, helical blued steel hairspring.

231 Marine chronometer with eight-day movement, ca. 1835, signed "Parkinson & Frodsham, 5 Budge Row, London, No. 355"; second dial near the XII, winding indicator in small curved window in the dial under the center; movement with chain and fusee, chronometer spring detent escapement, two-armed bimetallic balance and cylindrical hairspring with terminal curves.
Ineichen, Zürich

232 Movement of a marine chronometer, signed "Charles Frodsham. No. 2545", ca. 1845; rosewood case with brass corners, dial diameter 9.25 cm, two-armed bimetallic balance, blued helical hairspring, chronometer spring detent escapement, two-day going.
Asprey, London

233 Back of the movement of a pocket chronometer with silver case, signed "Barwise No. 159, London", 1847; white enameled dial with 53 mm diameter and blued steel hands; gilded movement with full plates, three-armed compensated balance with middle error compensation, helical blued steel hairspring, chronometer spring detent escapement.
Privately owned in Britain

234b Movement of M. Tobias' marine chronometer, with chain and reversed fusee, eight-day movement, additional small plate for balance and escapement, two-armed bimetallic compensated balance, chronometer spring detent escapement, helical hairspring.

234a Marine chronometer in mahogany case with gimbal mounting, signed "Morris Tobias, No. 2453", ca. 1840; dial with second ring near the VI and winding indication near the XII, dial diameter 11.4 cm.
Asprey, London

235a Marine chronometer in mahogany case, signed "James Sweetman Eiffe, London", ca. 1845; dial diameter 117 mm, gold hands, second dial and up-and-down indication.
Privately owned in Switzerland

235b Movement of J. S. Eiffe's marine chronometer, eight-day movement with chain and reversed fusee, separate escapement and balance structure with two plates and protection for the two-armed compensated balance, helical blued steel hairspring with terminal curves, escape wheel with cap jewel bearings on both sides, diamond cap jewel for the balance arbor.

236 Back of the movement of a marine chronometer, signed "Dent No. 1633", ca. 1840-45; driven by three barrels with single winding from the rear, four-armed compensated balance on a separate plate with detent, Savage-type lever system with two pins as Remontoire.
Photo by V. Mercer

237a Dial of a marine chronometer in mahogany case, signed "John Poole No. 684, London", ca. 1835.
Stender, Saint Michielsgestel, Holland

237b Side view of movement of J. Poole's marine chronometer, eight-day movement with chain and fusee, blued helical hairspring.

238a Marine chronometer in new mahogany case with gimbal mounting, dial with separate second ring and winding indicator near the XII, signed John Poole, London, No. 2008, ca. 1845.
Stender, Saint Michielsgestel, Holland

238b Side view of the movement capsule of J. Poole's marine chronometer removed from its wooden case.

237a △

237b △

238a ▽

238b

237c Rear view of the movement of J. Poole;s marine chronometer, two-armed bimetallic compensated balance, chronometer spring detent escapement.

238c Rear view of the movement of J. Poole's marine chronometer, bimetallic compensated balance with Poole's middle error compensation, eight-day movement.

239a Marine chronometer in mahogany case with gimbal mounting, signed "John Poole, London, No. 3274", post-1850, silvered dial, 100 mm diameter, with second dial by the VI and winding indication by the XII.
Clockmakers Company Museum, Guildhall, London

239b Back of the movement of J. Poole's marine chronometer, two-day movement with chain and fusee, two-armed compensated balance with middle error compensation, probably the very first, originally used by John Pennington, generally named after Poole because he made it known.

7c △

239a △

238c ▽

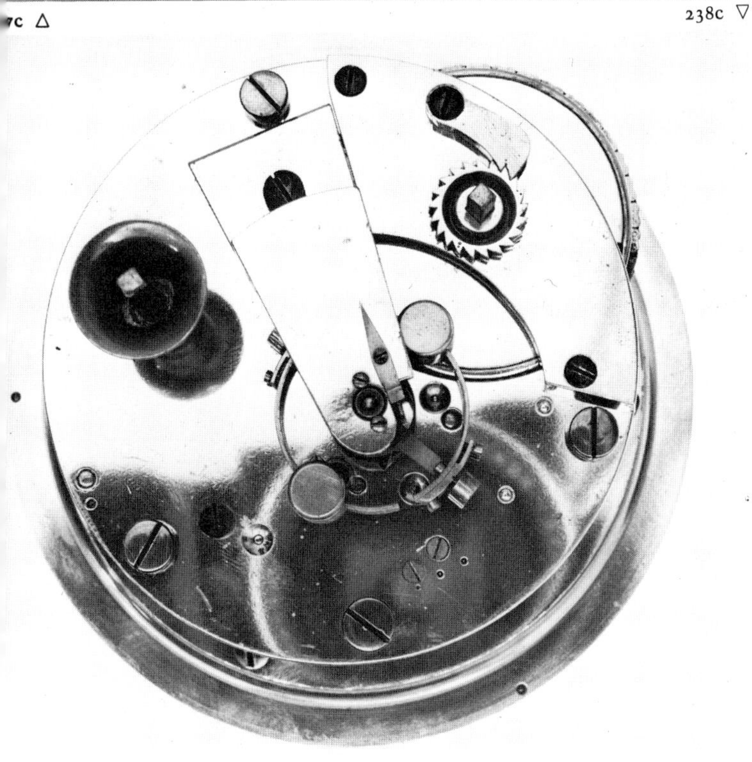

239b ▽

240a Marine chronometer without gimbal mounting, signed "Hri Motel No. 276", Paris, ca. 1845; dial diameter 80 mm.

240b Movement of Motel's marine chronometer, 36-hour movement with chain and fusee, four-armed bimetallic balance, conical hairspring, detent, three-step balance assembly construction.

240c Side view of the movement of Motel's marine chronometer, showing the front of the balance assembly.

240d Side view of the movement of Motel's marine chronometer, showing the side of the balance assembly.

241a Marine chronometer in mahogany case with gimbal mounting, signed "Vissière, Paris, No. 244", ca. 1850; 49-hour movement, minute hand from the center, hour and second indication on two smaller dials.

241b Movement of a marine chronometer, signed "Vissière, Paris, No. 216", 49-hour movement with chain and reversed fusee, two-armed compensated balance with helical hairspring, chronometer spring detent escapement.

242a Dial of a marine chronometer, signed "Loseby, London, No. 126", ca. 1850; mahogany case 16 cm long, 24-hour dial, silvered, with second dial by the 12 and winding indicator by the 6, gold hands.
British Museum, London, ex Ilbert collection

242b Movement of Loseby's marine chronometer no. 126, especially fine workmanship, two-day movement, Loseby's compensated balance with two mercury thermometer tubes for extra temperature compensation.

243a Marine chronometer in mahogany case, large gimbal mount, signed "Winnerl No. 336, Paris", ca. 1850.
Th. Beyer Museum of Timekeeping, Zürich

243b Movement of Winnerl's chronometer, in typical form with reversed fusee.

243c Enlarged view of the escapement of chronometer no. 336 by Winnerl.

244a Dial of a marine chronometer signed "Winnerl, no. 509", Paris, ca. 1850, with second dial by the XII and winding indicator by the VI, dial diameter 68 mm.
French Navy

243a △

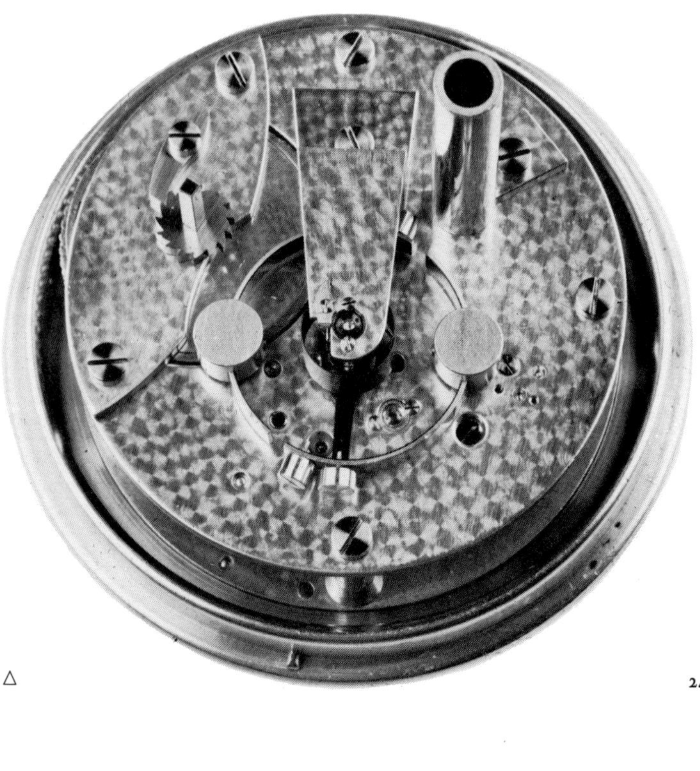

243b △

243c ▷

244b Movement of the marine chronometer by J. H. Winnerl, no. 509, 49-hour movement, two-armed bimetallic balance, helical hairspring, chronometer spring detent escapement.

245a Dial of a chronometer signed "Ame Jacob, St. Nicolas (près Diéppe), No. 272", ca. 1853; diameter 67 mm, with second dial by the VI and winding indicator by the III.
French Navy

245b Back of the movement of Jacob's marine chronometer, with two-armed bimetallic balance, chronometer spring detent escapement.

245c Side view of the movement of Jacob's chronometer, 36-hour movement with chain and fusee, helical hairspring.

245b △

a △

245c ▷

246c Movement of Breguet's chronometer, 36-hour movement, two-armed bimetallic balance, helical hairspring, chronometer spring detent escapement.

246a Marine chronometer in wooden case with gimbal mount, signed "Breguet No. 816", ca. 1852; dial diameter 65 mm, second dial by the VI, winding indicator by the XII.
French Navy

246d Side view of the movement of Breguet's chronometer, with chain and fusee.

246b Dial of Breguet's marine chronometer.

247a

247b

247a Marine chronometer in mahogany case with gimbal mounting, signed "Krille, Altona, No. 1399", ca. 1850; second dial by the XII, winding indicator by the VI, going time about 30 hours.
Bamberg Observatory

247b Movement of Krille's marine chronometer with bimetallic two-armed balance and blued helical hairspring.

248a Marine chronometer in mahogany case, partly with brass corner irons and inlaid brass strips, with gimbal mounting, signed "Charles Frodsham No. 3311", London, 1852; dial with second ring by the VI and winding indicator by the XII (winding mark at 24), dial diameter 8.5 cm.
Asprey, London

248b Movement of Frodsham's chronometer no, 3311, two-day movement with chain and reversed fusee, unusual two-armed bimetallic balance. Besides the normal Earnshaw arrangement of semicircular bimetallic strips with segment shaped weights, there are two bimetallic strips underneath with weight screws on their ends (flat bar) for middle temperature error compensation, but no test results have survived (see also #278b).

248a

248b

249a Marine chronometer in mahogany case with gimbal mounting, signed "F. Tiede, Berlin No. 376", ca. 1850-60, with second dial by the VI and winding indicator by the XII.
Babelsberg Observatory

249b Movement of F. Tiede's chronometer, with two-armed bimetallic balance and helical hair-spring, typical English style, See. #278b.

250a Marine chronometer in mahogany case, firmly mounted, signed "Winnerl No. 261", ca. 1855; enameled dial 5.5 cm in diameter, second ring by the XII, case dimensions 10.5 x 8.5 x 4.0 cm.
Privately owned, Bremen

250b Back of the movement of Winnerl's chronometer no. 261, two-armed bimetallic balance with middle temperature error compensation, helical hairspring, movement about 1 cm high, no chain and fusee, geared barrel.

251 Table chronometer in richly chased bronze case, gilded, gimbal mounted, signed "James McCabe, London, No. 448", ca. 1856; dial diameter 12 cm, winding indicator by the XII, eight-day movement, chronometer spring detent escapement, bimetallic balance.
Asprey, London

249a △

249b △

250a ▽

250b

WIND
UP
James McCabe
Royal Exchange
London No. 148

252a Marine chronometer in wooden case with gimbal mounting, signed "Dumas (successeur) de Gannery No. 34", with winding indicator by the XII.
Privately owned

252b Part of the movement of Dumas' no. 34, two-armed bimetallic balance, mainspring signed "Bourquin, 1851", reversed fusee.

253a Marine chronometer in wooden case with gimbal mounting, signed "Dumas, Elève Berthoud Prof., no. 295"; dial with minute from the center and separate small hour and second dials.
Arts et Métiers, Paris

253b Movement of Dumas' chronometer no. 295, with four-armed bimetallic balance, helical hairspring without terminal curves and unusual bridge for balance arbor bearing.

252a △

252b ▽

253a △

253b ▽

254a Marine chronometer in square mahogany case 152 mm long, signed "Callier (Paris) No. 763", ca. 1850; with winding indicator by the VI.
British Museum, London, ex Ilbert collection

254b Movement of Callier's chronometer no, 763, 50-hour movement, spring detent, very unusual balance with a two-part bimetallic strip system by which the screw bolts parallel to the axis are moved (no adjusting weights on the bolts). Probably made in collaboration with Winnerl, whose successor was Callier.

255 Back of the movement of a marine chronometer by Callier (Paris), no. 5, with typically French details (screws, locking ball and bridge), see Winnerl, #243b.

256a Table chronometer in wooden case, dial with special rings for second and winding indication, signed "Chas Frodsham, London No. 2412", ca. 1850.
Privately owned, Munich

256b Movement of the table chronometer by Chas. Frodsham, no. 2412, made with rectangular plates, the back plate of an unusual pattern; two-armed bimetallic compensated balance with helical hairspring, the balance arbor is perpendicular to the hand axis, 8-day movement.

257a Table chronometer in mahogany case, silvered dial, diameter 310 mm with separate rings for hours, minutes and seconds, plus winding indication by the XII of the small hour dial.
Sotheby, London

257b Side view of the movement of the table chronometer, eight-day movement with chain and fusee, balance assembly perpendicular to the hand axis as a platform, bimetallic compensated balance, helical blued hairspring with terminal curves (beat 1 second), spring detent chronometer escapement.

Opposite page

258a Marine chronometer in wooden case with gimbal mounting, signed "Tiede, Berlin, No. 257", ca. 1850-60; dial with separate rings for seconds and winding indication.
Privately owned, Stuttgart

258b Movement of Tiede's chronometer no. 257, 56-hour movement, chain and fusee, two-armed bimetallic balance.

258c Side view of movement of Tiede's chronometer no. 257, raw movement from Prescot, Preston.

259a Marine chronometer in wooden case with gimbal mounting, signed "B. Scharf A St Nicolas", no. 42, ca. 1860; dial with separate rings for second and winding indication.
Privately owned, Stuttgart

259b Back of the movement of Scharf's chronometer no. 42, 2-day movement with chain and reversed fusee, two-armed bimetallic balance, See. #252b.

259c Side view of the movement of Scharf's chronometer no. 42, helical steel hairspring.

8a △

259a △

258b ▽

259b ▷

258c ▽

259c ▽

260a Pocket chronometer in silver case, signed "Breguet et Fils, No. 4365", Paris, ca. 1860; dial diameter 51 mm.
French Navy

260b Movement of Breguet et Fils pocket chronometer no. 4365, 30-hour movement with chain and fusee, two-armed bimetallic compensated balance with helical steel hairspring, chronometer spring detent escapement.

261a Marine chronometer in mahogany case with gimbal mounting, signed "Viktor Kullberg No. 199", with separate dials for second and winding indication, ca. 1862.
Wood, Edinburgh

261b Enlarged view of the balance of Kullberg's chronometer no. 188, wide balance (flat rim) with sector-shaped weights.

262a Marine chronometer in rosewood case with gimbal mounting, signed "George Blackie No. 822", London 1862; dial with winding indicator by the XII and second ring by the VI.
Clockmakers Company Museum, Guildhall, London

262b Back of the movement of Blackie's chronometer no. 822, eight-day movement with chain and fusee, balance assembly on special plate, two-armed bimetallic balance with two short additional compensating arms perpendicular to the plane of the balance.

263a Marine chronometer in wooden case with gimbal mounting, signed "Andr. Hohwü, Amsterdam, No. 395", ca. 1860; dial with separate second and winding indication rings.
Bamberg Observatory

263b Movement of chronometer no. 395 by Hohwü; 56-hour movement with chain and fusee, two-armed bimetallic balance without secondary compensation.

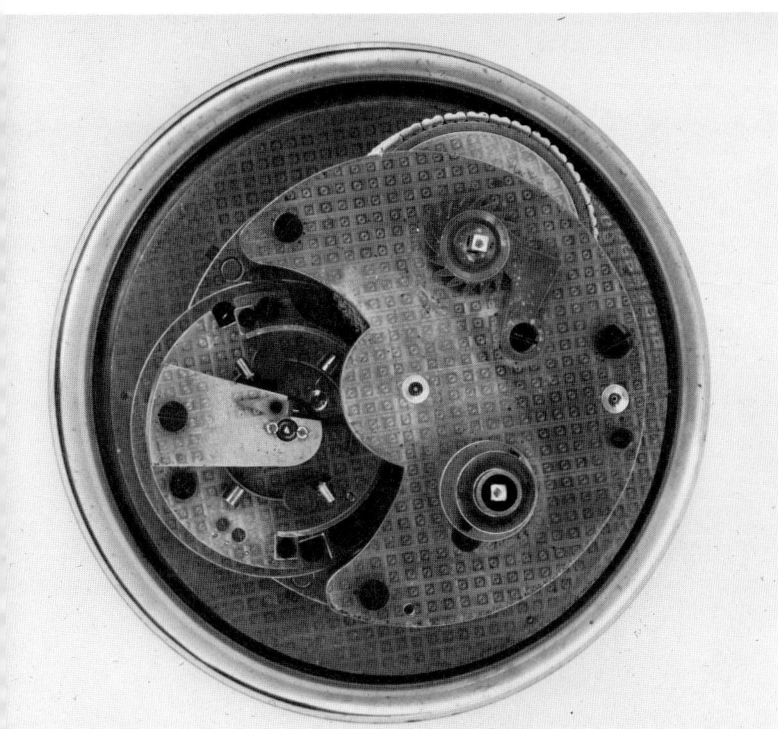

264a Dial of a pocket chronometer signed "Charles Fasoldt, Albany, New York", ca. 1865; dial diameter 4.5 cm.
Smithsonian Institution, Washington D.C.

264b Movement of Fasoldt's pocket chronometer no. 484, movement diameter 4.5 cm, one-day movement, direct barrel drive, unusual construction with several bridges, crown winding, engraved patent data, split bimetallic balance, Breguet hairspring fine regulation, Fasoldt's patented chronometer escapement, see Chamberlain, "It's About Time", pp. 92ff & 428ff.

265a Marine chronometer in mahogany case with gimbal mounting, signed "Julius Albers, Bremerhaven, No. 45", ca. 1870; dial diameter 10 cm, small dials for second and winding indication.
Smithsonian Institution, Washington D.C.

265b Movement of Albers' chronometer no. 45, movement diameter 9 cm, 56-hour running, chain and fusee, bimetallic compensated balance with helical hairspring and chronometer spring detent escapement, probably a British raw movement.

266a Dial of a marine chronometer, signed "Ate (Auguste) Louis Berthoud No. 55", Paris, ca. 1870; dial diameter 77 mm, with separate dials for second and winding indication.
French Navy

266b Movement of Berthoud's chronometer no. 55, 49-hour movement, two-armed bimetallic balance, gold hairspring without terminal curves, spring detent chronometer escapement, reversed fusee.

267a Movement of an observation watch in silver case, signed "Jules Jürgensen, Copenhagen, No. 7932", ca. 1850-55; dial diameter 4 cm, unusual bridge structure, Breguet hairspring, bimetallic compensated balance, chronometer pivoted detent escapement, one-day movement, direct barrel drive, key winding.
Smithsonian Institution, Washington, D.C.

267b Inner lid of Jürgensen's silver observation watch no. 7932.

6a △

267a ▽

266b △

267b ▽

269a Marine chronometer (mounted as compteur marine) signed "Leroy No. 235", ca. 1863; dial diameter 63 mm, with small dials for second and winding indication.
French Navy

269b Movement of Leroy's chronometer no. 235; movement diameter 60 mm (Ebauche Dumas No. 605), 40-hour movement, reversed fusee, two-armed bimetallic balance.

270a Silver pocket chronometer, signed "Philibert Perret, No. 401, Geneve", ca. 1865; small dials for second and winding indication.
Crott & Schmelzer Auction House, Aachen

270b Movement of Perret's pocket chronometer no. 401; key winding, chronometer escapement with spring, helical hairspring, two-armed bimetallic balance, interesting anti-"galloping" apparatus, chain and fusee.

268 Back of movement of a marine chronometer, signed "Henry P. Isaac, No. 1688", ca. 1870, two-day movement, diameter 10.6 cm, two-armed bimetallic balance with middle temperature error compensation after Molyneux, helical hairspring.
National Maritime Museum, Greenwich

270a △

270b

268 ▷

271 aMarine chronometer, mounted as "compteur marine", signed "S. Vissière", ca. 1870, with small dials for second and winding indication.
Antiquorum, Geneva

271b Movement of Vissière's chronometer with two-piece compensated balance, see #269b.

272a Silver pocket chronometer in wooden case, signed "Brüder Klumak, Wien, No. 4702", ca. 1870, dial diameter 5 cm.

272b Movement of Brüder Klumak chronometer no. 4702; two-armed bimetallic balance with 18 adjusting screws, chronometer pivoted detent escapement, patented by M. Petersen, Breguet hairspring with regulation, 36-hour going, geared barrel, see also #279.

273a Pocket chronometer in gold savonette case with colorfully enameled coat of arms, signed "Henry Dent Gardner, London" (hallmark 1873); dial diameter 51 mm, second out from the center and winding indicator by the XII.
Clockmakers Company Museum, Guildhall, London

273b Back of the movement of H. Dent Gardner's chronometer, two-armed bimetallic compensated balance, chronometer spring detent escapement, 30-hour movement, chain and fusee, key winding.

274 Marine chronometer movement signed "Henry P. Isaac, London No. 1129", ca. 1873; two-day movement, Kullberg's balance with flat rim and adjusting weights, crossarm and two beams combined, chronometer spring detent escapement, helical hairspring, movement diameter 10.3 cm.
National Maritime Museum, Greenwich

276a △

276b

275

Opposite page:

275 Rear view of a marine chronometer movement by Dent, no. 3097; eight-day movement with chain and fusee, special balance assembly construction with two plates and a bridge plus balance protection in case of chain breakage; the balance is the "staple balance" patented by Dent.
Photo by Dr. R. V. Mercer

276a Marine chronometer in mahogany case, 146 mm long, with gimbal mounting, signed "Auguste Berthoud, No. 293", ca. 1875.
British Museum, London, ex Ilbert collection

276b Movement of Berthoud;s chronometer no. 293, one-day movement, two-armed bimetallic balance, helical hairspring, reversed fusee, See. #243b.

277 Movement of a marine chronometer signed "H. P. Isaac, London, No. 1188", ca. 1876; two-day movement, Hartnup's balance with helical hairspring, chronometer spring detent escapement, movement diameter 10.2 mm.
National Maritime Museum, Greenwich

278a Marine chronometer in wooden case with gimbal mounting, signed "Tiede, Berlin No. 345", ca. 1870-80.
Babelsberg Observatory

278b Movement of chronometer no. 345, signed "Tiede, Nr. 20 Jägerstr. Berlin"; movement structure typically French (See #242b & 249b) with chain and reversed fusee, chronometer spring detent escapement, compensated balance with middle temperature error compensation (See #248b).

277

278a

278b

280a Marine chronometer in mahogany case with inlaid brass strips, gimbal mounting, signed "Thomas Mercer, London, No. 2696", ca. 1876.
Mercer, St. Albans.

280b Back of movement of Mercer's chronometer no. 2696 with chain and fusee, two-armed bimetallic balance with Mercer's compensation for the middle temperature error.

279a Marine chronometer in wooden case with gimbal mounting, signed "Matthias Petersen, No. 79, Altona", 1878-79; Petersen's chronometers no. 78 and 80 won 7th and 25th places in the 1879-80 chronometer tests.
Privately owned, Stuttgart

279b Movement of Petersen's chronometer no. 79, two-day movement, chain and fusee, two-armed compensated balance, helical hairspring, Petersen's patented chronometer pivoted detent escapement, see also #272.

281a Pocket chronometer in silver case, signed "A. Johannsen & Co., London No. 6068", 1880, with steel hands.
Privately owned

281b Back of movement of Johannsen's chronometer no. 6068, movement with chain and fusee, key winding from the back, two-armed compensated balance with nine adjusting screws per segment, chronometer spring detent escapement, helical hairspring.

282a Marine chronometer in mahogany case with gimbal mounting, signed "Whyte, Thomson & Co, Glasgow No. 5050", ca. 1880; dial with gold main hands and small rings for second and winding indication, 11 cm diameter.
Privately owned

282b Movement of Whyte, Thomson & Co. chronometer no. 5050; eight-day movement with chain and fusee, balance assembly in special small two-level form with protection from chain breakage, two-armed bimetallic balance, helical blued hairspring, chronometer spring detent escapement.

283a Marine chronometer in mahogany case with gimbal mounting, signed "W. Bröcking, Hamburg No. 826", ca. 1885; dial with separate small dials for second and winding indication, gold main hands, dial diameter 12 cm.
Privately owned

283b Movement of Bröcking's chronometer no. 826, eight-day movement with chain and fusee, special two-level balance assembly structure, two-armed bimetallic balance and helical hairspring, chronometer spring detent escapement, protective casing against chain breakage.

284a Marine chronometer in mahogany case with gimbal mounting, signed "Dent London, No. 44000", ca. 1880; 24-hour dial with two small rings for second and winding indication, diameter 116 mm. *French Navy*

284b Movement of Dent's chronometer no. 44000, 1888, eight-day movement with chain and reversed fusee.

284c Dial of Dent's chronometer no. 44000, with 24-hour division.

284d Side view of movement of Dent's chronometer no. 44000 with special three-level construction for the balance assembly, helical hairspring, balance after Dent's "single staple" patent.

285a Movement of an unusual marine chronometer, unsigned but probably an experimental piece by Robert Gardner, ca. 1880-90, with three separate small enameled dials for hours, minutes and seconds.
Bobinet, London

285b-c Enlarged rear view of the balance bearing bridge of R. Gardner's experimental chronometer, undivided three-arm balance, flat hairspring, unique temperature compensation.

286a Marine chronometer in wooden case with gimbal mounting, signed "Knoblich & Dencker, Hamburg No. 1855", 1890-1900; dial with two small rings for second and winding indication, gold hour and minute hands.
Babelsberg Observatory

286b Back of movement of Knoblich & Dencker chronometer no. 1855; 56-hour movement with chain and reversed fusee, two-armed bimetallic balance, helical hairspring, chronometer spring detent escapement.

287a Marine chronometer in wooden case with gimbal mounting, signed "A. Eppner, Berlin & Laehn No. 25", ca. 1880; dial with two small rings for second and winding indication.
Altona Museum, Hamburg

287b Movement of Eppner chronometer, movement with chain and fusee, balance may have been rebuilt in the 20th Century, helical hairspring.

288a Dial of a marine chronometer in a wooden case with gimbal mounting, signed "E. Delépine, No. 1455, St. Nicolas" (près Dieppe), 1885; diam diameter 90 mm, small dials for second and winding indication.
French Navy

288b Movement of Delépine's chronometer no. 1455; two-armed bimetallic balance, helical palladium hairspring, chronometer spring detent escapement, special balance attachment (see Breguet).

289a Silver pocket chronometer, signed Albert H. Potter, Geneva, 1895.
Crott & Schmelzer Auction House, Aachen

289b Rear of movement of Potter's silver pocket chronometer, unusual bridge layout for bearings and arbors, Potter's patented barrel and chronometer pivoted detent escapement.

290a Marine chronometer in wooden case with carrying strap, gimbal mounting, signed "H. Conant, Tiffany & Co. Makers (New York)"; patented August 23, 1887, very unusual dial pattern with hour, minute and second hands from the center plus four additional small second dials whose purpose is not known.
Time Museum, Rockford, Illinois

290b Back of the movement of the Conant-Tiffany chronometer, with four spring windings and four balance assembly's on small plates with two-armed bimetallic balances and helical hairsprings, one balance (below) with Poole's middle temperature error compensation.

291a Marine chronometer in wooden case, fixed in place without gimbal, signed "Max Richter, Berlin", ca. 1890; dial with gold main hands and two small dials for second and winding indication.
Babelsberg Observatory

291b Movement of the chronometer by Richter, Berlin; eight-day movement with chain and reversed fusee, balance assembly on intermediate plate with two-armed bimetallic balance, chronometer spring detent escapement. See #198.

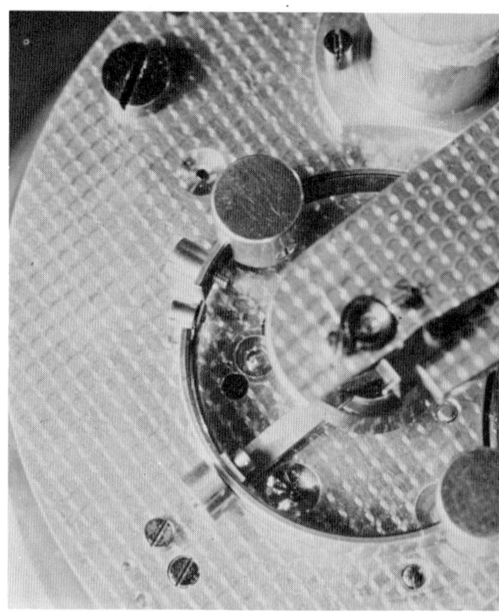

292a Marine chronometer in mahogany case with gimbal mounting, signed "Dobbie McInnes Ltd, Glasgow No. 5.4273", ca. 1890.
Wood, Edinburgh

292b Enlarged view of the two-armed bimetallic balance with middle temperature error compensation (similar to Mercer's system), with the opposite order of the metals to that in the main segments (movement probably made by Johannsen of London).

293a Marine chronometer in mahogany case, signed "T. S. & J. D. Negus, New York, No. 2245", ca. 1885; dial diameter 10 cm, with small second and winding indications.
Smithsonian Institution, Washington D.C.

293b Rear of the movement of Negus chronometer no. 2245; 56-hour movement, 9 cm diameter, two-armed bimetallic compensated balance, helical hairspring, chain and fusee, chronometer spring detent escapement.

294a Marine chronometer, signed "I. A. Seckel & Zonen, Rotterdam, No. 467", ca. 1885.
Privately owned, Bremen

294b Movement of the Seckel & Zonen chronometer no. 467; 56-hour movement with chain and fusee, two-armed bimetallic compensated balance with special device to compensate for the middle temperature error, Kullberg system.

295 Movement of a marine chronometer, signed "John Hume, Newcastle upon Tyne, No. 2050", 1890-91; two-day movement with chain and fusee, two-armed bimetallic balance with Kullberg's (see early version, #294b) added compensation for the middle temperature error, helical palladium hairspring, chronometer spring detent escapement, movement diameter 9.9 cm (chromed case, key and gimbal).

296 Movement of a marine chronometer, signed "Usher and Cole, London No. 6517", ca. 1890; eight-day movement with chain and fusee, two-armed bimetallic balance with added device to compensate for the middle temperature error, chronometer spring detent escapement.
Photo by Tony Mercer

No. 2
BULLETIN DE 1ère CLASSE
de l'Observatoire de Neuchatel
PAUL DITISHEIM
LA CHAUX-DE-FONDS
PREMIER
AUX CONCOURS CHRONOMETRIQUES
des Observatoires de
Washington
& NEUCHATEL
60
10
20
30
40

(1).—*continued*—WEEKLY SUMS OF

NAME OF MAKER.	Number.	Whether Pocket, 1, 2, or 8 Days.	Jan. 1—8	Jan. 8—15	Jan. 15—22	Jan. 22—29	Jan. 29—Feb. 5	Feb. 5—12	Feb. 12—19	Feb. 19—26
Widenham	1950	1	..	..	+17·4	+14·6	+13·5	+11·2	+ 8·3	+ 8·4
Litherland and Davies,	[illegible]	2	..	−17·9	−13·1	−10·5	− 3·3	+ 0·1	+ 4·8	+ 3·6
Dent	1659	2	..	..	..	..	..	..	..	..
Carter	397	2	..	..	+ 7·4	+ 7·9	+12·8	+15·0	+19·9	+18·3
Molyneux	2172	2	..	..	− 9·1	− 9·3	−10·4	− 9·5	−19·7	−15·2
Parkinson & Frodsham,	2496	2	..	− 9·0	− 6·3	− 8·1	− 5·0	− 2·6	+ 0·5	− 1·1
Brockbanks & Atkins,	1002	2	..	..	− 3·8	− 7·2	−10·8	−17·6	−23·6	−26·7
Birchall	327	P	..	..	−21·1	−20·3	−10·6	− 4·4	− 0·1	− 0·4
Molyneux	2184*	2	..	..	−7·5)	(−6·5)	(−8·1)	− 7·5	− 7·0	− 2·8
Molyneux	2165*	8	..	..	(−5·3)	(−11·8)	(−4·4)	− 2·1	+ 1·0	+ 3·2
Parkinson & Frodsham,	2451	2	..	− 2·2	+ 5·0	+ 4·7	− 0·7	− 2·5	0·0	+ 0·1
Muston	671	2	..	..	+ 1·9	+ 3·9	+ 6·9	+ 9·2	+ 6·8	+ 9·0
Dent	1584	2	..	..	−28·9	−28·8	−24·0	−23·7	−21·1	−19·2
Muston	656	8	..	..	−23·3	−22·8	−19·8	−17·1	−17·6	−15·2
Parkinson & Frodsham,	2452	2	..	− 4·3	− 0·9	− 0·8	+ 3·7	+ 5·4	+ 8·6	+ 5·9
C. Frodsham	1960	2	..	− 9·1	− 1·7	− 0·5	+ 4·2	+ 3·5	+ 3·4	+ 2·3
Connell	299	2	..	..	−41·1	−38·3	−31·1	−21·9	−21·5	−26·8
Arnold	1157	2	..	..	−23·9	−26·1	−29·5	−27·9	−32·4	−28·3
Murray	577	P	..	+ 8·3	+ 2·6	+ 2·1	+ 6·5	+ 4·5	+ 5·9	+ 3·8
Hewitt	1044	2	..	−20·4	−16·2	− 7·8	− 7·0	− 8·4	− 8·5	− 9·8
Dent	1656	2	..	..	..	..	..	..	..	..
Appleton	404	2	..	..	−15·6	−18·0	−15·4	−14·8	−12·9	−12·8
Carter	202	1	..	..	−15·8	−13·2	− 7·6	− 5·8	+ 0·4	− 0·2
C. Frodsham	1917	2	..	−10·1	+ 2·0	+ 0·9	+ 4·3	+ 5·5	+ 1·0	− 2·8
Connell	297	2	..	..	−39·7	−38·2	−23·5	−23·5	−17·8	−19·4
Carter	345	2	..	..	+42·9	+42·3	+44·3	+44·0	+44·8	+45·9
Porthouse	65209	2	..	..	−22·1	−19·0	−12·7	−15·1	−10·1	−10·7
Parkinson & Frodsham,	2467	2	..	− 2·9	− 0·6	− 7·2	− 8·6	−10·1	−10·5	− 8·7
Hewitt	1019	2	..	−29·1	−26·1	−25·3	−20·4	−15·4	−13·9	−12·0
Dent	1559	2	..	..	−22·4	−22·3	−22·8	−20·7	−21·1	−18·5
Parkinson & Frodsham,	2269	2	..	− 2·0	+ 3·3	+ 3·8	+ 1·0	− 2·2	− 5·7	− 3·6
Carter	410	2	..	..	+23·7	+23·6	+25·0	+22·9	+22·2	+22·2
Hewitt	973*	2	..	(−17·9)	(−15·9)	(−19·2)	(−14·2)	−15·0	−19·5	−14·6
Hewitt	1022	2	..	..	+12·9	+ 6·2	+ 4·0	+ 4·8	− 6·3	− 8·4
E. J. Massey	104	2	..	..	−44·4	−47·4	−42·5	−43·4	−38·6	−44·0
Dent	1645	2	..	..	..	..	..	..	..	..
Dent	1591	2	..	..	−31·1	[illegible]	−25·1	−23·3	−20·3	−21·6
Roskell	[illegible]	2	..	..	−29·2	[illegible]	−13·2	−13·6	− 9·9	−11·9
Dent	1599	2	..	..	+18·7	+19·8	+21·1	+18·2	+11·9	+15·9
Pennington	252	1	..	..	+ 14·1	+ 6·0	− 1·5	− 4·8	− 7·0	− 4·3
Chronometrical Thermometer.			[illegible]	+2105·6	+1923·4	[illegible]	+1465·0	+1397·6	+1026·9	+1358·8

309 Going rates from 40 chronometer tests in 1841 and 1842 at the Royal Observatory in Greenwich (2 pocket and 38 marine chronometers), including three with one-day and two with eight-day movements, extending over half a year (the *weekly* rates are stated).
Greenwich Archives

of the oil, and cause an insufficient isochronism of the oscillating system. This second type of deviation is particularly disturbing in long trips, but is of little significance in short ones."

It cannot be ascertained whether Airy suggested a different trial number for other types of use than that in the Royal Navy. In any case, Airy's suggestion introduced a classification system that was used and gave good service for decades. Year after year, into the 20th Century, chronometer tests were carried out on this basis at Greenwich. The weekly rates recorded for chronometers on hand for testing during the first half of the 1841-1842 observation period are shown in #309. There were forty chronometers, submitted by various chronometer manufacturers, some of them very well known.

Anyone not familiar with statistical work will probably be overwhelmed at first glance by the mass of data at hand, and wonder why such tables could go on for many decades. Upon closer observation, though, it is realized that mass after mass of great or small weekly rates can be seen to give the impression of well-defined periods in which the difference between small and large rates often attains a considerable range and even exceeds it. Here, among other things, the middle temperature error comes into view.

In the table in #310 there is a list of the best qualified chronometers tested in Greenwich in the 1840-1914 period, with the names of their makers and the attained statistics of their rates. One sees that over the years the rate deviations have become smaller, which proves that

Y RATES, IN THE ORDER OF TIME.

2	† Mar. 12—19	† Mar. 19—26	† Mar. 26–April 2	April 2—9	April 9—16	April 16—23	April 23—30	Apr. 30–May 7	May 7—14	May 14—21	May 21—28	May 28–June 4	June 4—11	June 11—18	† June 18—25	† June 25–July 2	† July 2—9	† July 9—16
8	+23·9	+29·5	+27·4	+25·8	+25·7	+25·2	+24·7	+26·3	+25·6	+25·5	+24·0	+23·0	+20·6	+22·5	+19·6	+22·5	+24·6	+24·9
5	− 8·7	− 5·2	− 9·3	− 3·7	− 5·2	− 1·1	+ 1·9	+ 5·3	+ 6·8	+ 4·3	+ 4·4	+ 3·5	− 1·8	− 1·6	− 9·9	−10·6	− 7·9	−10·6
1	−23·8	−19·0	−21·0	− 7·9	− 6·4	− 4·3	− 6·7	− 5·8	− 3·5	− 7·1	− 5·0	− 6·1	−10·2	−10·8	−14·8	−14·6	−12·2	−14·4
4	+11·2	+19·3	+14·5	+19·1	+11·5	+17·0	+26·0	+30·1	+31·5	+33·1	+34·8	+31·5	+28·5	+27·9	+18·7	+19·1	+22·2	+21·3
5	−16·1	− 9·8	−12·2	− 4·2	− 7·1	− 4·2	− 2·5	+ 2·3	+ 0·4	− 2·8	− 0·6	− 2·4	− 3·9	− 2·7	−11·8	−14·8	−13·3	−15·9
0	+11·5	+14·5	+13·4	+ 5·5	+ 5·0	+ 9·2	+15·0	+15·3	+14·3	+14·9	+15·3	+13·9	+13·7	+16·4	+13·4	+12·5	+13·7	+12·3
0	− 8·3	− 2·3	− 4·7	− 3·1	− 5·3	− 5·0	− 6·0	− 7·0	−10·7	−10·6	− 9·5	−10·4	−10·8	− 6·2	− 9·0	− 7·1	−10·7	−13·0
3	−12·0	− 6·1	+ 1·1	− 1·3	+ 1·9	+ 1·9	+ 5·9	+ 6·7	− 1·4	− 0·5	+ 0·8	+ 3·1	− 0·3	− 3·2	− 7·0	− 7·4	− 6·7	−10·5
1	+ 0·1	+ 3·6	+ 1·5	+ 1·1	+ 1·1	− 1·1	− 0·8	+ 1·9	+ 0·6	− 1·0	− 0·5	+ 0·9	− 1·9	+ 1·3	[−13·7]	[−10·0]	[−10·3]	[−11·5]
9	− 7·3	− 3·7	− 0·8	−10·6	− 9·0	− 6·3	− 1·2	− 0·8	− 1·2	− 2·2	+ 1·1	− 1·8	+ 0·7	+ 4·1	[−7·3]	[−11·3]	[−20·5]	[−19·7]
6	− 0·8	+11·3	+ 7·9	+10·1	+ 6·7	+10·0	+17·6	+18·4	+16·0	−18·0	+17·7	+18·4	+14·7	+14·5	+13·2	+16·1	+16·7	+14·5
3	− 7·6	+ 3·0	+ 2·9	+ 8·4	+ 9·0	+13·2	+13·1	+17·9	+18·4	+17·7	+19·3	+14·1	+13·4	+15·1	+10·3	+ 4·8	+ 5·2	+ 2·0
2	−16·1	− 9·7	− 8·5	− 7·0	− 7·5	− 3·2	− 0·3	+ 0·8	− 0·4	+ 0·6	+ 1·2	− 0·1	+ 0·6	+ 2·8	+ 0·7	+ 3·2	+ 5·5	+ 6·4
8	−18·3	−15·0	−15·5	− 5·9	− 5·7	− 5·1	− 6·3	− 1·0	− 2·3	− 1·3	− 3·4	− 2·1	− 6·9	− 9·0	−13·4	−13·8	−10·8	−12·6
0	+15·9	+17·8	+16·6	+11·4	+ 9·5	+15·4	+20·7	+20·4	+18·3	+21·0	+22·6	+24·5	+25·7	+28·5	+26·9	+22·4	+18·4	+16·6
9	+ 5·8	+11·0	+10·4	+15·5	+15·9	+19·0	+20·8	+22·5	+21·7	+22·8	+24·5	+24·4	+19·3	+19·9	+16·1	+18·4	+20·3	+20·1
3	−25·1	−16·7	−22·6	−30·3	−35·6	−27·3	−17·7	−13·5	−21·3	−16·2	−17·0	−11·1	−15·9	−17·6	−24·7	−25·8	−24·4	−23·8
9	−27·8	−23·9	−21·7	−15·2	−11·9	− 8·5	− 6·9	− 3·9	− 1·9	+ 0·5	+ 1·7	+ 5·8	+ 3·6	+ 3·9	+ 6·1	+ 4·3	+ 2·5	+ 0·9
0	− 4·5	− 0·2	− 7·3	+ 2·0	+ 1·7	+ 9·7	+ 3·7	+ 2·9	− 9·8	− 3·2	− 3·0	− 4·2	−11·1	−10·8	−14·7	−21·2	−18·6	−23·6
0	−11·7	− 9·9	−10·3	− 0·3	+ 0·7	+ 4·5	+ 7·8	+ 9·0	+10·3	+10·3	+11·8	+ 8·4	+ 4·9	+ 3·9	− 2·4	− 0·5	+ 2·7	+ 2·7
4	−12·2	− 3·5	− 2·8	+10·1	+12·3	+12·4	+ 9·2	+13·0	+13·5	+12·6	+12·9	+11·5	+10·7	+11·6	+ 9·8	+11·1	+15·3	+15·1
3	−24·1	−18·7	−21·1	− 4·7	− 3·5	− 3·6	− 5·9	− 3·8	− 2·6	− 4·2	− 4·1	− 6·3	− 9·3	−10·8	−21·0	−21·0	−18·9	−20·0
7	+ 4·2	+ 4·6	+ 9·2	+15·0	+15·1	+16·[illegible]	+20·1	+21·4	+23·7	+24·2	+26·6	+25·5	+21·8	+22·8	(A)	(G)	(G)	+ 0·7
9	−20·9	−16·0	−13·5	− 1·1	+ 1·9	+ 7·6	− 3·6	− 4·1	− 4·7	− 6·3	+ 0·1	− 7·4	− 5·1	− 9·5	−13·4	−19·7	−20·9	−21·2
0	−27·8	−21·8	−24·7	−20·3	−20·5	−16·7	−13·2	−14·3	−14·7	−12·5	−12·6	−12·5	−15·4	−16·1	−23·3	−21·2	−18·6	−19·5
5	+33·7	+40·8	+40·7	+55·9	+53·7	+58·5	+56·7	+56·9	+57·5	+57·8	+60·5	+57·3	+53·6	+53·3	+47·2	+52·9	+53·9	+53·3
4	− 0·2	+ 3·4	+ 2·2	− 2·3	− 2·0	+ 4·1	+12·7	+14·8	+14·8	+16·3	+18·0	+16·7	+13·6	+12·9	+ 8·6	+ 9·8	+11·9	+10·6
8	+ 3·0	+ 9·6	+ 9·4	+15·2	+15·7	+18·8	+19·1	+19·1	+18·4	+14·7	+12·7	+ 6·3	+ 9·7	+12·7	+ 7·2	+ 9·0	+ 7·8	+ 8·0
8	−19·1	−11·1	−13·1	+ 0·8	− 1·9	− 2·8	− 3·6	− 1·3	+ [illegible]·3	− 1·7	− 2·5	− 5·9	− 9·0	− 7·5	−15·7	−14·0	−11·1	−12·9
9	−33·2	−25·5	−23·6	− 8·9	− 7·0	− 4·5	− 8·7	− 6·5	− 4·7	− 5·9	− 5·0	− 7·5	− 9·3	− 9·8	−16·3	−15·5	−15·2	−16·1
1	[illegible] 0·2	[illegible] 2·0	− 4·1	+10·7	+11·5	+15·2	+13·2	+15·1	+15·6	+13·2	+15·2	+11·4	+ 7·7	+ 7·4	+ 2·5	+ 2·6	+ 4·8	+ 1·5
0	+ 0·5	+ 5·8	+ 3·7	+18·9	+16·9	+19·6	+19·1	+20·5	+23·0	+21·3	+22·6	+18·3	+14·6	+14·9	+ 5·1	+ 6·4	+ 8·7	+ 7·7
4	−37·7	−30·8	−22·0	−10·0	− 9·6	−12·6	−17·0	−19·8	−16·0	−19·9	−19·2	−22·3	−20·5	−26·1	[−17·4]	[−17·4]	[−11·1]	[−2·6]
5	− 3·1	− 7·4	− 9·3	+ 5·4	+ 9·9	+14·1	+12·9	+11·9	+14·1	+14·0	+15·4	+13·1	+10·0	+10·0	+ 6·5	+ 4·4	+ 6·6	+ 5·9
6	−25·5	−21·8	−21·6	−35·1	−39·9	−35·8	−27·3	−27·6	−[illegible]·6	−30·0	−30·0	−28·4	−27·4	−25·9	−25·5	−26·1	−25·8	−26·6
1	− 0·6	+10·7	+10·1	+24·1	+25·0	+25·[illegible]	+20·0	+23·6	+23·8	+20·4	+19·1	+15·4	+15·3	+18·6	+17·5	+22·1	+26·8	+26·8
9	− 7·2	− 2·4	− 3·4	− 8·8	−10·4	− 3·2	+ 2·7	+ 3·5	+ 2·8	+ 6·7	+ 8·1	+10·3	+10·7	+12·6	+10·9	+12·6	+14·0	+15·2
1	− 9·8	− 3·4	− 6·1	+ 1·2	+ 2·6	+ 4·5	+ 5·5	+ 6·2	+ 6·0	+ 4·8	+ 5·0	+ 2·8	+ 0·8	+ 1·1	− 5·4	− 1·7	− 0·2	− 1·5
3	+ 2·5	+ 6·8	+ 1·9	+21·5	+22·9	+19·2	+12·2	+11·8	+16·2	+14·5	+14·7	+13·1	+10·5	+11·3	+ 6·6	+12·8	+11·0	+10·9
6	+ 8·3	+16·9	+16·1	+29·1	+30·5	+29·9	+23·1	+26·1	+25·8	+20·2	+19·0	+1[illegible]·8	+10·0	+11·3	+ 7·1	+11·8	+14·4	+11·8
[illegible]·9	−1389·7	−1123·6	−1384·8	+[illegible]	+1169·5	[illegible]	[illegible]	+42·0	+[illegible]	+107·[illegible]	+59·9	[illegible]	−724·6	929·8	−1697·0	−1651·0	−1590·5	−1757·0

the quality has improved gradually.

To add to this table, which illustrates the development of the fine mechanical period, #311 shows the rates and trial numbers of the first marine chronometers of the heroic epoch, which Gould calculated on the basis of the old tests — which did not cover a whole year. The data in these tables show in an impressive manner what progress has been made in chronometer quality over the years.

The second official clock-testing center in Britain was set up at Kew in 1884 to assist Greenwich; its activities were later taken over by the NPL (National Physical Laboratory) in Teddington. This Kew-Teddington service was created with regulations essentially corresponding to those of the Geneva Observatory.

The range of responsibilities met at Teddington includes, as D. De Carle[1] has set forth in detail in three pages, not only chronometers but also ordinary (time of day) pocket watches. Numerous other conditions concerning length of testing, temperature ranges and positions for various quality classes (grades 1a-1d, etc.) are presented in detail.

During the 19th Century, various laboratories and observatories in Europe and elsewhere have taken up the testing of chronometers. As early as 1790, a competition for clocks took place at the Geneva Observatory, followed by a second in 1817. But only as of 1842 were the Geneva competitions extended to include chronometers. The Geneva tests are limited to 44 days. There, according to De Carle, the following categories are differentiated: (a) pocket chronometers (deck watches), and (b) and (c) for

GREENWICH TRIALS, 1840—1914.

Year.	Maker's Name and No.	Remarks.	*a*.	*b*.	Trial No (*a*—2*b*).
1840	Molyneux & Sons	No details available	11.4	4.3	20.0
1841	Litherland & Davies, 915/18279	No details available	15.5	4.1	23.7
1842	Molyneux, 2166	No details available	10.0	4.6	19.2
1843	Molyneux, 2185	No details available	11.1	5.1	21.3
1844	Appleton, 468	No details available	13.3	5.0	23.3
1845	Poole, 1155	No details available	7.6	6.7	21.0
1846	Hutton, 138	No details available	11.7	7.6	26.9
1847	Massey, 123	No details available	7.2	5.6	18.4
1848	Loseby, 115	No details available	8.7	4.6	17.9
1849	Eiffe, 662	No details available	13.2	10.3	33.8
1850	Loseby, 123	Loseby's mercurial balance	12.7	4.7	22.1
1851	Loseby, 127	Loseby's mercurial balance	16.5	4.4	25.3
1852	Loseby, 125	Loseby's mercurial balance	11.7	9.4	30.5
1853	Lister & Son	Auxiliary compensation	16.4	11.6	39.6
1854	Poole, 1585	Poole's auxiliary	8.6	4.5	17.6
1855	Lawson, 1163	Auxiliary compensation (Poole's)	17.4	6.4	30.2
1856	J. Muirhead, 2119	Poole's auxiliary	23.1	6.8	36.7
1857	Hornby, 1300	Hartnup's balance	23.9	7.3	38.5
1858	Blackie, 497	Auxiliary compensation (unpublished)	14.3	6.2	26.7
1859	Campbell, 837	Ordinary balance with slight alteration	9.6	4.7	19.0
1860	Birchall, 642	Ordinary balance with slight alteration	18.1	3.6	25.3
1861	McGregor, 2776	Poole's auxiliary	14.1	5.2	24.6
1862	Simpson & Roberts, 9/1161	Kullberg's "flat rim" balance	10.9	5.7	22.3
1863	Fletcher, 2690	Auxiliary compensation	6.8	3.6	14.0
1864	Kullberg, 760	Kullberg's "flat-rim" balance	11.2	3.7	18.6
1865	Webb, 5388	Auxiliary compensation	8.7	5.0	18.7
1866	McGregor, 3795	Auxiliary compensation	7.4	5.8	19.0
1867	Sewill, 2263	Auxiliary compensation	9.0	3.6	16.2
1868	Birchall, 1069	Ordinary balance with slight alteration	8.3	4.1	16.5
1869	Fletcher, 2972	Auxiliary compensation	7.9	4.7	17.3
1870	M. F. Dent, 2533	Ordinary balance with extra long rims	5.5	3.8	13.1
1871	Frodsham, 3423	Frodsham's reversed balance	7.8	3.7	15.2
1872	Kullberg, 1799	Kullberg's "flat-rim" balance	10.6	2.8	16.2
1873	Weichert, 2,300	Kullberg's "flat-rim" balance	5.1	3.0	11.1
1874	Sewill, 3084	Auxiliary compensation	9.9	5.7	21.3
1875	Highley, 5435	Auxiliary acting in cold	6.2	3.3	12.8
1876	Graham & Parkes, 1283/21493	Poole's auxiliary	7.0	2.9	12.8
1877	Isaac, 1612	Unpublished auxiliary	9.0	3.5	16.0
1878	Pyott, 478	Ordinary balance with slight addition	11.0	5.5	22.0
1879	Keys, 282	Ordinary balance with slight alteration	6.0	2.9	11.8
1880	Cornell, 5658	No details available	16.3	3.7	23.7
1881	Kullberg, 3972	Reversed detent with short spring	8.4	2.3	13.0
1882	Kullberg, 4066	Reversed detent with short spring	5.1	2.0	9.1
1883	Kullberg, 4067	Reversed detent with short spring	8.2	3.4	15.0
1884	Brunner, 3777	No details available	8.0	3.2	14.4
1885	Kullberg, 4139	Reversed detent with short spring	10.3	4.6	19.5
1886	Uhrig, 453	Continuously acting auxiliary	4.0	2.8	9.6
1887	Uhrig, 452	No details available	6.8	3.2	13.2
1888	Kullberg, 4915	Auxiliary: palladium spring: reversed detent	5.3	3.7	12.7
1889	Kullberg, 4739	Auxiliary: palladium spring: reversed detent	3.7	2.7	9.1
1890	Uhrig, 514	Continuously acting auxiliary	8.4	2.6	13.6
1891	Kullberg, 5162	Reversed detent with short spring	8.8	4.7	18.2
1892	Uhrig, 502	Continuously acting auxiliary	10.7	3.6	17.9
1893	Isaac, 1933	Auxiliary compensation	8.5	3.3	15.1
1894	Kullberg, 5433	Auxiliary compensation: reversed detent	3.7	3.4	10.5
1895	Kullberg, 5512	Ordinary balance: palladium balance spring	5.6	3.6	12.8
1896	Kullberg, 6059	Auxiliary compensation: palladium balance-spring	5.1	3.7	12.5
1897	Gardner, 5/4162	Auxiliary compensation: palladium balance-spring	6.0	2.8	11.6
1898	Usher & Cole, 8237	Ordinary balance: palladium balance-spring	6.9	2.5	11.9
1899	Kullberg, 6564	Auxiliary compensation: palladium balance-spring: reversed detent	8.0	3.2	14.4
1900	Kullberg, 6673	Auxiliary compensation: palladium balance-spring: reversed detent	6.7	2.5	11.7
1901	Johannsen & Co., 5219	Auxiliary acting in heat: palladium balance-spring	5.3	2.5	10.3
1902	Johannsen & Co., 5217	Auxiliary acting in heat: palladium balance-spring	7.0	3.2	13.4
1903	Kullberg, 6620	Auxiliary compensation: palladium balance-spring: reversed detent	5.8	3.0	11.8
1904	Kullberg, 7273	Auxiliary compensation: palladium balance-spring	9.0	3.0	15.0
1905	Kullberg, 7583	Auxiliary compensation: palladium balance-spring: reversed detent	8.6	4.6	17.8
1906	Kullberg, 7612	Auxiliary compensation: palladium balance-spring	7.5	3.3	14.1
1907	Johannsen & Co., 6150	Auxiliary acting in heat: palladium balance-spring	14.1	5.5	25.1
1908	Kullberg, 8042	Auxiliary compensation: palladium balance-spring: reversed detent	6.7	4.4	15.5
1909	Kullberg, 8075	Auxiliary compensation: palladium balance-spring	5.0	2.6	10.2
1910	Kullberg, 8159	Auxiliary compensation: palladium balance-spring	6.6	2.3	11.2
1911	T. Mercer, 8306	Mercer's auxiliary: palladium balance-spring	5.0	2.6	10.2
1912	Kullberg, 8361	Auxiliary compensation: palladium balance-spring	5.2	2.4	10.0
1913	Kullberg, 8472	Auxiliary compensation: palladium balance-spring: reversed detent	4.9	2.5	9.9
1914	Kullberg, 8734	Auxiliary compensation: palladium balance-spring	6.2	3.4	13.0

310 List of all the chronometers ranked in first place between 1840 and 1914 at the Royal Observatory, Greenwich, with their test values a, b in 29-week tests and comments on special details of construction, particularly of temperature compensation.
(Gould)

Astronomic and Chronometric Observatory of Neuchatel, Switzerland

CHRONOMETER NO. 4186 — Year 1942

by Ulysse Nardin S.A. **FIRST CLASS TEST FOR MARINE CHRONOMETERS** No. of Depot: M I 22

at Le Locle (Suisse) — Category A

ABSTRACT FROM THE REGISTER

Periods	Date	Daily rate	Deviation from average of period	Temp.	Barom.
1	Jan. 1942 12-13	(+1.92)	—	(31.8)	(716)
	13-14	+1.94	-0.05	31.8	711
	14-15	+1.61	+0.28	32.0	716
	15-16	+1.87	+0.02	32.0	719
	16-17	+1.95	-0.06	32.0	716
	17-18	+1.99	-0.10	31.9	719
	18-19	+2.00	-0.11	31.9	720
	Average	+1.89	±0.10	31.9	717
2	19-20	(+1.60)	—	(25.1)	(718)
	20-21	+1.65	+0.02	25.1	718
	21-22	+1.62	+0.05	24.9	721
	22-23	+1.69	-0.02	25.0	725
	23-24	+1.61	+0.06	24.9	721
	24-25	+1.80	-0.13	25.0	709
	25-26	+1.66	+0.01	25.0	708
	Average	+1.67	±0.05	25.0	717
3	26-27	(+1.52)	—	(18.2)	(715)
	27-28	[illegible]60	[illegible]28	[illegible]	71[illegible]
	28-29	+1.79	+0.09	18.1	713
	29-30	+1.92	-0.04	17.9	712
	30-31	+1.92	-0.04	18.0	714
	31-1	+1.95	-0.07	18.0	712
	Fév. 1942 1-2	+2.08	-0.20	18.0	718
	Average	+1.88	±0.12	18.0	714
4	2-3	(+1.85)	—	(10.5)	(719)
	3-4	+1.95	+0.06	10.5	719
	4-5	+1.93	+0.08	10.6	712
	5-6	+1.98	+0.03	10.8	713
	6-7	+2.05	-0.04	10.8	714
	7-8	+2.10	-0.09	10.5	717
	8-9	+2.05	-0.04	10.4	719
	Average	+2.01	±0.06	10.6	716
5	9-10	(+2.43)	—	(3.8)	(716)
	10-11	+2.44	-0.09	3.9	71[illegible]
	11-12	+2.44	-0.09	3.9	718
	12-13	+2.31	+0.04	3.9	720
	13-14	+2.30	+0.05	3.[illegible]	720
	14-15	+2.27	+0.08	4.0	72[illegible]
	15-16	+2.37	-0.02	4.0	725
	Average	+2.35	±0.06	3.9	720
6	Fév. 1942 16-17	(+2.11)	—	(10.6)	(723)
	17-18	+2.16	+0.20	10.5	723
	18-19	+2.20	+0.16	10.5	726
	19-20	+2.30	+0.06	10.8	724
	20-21	+2.48	-0.12	10.8	719
	21-22	+2.52	-0.16	10.4	714
	22-23	+2.52	-0.16	10.8	709
	Average	+2.36	±0.14	10.6	722
7	23-24	(+2.66)	—	(18.3)	(704)
	24-25	+2.72	-0.10	18.0	704
	25-26	+2.56	+0.06	17.9	708
	26-27	+2.54	+0.08	18.2	712
	27-28	+2.66	-0.04	18.2	716
	28-1	+2.58	+0.04	18.0	717
	Mars 1942 1-2	+2.69	-0.07	18.1	718
	Average	+2.62	±0.06	18.1	712
8	2-3	(+2.19)	—	(25.2)	(721)
	3-4	[illegible]24	[illegible]12	25.[illegible]	72[illegible]
	4-5	+2.30	-0.04	25.0	713
	5-6	+2.22	+0.04	25.1	712
	6-7	+2.32	-0.06	25.2	709
	7-8	+2.39	-0.13	25.1	715
	8-9	+2.18	+0.08	25.1	721
	Average	+2.26	±0.08	25.1	715
9	9-10	(+1.96)	—	(32.0)	(719)
	10-11	+2.02	+0.32	32.0	721
	11-12	+2.21	+0.13	32.0	719
	12-13	+2.28	+0.06	32.0	714
	13-14	+2.35	-0.01	32.0	715
	14-15	+2.39	-0.05	32.0	717
	15-16	+2.78	-0.44	32.0	720
	Average	+2.34	±0.17	32.0	717

Supplementary test

RÉSUMÉ

Periods	Average rate	Average deviation	Total deviation	Temp.	Barom.
1	+1.89	±0.10	0.62	31.9	717
2	+1.67	0.05	0.29	25.0	717
3	+1.88	0.12	0.72	18.0	714
4	+2.01	0.06	0.34	10.6	716
5	+2.35	0.06	0.37	3.9	720
6	+2.36	0.14	0.86	10.6	722
7	+2.62	0.06	0.39	18.1	712
8	+2.26	0.08	0.47	25.1	715
9	+2.34	0.17	1.01	32.0	717

COMPENSATION

Periods	t	Observed rate m	$C' = -\frac{1}{5}\left(4\frac{m_{32}-m_4}{t_{32}-t_4}+\frac{m_{25}-m_{11}}{t_{25}-t_{11}}\right)$	$S' = -\frac{1}{4}\left[(m_{32}-m_{18})+(m_4-m_{18}) + (m_{25}-m_{18})+(m_4-m_{11})\right]$
5	3.9	$m_4 = +2.35$	$\frac{m_{32}-m_4}{t_{32}-t_4} = \frac{-0.24}{28.0}$	$m_4 - m_{18} = +0.10$
$\frac{1}{2}(4+6)$	10.6	$m_{11} = +2.18$	$= -0.009$	$m_4 - m_{11} = +0.17$
$\frac{1}{2}(3+7)$	18.0	$m_{18} = +2.25$	$\frac{m_{25}-m_{11}}{t_{25}-t_{11}} = \frac{-0.22}{14.4}$	
$\frac{1}{2}(2+8)$	25.0	$m_{25} = +1.96$	$= -0.015$	$m_{25} - m_{18} = +0.15$
$\frac{1}{2}(1+9)$	31.9	$m_{32} = +2.11$	$4\frac{m_{32}-m_4}{t_{32}-t_4} = -0.036$	$m_{32} - m_{18} = -0.14$
Total		+10.85	-0.051	$p = +0.28$
Average		$m' = +2.17$	$C' = -0.010$	$S' = +0.07$

Daily rate at the temperature t:

$$m_t = m_o + C'(t - 18.0°) + d(t - 18.0°)^2$$

$$= +2.13 - 0.010\,(t - 18.0°) + 0.0004\,(t - 18.0°)^2$$

$$m_o = m' - \frac{p}{7} = +2.17 - 0.04 = +2.13 \qquad d = \frac{p}{686} = +0.0004$$

1. Average deviation of the daily rate . . . . E = ±0.09
2. Thermal coefficient . . . . C′ = -0.010
3. Proportional deviation . . . . S′ = +0.07

Average rate of the 1st period . . . . = +1.89

Average rate of the 9th period . . . . = +2.34

4. Recovery of rate . . . . R = +0.45

312 Test table of 1942 for a Ulysse Nardin marine chronometer at the Neuchâtel Observatory over nine weeks, with stated daily rates, prevailing temperatures, daily deviation compared to weekly mean, and barometer reading. Also a summary and an evaluation of the compensation. *(M. Whitney)*

The Early Trials.

The trial numbers given in the following table have been computed from the MS. records at Greenwich, the selection of a period of 29 weeks being governed by the circumstances of each case. The period is always a continuous one.

The range of temperature may be taken, except in the case of Harrison's No. 4, as 40°-70° Fahr.

The trial numbers have been calculated from the formula $(a + 2b)$.

Machine	Period of Trial	a	b	Trial No.
Harrison No. 1	7.11.66 to 1.6.67	668·0	235·5	1139·0
Harrison No. 2	4.11.66 to 25.5.67	289·5	200·7	690·9
Harrison No. 3	9.10.66 to 7.1.67	58·5	36·0	130·5
	3.2.67 to 25.5.67	290·8	68·9	428·6
Harrison No. 4	7.7.66 to 22.3.67	59·8	32·3	124·4
Kendall No. 1	3.5.70 to 22.11.70	82·0	28·0	138·0
Kendall No. 2	12.9.72 to 10.3.73	57·0	49·0	155·0
Mudge No. 1	8.4.77 to 29.10.77	5·97	2·88	11·73
Arnold No. 36	29.3.79 to 31.10.79	21·17	11·87	44·91
Mudge "Green"	29.6.89 to 17.1.90	31·46	13·33	58·12
Mudge "Blue"	28.6.89 to 17.1.90	22·87	8·41	39·39
Earnshaw No. 1	27.3.00 to 15.10.00	14·95	5·31	25·57
Earnshaw No. 2	27.3.00 to 15.10.00	17·18	6·62	30·42

311 Test and quality figures of several important chronometers of the heroic epoch, calculated from contemporary 29-week tests at Greenwich.
(Gould)

I Laufende Nr.	II Name und Wohnort des Fabrikanten	III Fabrik-Nr.	1916 Okt. 23 bis Nov. 5 30°	191[illegible] Nov. bis Nov. 25°
	Klasse I.		sek	sek
1	A. Lange & Söhne, Glashütte i. Sa.	184	+ 0.04	— 0.0
2	Chronometer-Werke, Hamburg	827	+ 1.11	+ 1.0
3	Chronometer-Werke, Hamburg	829	+ 1.10	+ 1.2
4	C. Wiegand, Peine	63	— 0.31	— 0.3
5	Joh. Raabe, Glashütte i. Sa.	1195	+ 0.58	+ 0.4
6	A. Lange & Söhne, Glashütte i. Sa.	181	+ 0.16	+ 0.2
7	Chronometer-Werke, Hamburg	828	+ 0.52	+ 0.4
8	Chronometer-Werke, Hamburg	838	+ 0.29	+ 0.0
9	A. Lange & Söhne, Glashütte i. Sa.	222	— 0.30	— 0.2
10	A. Lange & Söhne, Glashütte i. Sa.	182	+ 0.04	— 0.0
11	Chronometer-Werke, Hamburg	810	+ 0.23	+ 0.0
12	A. Lange & Söhne, Glashütte i. Sa.	183	+ 0.34	+ 0.3
13	C. Wiegand, Peine	58	— 0.45	— 0.7
14	Chronometer-Werke, Hamburg	839	+ 1.0[illegible]	+ 0.8
15	Chronometer-Werke, Hamburg	822	— 0.30	— 0.5
16	Chronometer-Werke, Hamburg	790	+ 0.31	— 0.0
17	A. Lange & Söhne, Glashütte i. Sa.	254	— 0.93	— 0.9
18	F. Lidecke, Geestemünde	442	+ 0.37	+ 0.2
19	A. Lange & Söhne, Glashütte i. Sa.	255	— 0.43	— 0.7
20	A. Lange & Söhne, Glashütte i. Sa.	256	+ 0.23	+ 0.0
21	A. Lange & Söhne, Glashütte i. Sa.	259	— 0.70	— 0.9
22	A. Lange & Söhne, Glashütte i. Sa.	242	— 1.12	— 1.3
23	Chronometer-Werke, Hamburg	813	+ 0.13	— 0.1
24	A. Lange & Söhne, Glashütte i. Sa.	189	+ 0.56	+ 0.4
25	Chronometer-Werke, Hamburg	823	+ 1.01	+ 0.9
26	Chronometer-Werke, Hamburg	826	+ 0.12	— 0.2
27	Chronometer-Werke Hamburg	831	+ 0.04	— 0.2
28	Chronometer-Werke, Hamburg	830	+ 0.30	— 0.0
29	Chronometer-Werke, Hamburg	820	— 0.54	— 0.8
30	Chronometer-Werke, Hamburg	763	+ 0.04	— 0.1
31	Chronometer-Werke, Hamburg	835	+ 0.13	— 0.2
32	C. Wiegand, Peine	64	— 2.02	— 2.3
33	Chronometer-Werke, Hamburg	837	+ 0.61	+ 0.2
34	A. Lange & Söhne, Glashütte i. Sa.	185	+ 0.23	+ 0.3
35	A. Lange & Söhne, Glashütte i. Sa.	245	— 0.45	— 0.7
36	F. Lidecke, Geestemünde	456	— 1.22	— 1.5
37	Chronometer-Werke, Hamburg	812	— 0.93	— 1.0
38	A. Lange & Söhne, Glashütte i. Sa.	251	— 0.56	— 0.8
39	A. Lange & Söhne, Glashütte i. Sa.	257	— 0.52	— 0.9
40	Chronometer-Werke, Hamburg	836	+ 0.[illegible]	+ 0.3
41	A. Lange & Söhne, Glashütte i. Sa.	250	— 0.82	— 0.5
42	C. Wiegand, Peine	62	— 0.09	— 0.48
43	F. Lidecke, Geestemünde	469	— 0.77	— 1.06
44	A. Lange & Söhne, Glashütte i. Sa.	253	— 0.52	— 0.90
45	L. Kurtz, Münster i. W.	118	— 4.41	— 4.72
46	Chronometer-Werke, Hamburg	824	+ 0.22	+ 0.18
47	A. Lange & Söhne, Glashütte i. Sa.	252	— 0.[illegible]	— 1.22
48	C. Wiegand, Peine	60	+ 0.01	0.00
49	C. Wiegand, Peine	57	— 0.47	— 0.92
50	C. Wiegand, Peine	53	— 1.06	— 0.96

40. Chronometer-Wettbewerb-Prüfung 1916/17. **Gang-Tabelle.**

IV Tägliche Gänge: 1916 Nov. 21 bis Nov. 29, 15°	1916 Nov. 29 bis Dez. 7, 10°	1916 Dez. 7 bis Dez. 15, 10°	1916 Dez. 15 bis Dez. 23, 15°	1916 Dez. 23 bis Dez. 31, 20°	1916/17 Dez. 31 bis Jan. 8, 25°	1917 Jan. 8 bis Jan. 16, 30°	V Auf die Mitte der Untersuchungszeit bezogene mittlere tägliche Gänge: 30°	25°	20°	15°	10°	VI A	VII B	VIII C	IX A+2B+C
sek	sek	sek	sek	sek	sek	sek	sek	sek	sek	sek	sek	sek	sek	sek	sek
+ 0.03	— 0.01	+ 0.02	+ 0.11	+ 0.12	+ 0.14	+ 0.18	+ 0.11*	+ 0.02	+ 0.03	+ 0.07	+ 0.01*	+ 0.10	0.10	+ 0.003	0.30
+ 0.97	+ 0.81	+ 0.94	+ 0.90	+ 1.04†	+ 1.02	+ 1.04	+ 1.08*	+ 1.04	+ 1.01	+ 0.93	+ 0.88*	+ 0.20	0.09	— 0.001	0.38
+ 1.02	+ 1.02	+ 0.92	+ 1.00	+ 0.97	+ 0.89	+ 0.73†	+ 0.92*	+ 1.05*	+ 1.01	+ 1.01	+ 0.97	— 0.13	0.13	— 0.005	0.39
— 0.33	— 0.25	— 0.45†	— 0.53	— 0.43	— 0.51	— 0.48	— 0.39	— 0.43	— 0.45*	— 0.43	— 0.35*	— 0.10	0.20	— 0.002	0.50
+ 0.42	+ 0.34	+ 0.21	+ 0.09	+ 0.30†	+ 0.19	+ 0.33	+ 0.45*	+ 0.31	+ 0.39	+ 0.25*	+ 0.28	+ 0.20	0.16	— 0.004	0.52
+ 0.30	+ 0.06†	+ 0.10	+ 0.09	+ 0.04	— 0.03	+ 0.18	+ 0.17	+ 0.09	+ 0.14	+ 0.20*	+ 0.08*	+ 0:12	0.21	— 0.002	0.54
+ 0.49	+ 0.68	+ 0.58	+ 0.36†	+ 0.55	+ 0.59	+ 0.58	+ 0.55	+ 0.50	+ 0.47	+ 0.42*	+ 0.63*	— 0.21	0.17	+ 0.002	0.55
— 0.10	— 0.10	— 0.06	+ 0.02	0.00	+ 0.10	+ 0.29	+ 0.29*	+ 0.10	+ 0.03	— 0.04	— 0.08*	+ 0.37	0.11	0.000	0.57
— 0.37	— 0.60†	— 0.65	— 0.82	— 0.65	— 0.56	— 0.53	— 0.41*	— 0.42	— 0.46	— 0.59	— 0.62*	+ 0.21	0.18	— 0.004	0.57
— 0.24	— 0.17	— 0.20	— 0.21	— 0.01†	+ 0.10	+ 0.24	+ 0.14*	+ 0.05	— 0.06	— 0.22*	— 0.18	+ 0.36	0.11	+ 0.002	0.58
+ 0.01	— 0.13	— 0.17	+ 0.03	+ 0.18	+ 0.14	— 0.11†	+ 0.06	+ 0.09*	+ 0.09	+ 0.02	— 0.15*	— 0.24	0.19	— 0.001	0.62
+ 0.22	— 0.03	— 0.03	— 0.21	— 0.15	— 0.13	+ 0.13†	+ 0.24*	+ 0.11	+ 0.01	0.00	— 0.03*	+ 0.27	0.19	— 0.006	0.65
— 0.77	— 0.73	— 0.83	— 0.81	— 0.81	— 0.82	— 0.77	— 0.61*	— 0.78	— 0.83*	— 0.79	— 0.78	+ 0.22	0.23	— 0.003	0.68
+ 0.90	+ 0.70	+ 0.59	+ 0.84†	+ 1.03	+ 0.92	+ 0.97	+ 1.02*	+ 0.88	+ 0.92	+ 0.87	+ 0.65*	+ 0.37	0.16	0.000	0.69
— 0.58	— 0.43	— 0.51	— 0.46	— 0.56	— 0.73	— 1.04†	— 0.67*	— 0.63	— 0.66	— 0.52	— 0.47*	— 0.20	0.26	— 0.007	0.73
+ 0.20	+ 0.30	+ 0.28	+ 0.15	+ 0.24	+ 0.37	+ 0.31	+ 0.31*	+ 0.17	+ 0.17*	+ 0.18	+ 0.29	+ 0.14	0.30	+ 0.004	0.74
— 1.17	— 1.25	— 1.37	— 1.42	— 1.20	— 1.11	— 0.83†	— 0.91*	— 1.04	— 1.16	— 1.29	— 1.31*	+ 0.40	0.18	0.000	0.76
+ 0.40	+ 0.69†	+ 0.55	+ 0.48	+ 0.40	+ 0.33	+ 0.29	+ 0.33	+ 0.28*	+ 0.38	+ 0.44	+ 0.62*	— 0.34	0.21	0.000	0.76
— 0.72	— 0.86	— 1.01	— 0.91	— 0.72	— 0.72	— 0.64	— 0.56*	— 0.75	— 0.79	— 0.81	— 0.93*	+ 0.37	0.20	— 0.001	0.77
— 0.32	— 0.45	— 0.64	— 0.77	— 0.60	— 0.46	— 0.21†	+ 0.01*	— 0.22	— 0.41	— 0.54	— 0.54*	+ 0.55	0.11	— 0.007	0.78
— 1.21	— 1.32	— 1.44	— 1.42	— 1.23	— 1.14	— 1.09	— 0.89*	— 1.06	— 1.17	— 1.31	— 1.38*	+ 0.49	0.16	— 0.004	0.81
— 1.57	— 1.63	— 1.42	— 1.11†	— 0.96	— 1.14	— 1.14	— 1.13*	— 1.25	— 1.18	— 1.34	— 1.52*	+ 0.39	0.21	+ 0.002	0.81
— 0.12	+ 0.07	+ 0.05	— 0.05	— 0.06	— 0.18	— 0.33	— 0.08	— 0.14	— 0.15*	— 0.08	+ 0.06*	— 0.21	0.30	— 0.004	0.81
+ 0.20	— 0.10†	— 0.01	+ 0.11	+ 0.29	+ 0.27	+ 0.25	+ 0.40*	+ 0.38	+ 0.27	+ 0.15	— 0.05*	+ 0.45	0.19	— 0.004	0.83
+ 1.12	+ 1.24	+ 1.20	+ 1.09	+ 1.01	+ 0.95	+ 0.63†	+ 0.82*	+ 0.93	+ 1.00	+ 1.11	+ 1.22*	— 0.40	0.22	— 0.002	0.84
— 0.32	— 0.30	— 0.38	— 0.53	— 0.54	— 0.53	— 0.58	— 0.23*	— 0.38	— 0.49*	— 0.43	— 0.34	+ 0.26	0.29	— 0.007	0.85
— 0.21	— 0.19	— 0.20	— 0.31	— 0.47	— 0.48	— 0.68	— 0.32	— 0.36	— 0.41*	— 0.26	— 0.19*	— 0.22	0.33	— 0.007	0.88
— 0.47	— 0.47	— 0.42	— 0.40	— 0.41	— 0.27	— 0.21	+ 0.05*	— 0.15	— 0.28	— 0.43	— 0.44*	+ 0.49	0.21	— 0.006	0.92
— 0.00	— 0.05	0.00	0.77	— 0.71	0.00	1.00	0.77	0.05*	0.70	— 0.00	— 0.03*	— 0.22	0.35	— 0.003	0.92
+ 0.38	+ 0.44	+ 0.37	+ 0.27	+ 0.32	+ 0.22	+ 0.41	+ 0.23	+ 0.02*	+ 0.25	+ 0.33	+ 0.41*	— 0.39	0.27	+ 0.006	0.93
— 0.48	— 0.34	— 0.18	— 0.25	— 0.26	— 0.22	— 0.19	— 0.03*	— 0.23	— 0.35	— 0.36*	— 0.26	+ 0.33	0.30	— 0.002	0.93
— 2.66	— 2.75	— 2.70	— 2.64	— 2.34	— 2.49	— 2.46	— 2.24*	— 2.43	— 2.43	— 2.65	— 2.72*	+ 0.48	0.23	— 0.004	0.94
— 0.03	— 0.14	— 0.33	— 0.37	— 0.37	— 0.26	— 0.18	+ 0.21*	— 0.01	— 0.11	— 0.20	— 0.23*	+ 0.44	0.26	— 0.010	0.97
+ 0.72	+ 0.47†	+ 0.47	+ 0.35	+ 0.30	+ 0.24	+ 0.17	+ 0.21*	+ 0.30	+ 0.40	+ 0.53*	+ 0.47	— 0.32	0.33	— 0.002	0.98
— 0.81	— 1.14	— 1.10	— 0.90	— 0.67	— 0.74	— 0.76	— 0.58*	— 0.76	— 0.71	— 0.85	+ 1.12*	+ 0.54	0.23	— 0.002	1.00
— 2.05	— 2.08	— 2.22	— 2.36	— 2.18	— 2.13	— 2.03	— 1.62*	— 1.86	— 2.06	— 2.20*	— 2.15	+ 0.58	0.22	— 0.010	1.03
— 0.42†	— 0.36	— 0.35	— 0.25	— 0.42	— 0.56	— 0.88	— 0.91*	— 0.78	— 0.61	— 0.33*	— 0.35	— 0.58	0.23	+ 0.004	1.04
— 0.96	— 0.67†	— 0.81	— 0.97	— 0.91	— 0.83	— 0.74	— 0.65*	— 0.83	— 0.96	— 0.96*	— 0.74	+ 0.31	0.37	— 0.001	1.05
— 1.30	— 1.35	— 1.36	— 1.31	— 1.28	— 1.35	— 1.44	— 0.98*	— 1.15	— 1.20	— 1.31	— 1.35*	+ 0.37	0.35	— 0.010	1.08
+ 0.29	+ 0.52	+ 0.60	+ 0.43	+ 0.33	+ 0.30	— 0.15†	+ 0.20*	+ 0.31	+ 0.27	+ 0.36	+ 0.56*	— 0.36	0.36	— 0.005	1.08
— 1.18†	— 1.41	— 1.43	— 1.30	— 1.09	— 0.98	— 0.82	— 0.57*	— 0.76	— 0.97	— 1.24	— 1.42*	+ 0.85	0.12	— 0.007	1.10
— 0.74	— 1.04	— 0.92	— 0.98	— 0.82	— 0.76	— 0.44	— 0.26*	— 0.62	— 0.69	— 0.86	— 0.98*	+ 0.72	0.21	— 0.005	1.14
— 1.54†	— 1.84	— 1.83	— 1.56	— 1.45	— 1.06	— 1.35	— 1.06*	— 1.06	— 1.30	— 1.55	— 1.83*	+ 0.77	0.19	— 0.004	1.15
— 1.24	— 1.55	— 1.55	— 1.38	— 1.16	— 1.02	— 0.98	— 0.75*	— 0.96	— 1.13	— 1.31	— 1.55*	+ 0.80	0.18	— 0.004	1.16
— 4.64	— 4.49	— 4.51	— 4.72	— 4.91	— 5.11	— 5.18	— 4.81	— 4.92*	— 4.84	— 4.68	— 4.50*	— 0.42	0.38	— 0.009	1.19
+ 0.88	+ 0.71	+ 0.41	+ 0.14	+ 0.18	+ 0.13	+ 0.10	+ 0.16	+ 0.15*	+ 0.43	+ 0.51	+ 0.56*	— 0.41	0.41	— 0.001	1.23
— 1.61	— 2.01†	— 2.03	— 1.70	— 1.49	— 1.47	— 1.31	— 1.13*	— 1.34	— 1.42	— 1.65	— 2.02*	+ 0.89	0.18	— 0.005	1.25
+ 0.56†	+ 0.60	+ 0.79	+ 0.51	+ 0.23	+ 0.13	+ 0.13	+ 0.07	+ 0.07*	+ 0.16	+ 0.53	+ 0.70*	— 0.63	0.31	+ 0.002	1.25
— 1.37	— 1.59	— 1.72	— 1.55	— 1.32	— 1.32	— 1.29	— 0.88*	— 1.12	— 1.22	- 1.46	— 1.65*	+ 0.77	0.25	— 0.009	1.28
— 0.55	— 0.77	— 0.50	— 0.63	— 0.77	— 1.26†	— 1.51	— 1.28*	— 1.11	— 0.82	— 0.59*	— 0.63	— 0.69	0.32	— 0.006	1.34

I Laufende Nr.	II Name und Wohnort des Fabrikanten	III Fabrik-Nr.	IV Tägliche Gänge 1916 Okt. 28 bis Nov. 5	1916 Nov. 5 bis Nov. 13	1916 Nov. 13 bis Nov. 21	1916 Nov. 21 bis Nov. 29	1916 Nov. 29 bis Dez. 7	1916 Dez. 7 bis Dez. 15	1916 Dez. 15 bis Dez. 23	1916 Dez. 23 bis Dez. 31	1916/17 Dez. 31 bis Jan. 8	1917 Jan. 8 bis Jan. 16	
			30°	25°	20°	15°	10°	10°	15°	20°	25°	30°	30°
			sek	sek	sek	sek	sek	sek	sek	sek	sek	sek	sek
51	F. Vetterlein, Glashütte i. Sa.	10	— 2.44	— 2.62	— 2.65	— 2.18	— 1.92	— 1.87	— 2.36†	— 2.58	— 2.73	— 2.79	— 2.61
52	F. Lidecke, Geestemünde	478	+ 0.52	+ 0.28	— 0.21†	— 0.55	— 0.34	— 0.75	— 0.60	— 0.26	— 0.16	— 0.09	+ 0.21*
53	C. Wiegand, Peine	61	— 0.63	— 0.86	— 1.13	— 1.37	— 1.85†	— 1.88	— 1.97	— 1.52	— 1.45	— 1.25	— 0.94*
54	L. Kurtz, Münster i. W.	119	— 2.91	— 3.27	— 3.09	— 3.06	— 2.93	— 3.09	— 3.21	— 2.54†	— 2.92	— 3.30	— 3.11
55	F. Lidecke, Geestemünde	431	+ 0.96	+ 1.17	+ 1.46	+ 1.47	+ 1.74	+ 1.28	+ 0.96	+ 0.82	+ 0.98	+ 0.41†	+ 0.69*
56	F. Lidecke, Geestemünde	468	+ 0.53	+ 0.01†	— 0.44	— 0.45	— 0.83	— 0.98	— 0.59	— 0.25	— 0.06	— 0.02	+ 0.25*
57	F. Lidecke, Geestemünde	451	+ 0.52	— 0.21†	— 0.51	— 0.89	— 1.48	— 1.43	— 0.89	— 0.33	+ 0.01	— 0.14	+ 0.19*
	Klasse II.												
1	F. Lidecke, Geestemünde	459	— 0.68	— 1.21†	— 1.69	— 2.12	— 2.30	— 2.35	— 2.26	— 1.94	— 1.85	— 1.83	— 1.25*
2	F. Lidecke, Geestemünde	476	+ 0.14	— 0.66†	— 1.10	— 1.81	— 2.27	— 2.34	— 1.93	— 1.34	— 1.05	— 0.71	— 0.28*
3	F. Lidecke, Geestemünde	475	+ 0.28	— 0.58†	— 1.28	— 1.97	— 2.75	— 2.85	— 2.15	— 1.53	— 1.16	— 1.03	— 0.37*
4	Union, Glashütte i. Sa.	24	+ 1.72	+ 0.84†	+ 0.28	— 0.49	— 0.93	— 1.11	— 0.60	+ 0.19	+ 0.67	+ 1.45	+ 1.59*
5	F. Lidecke, Geestemünde	460	+ 1.20	+ 0.45	+ 0.20	— 0.49	— 1.75†	— 1.84	— 1.56	— 0.42	+ 0.54	+ 0.98	+ 1.09*
6	Union, Glashütte i. Sa.	20	+ 1.33	+ 0.08	— 0.61	— 0.78	— 1.66	— 1.71	— 0.41	+ 0.30	+ 0.60	+ 1.07	+ 1.20*
7	F. Lidecke, Geestemünde	472	+ 4.21	+ 3.04	+ 1.94	+ 0.67†	— 0.31	— 0.36	+ 0.69	+ 1.86	+ 2.49	+ 2.95	+ 3.58*
	Klasse III.												
1	C. Wiegand, Peine	38	+ 0.20	+ 0.09	+ 0.32	+ 0.83†	+ 0.57	+ 0.10	— 0.29	— 0.61	— 1.07	— 0.90	— 0.35
2	L. Kurtz, Münster i. W.	117	— 0.35	— 0.66	— 0.71	— 1.16	— 1.42	— 2.17†	— 2.78	— 2.47	— 2.19	— 1.59	— 0.97*
3	A. Lange & Söhne, Glashütte i. Sa.	177	— 0.75	— 1.07	— 1.62	— 2.06	— 2.89†	— 2.95	— 2.70	— 2.27	— 2.09	— 1.71	— 1.23*
4	Union, Glashütte i. Sa.	33	+ 1.00	— 0.58†	— 1.93	— 2.80	— 3.48	— 3.48	— 2.82	— 1.95	— 1.21	— 0.32	+ 0.34*
5	Union, Glashütte i. Sa.	26	+ 2.96	+ 1.22†	— 0.45	— 1.83	— 3.01	— 2.72	— 1.20	— 0.32	+ 1.05	+ 2.54	+ 2.75*
	Außer Wettbewerb. **Klasse II.**												
1	F. Lidecke, Geestemünde	470	+ 2.75	+ 1.92	+ 1.36	+ 0.83	+ 0.23	— 0.04	+ 0.59	+ 1.60†	+ 2.12	+ 2.93	+ 2.84*
2	F. Lidecke, Geestemünde	477	+ 2.53	+ 1.21	+ 0.40	— 1.05	— 1.99	— 2.10	— 1.23	+ 0.10	+ 0.80	+ 1.65	+ 2.09*
	Eppner Chronometrische Thermometer		+ 153.09	+ 89.66	+ 27.99	— 43.77	— 100.12	— 101.94	— 43.45	+ 29.11	+ 81.04	+ 151.27	+ 152.18
	Tiede ohne Kompensation		+ 151.37	+ 92.86	+ 35.77	— 28.86	— 86.88	— 88.81	— 24.65	+ 37.79	+ 87.40	+ 152.75	+ 152.06
	Mittlere Dekadentemperatur, Cels.		30.0°	25.0°	20.0°	14.9°	10.0°	10.0°	15.0°	20.0°	24.9°	30.0°	30.0°
	Extreme der mittl. Tagestemperatur		30.9-27.0	25.5-24.2	21.1-19.5	15.8-14.2	10.9 -9.2	10.6 -9.5	15.7-14.5	21.0-18.0	25.7-23.5	30.7-29.2	
	Mittlere relative Feuchtigkeit in %		52	51	50	54	58	63	62	57	58	50	

313 Going results of the competition at the Deutsche Seewarte, Hamburg, in the winter of 1916-17, for 71 marine chronometers made by various German chronometer makers, with details of daily rates (weekly means) for various temperatures (10 to 30 degrees Celsius), plus their range of deviation, the prevailing conditions, mean humidity and the auxiliary powers A, B, C plus A + 2B + C for the class ratings of the Deutsche Seewarte.
Annalen der Hydrographie XXXXV (1917), No. VIII.

Das Chronometer Chronometer - Werke Nr.

ist in der Zeit vom 31. Mai 1939 bis zum 21. Juli

~~derxxxxWettbewerb~~ einer großen -Prüfung unterworfen worden und hat dabei folgende Ergebniss

Mittlere tägliche Gangschwankung	S :	0.16 Se
Größte Gangschwankung	S1:	0.34 "
Größte Gangänderung	S2:	0.63 "
Temperaturfehler	T :	0.63 "
Wärmefehler	W:	-0.01 "
Kältefehler	K :	-0.63 "
Langfristige Gangänderung	C :	0.92 "
Neigungsfehler	N :	"
Isochronismusfehler	Z :	"
Aufzugsfehler	Z':	"

Das Chronometer hat die Wettbewerb-Prüfung bestanden und entspricht daher Anforderungen der Deutschen Seewarte.

314 Test certificate of the Deutsche Seewarte for a chronometer, dated July 22, 1939.

V			VI	VII	VIII	IX
e der Untrsuchungszeit ittlere tägliche Gänge						A+2B+C
20°	15°	10°	A	B	C	
sek	sek	sek	sek	sek	sek	sek
— 2.62	— 2.27	— 1.89*	— 0.79	0.29	— 0.003	1.37
— 0.24	— 0.57*	— 0.54	+ 0.78	0.29	— 0.008	1.37
— 1.33	— 1.67	— 1.86*	+ 0.92	0.25	— 0.010	1.43
— 2.81*	— 3.13*	— 3.01	+ 0.32	0.59	0.000	1.50
+ 1.14	+ 1.22	+ 1.51*	— 0.82	0.36	— 0.006	1.55
— 0.35	— 0.52	— 0.90*	+ 1.15	0.23	— 0.004	1.61
— 0.42	— 0.89	— 1.45*	+ 1.64	0.32	— 0.003	2.28
— 1.82	— 2.19	— 2.32*	+ 1.07	0.26	— 0.013	1.60
— 1.22	— 1.87	— 2.30*	+ 2.02	0.29	— 0.009	2.61
— 1.41	— 2.06	— 2.80*	+ 2.43	0.25	— 0.014	2.94
+ 0.24	— 0.55	— 1.02*	+ 2.61	0.22	— 0.003	3.05
— 0.11	— 1.02	— 1.79*	+ 2.88	0.54	— 0.001	3.96
— 0.15	— 0.60	— 1.68*	+ 2.88	0.58	+ 0.003	4.04
+ 1.90	+ 0.68	— 0.33*	+ 3.91	0.27	— 0.014	4.46
— 0.15	+ 0.27	+ 0.34*	— 0.83	0.30	— 0.018	1.45
— 1.59	— 1.97*	— 1.79	+ 1.00	0.75	— 0.022	2.52
— 1.95	— 2.38	— 2.92*	+ 1.69	0.42	— 0.016	2.55
— 1.94	— 2.81	— 3.48*	+ 3.82	0.63	— 0.015	5.09
— 0.38	— 1.52	— 2.86*	+ 5.61	0.33	— 0.004	6.27
+ 1.48	+ 0.71	+ 0.09*	+ 2.75	0.32	+ 0.003	3.39
+ 0.25	— 1.14	— 2.04*	+ 4.13	0.40	— 0.010	4.94
+ 28.55	— 43.61	— 101.03				
+ 36.78	— 26.76	— 87.84				
20.0°	15.0°	10.0°				

ordinary pocket watches. Prescribed temperature ranges and various testing positions are likewise specified in detail by De Carle.

Another important clock testing center in Switzerland is the Official Bureau for Clock Testing, with branches in Biel, La Chaux-de-Fonds, Le Locle, St. Imier and Geneva. The tasks of the Official Bureau, though, seem to include only the testing of pocket watches and wristwatches. On the other hand, the Observatory of Neuchâtel, founded in 1858-59, includes the testing of chronometers. But only two positions for pocket chronometers are prescribed there: vertical hanging with pendant upward, and horizontal lying with dial upward. In addition, only two temperature ranges are called for: ordinary environment and sweat-box temperatures. The table in #312 shows a modern test document from the Observatory of Neuchâtel, issued in 1942.

Two official centers in Germany have dealt with chronometer testing, namely the Deutsche Seewarte in Hamburg, founded in 1875, and where in 1882 the first thorough examination — though essentially for scientific purposes such as polar explorations — was undertaken; and the Physicakalisch-Technische Reichsanstalt in Braunschweig (PTR). De Carle offers a brief description of these two test centers along with those of the Swiss centers, while Repsold[2] provides a thorough account of the chronometer and pocket-watch tests at the Deutsche Seewarte. Rate results for the chronometer competition tests of 1916-1917 at the latter site, also including the class divisions for the 71 marine chronometers, are shown in #313, while #314 shows a test certificate for a chronometer from the Deutsche Seewarte, which was given to the entrant when the chronometer was submitted.

The handling of chronometer testing and official activities in France was conducted in a very different manner from that in Britain and Germany. Guyot[3] writes of it: "Chronometer supplying in France originally took place on a basis of pure trust. Until 1832 they were supplied to the French Navy exclusively by the official naval clockmakers (Horlogers de la Marine). This privilege was ended in 1832 on the petition of A. Jacob of St. Nicolas d'Aliermont, who had sent a complaint to the renowned physicist F. Arrago. Through his influence, a general competition for the purchasing of chronometers by the state was introduced in 1833, and its administration was turned over to the Paris Observatory. From then on, the marine chronometers had to be tested first during two months at sea and thereafter at the observatory in Paris. Until 1838 the chronometers were kept in an icebox at the observatory, at a constant temperature of 12 to 13 degrees Celsius, and only very seldom exposed to a decrease to 0 degrees; the chronometers were generally kept this way for a whole year, without any notice being taken of any errors that might have occurred in the interim. Unfortunately, until now it has not been possible to obtain the organizational regulations used in the period up to 1855, since all dealings were to be made only orally between 1840 and 1855." Guyot also mentions that at that time, chronometers rejected after the first tests could be reworked by the chronometer makers and submitted again for testing. As of 1839 three temperature areas:

TABLE I.—RATES OF CHRONOMETERS ON COMPETITIVE TRIAL AT THE U.

Chronometer. Maker.	No.	Description of balance as given by the maker.	Dec. 31 to Jan. 7.	Jan. 7 to Jan. 14.	Jan. 16 *to Jan. 23.	Jan. 24 *to Jan. 31.	Feb. 1 *to Feb. 8.	Feb. 9 *to Feb. 16.	Feb. 16 *to Feb. 23.	Feb. 23 *to Mar. 1.	Ma * Ma
			s.	s.	s.	s.	s.	s.	s.	s.	
T. S. & J. D. Negus.	1615	Ordinary balance, with Negus' correction	—0.529	— .461	— .246	— .274	— .514	— .936	—1.077	— .666	—
Wm. Bond & Son	510	Ordinary balance	—0.493	— .533	—1.389	+ .047	+ .771	— .150	— .720	— .237	+
Jno. Bliss & Co.	2768	Bliss' auxiliary balance	—0.350	+ .217	— .246	+1.226	+1.986	+1.743	+1.637	+1.870	+
T. S. & J. D. Negus.	1655	Ordinary balance, with Negus' correction	+1.150	+1.431	+1.290	+1.083	+ .093	— .436	— .506	— .023	+
do.	1597	Same	+0.507	+ .789	+ .183	+ .654	+ .093	— .436	— .827	— .416	—
Wm. Bond & Son	529	Ordinary balance	—0.243	— .390	—1.603	+ .190	+1.164	+ .243	— .434	— .059	+
D. Eggert's Sons	863	Auxiliary balance, with Airy's supplementary	—0.314	— .140	— .603	+ .369	+ .414	—1.221	—1.756	—1.059	+
Wm. Bond & Son.	467	Ordinary balance	—0.100	— .033	—1.424	— .346	+1.450	+ .457	+ .209	+ .477	+
Jno. Bliss & Co.	2809	Flat rim balance	—0.814	— .426	+ .433	— .774	— .550	+ .207	+ .530	+ .477	±
T. S. & J. D. Negus.	1605	Ordinary balance, with Negus' correction	—0.921	— .747	+ .004	— .453	—1.750	—2.007	—1.863	—1.094	—
D. Eggert's Sons	867	Ordinary balance	+1.114	+1.146	+ .540	+1.476	+1.093	—1.186	—2.506	—1.487	+
Jno. Bliss & Co.	2803	Bliss' auxiliary balance	+1.043	+1.324	+1.183	+1.940	+1.164	— .864	— .149	— .201	+
do.	2804	Same	+1.293	+1.396	+ .933	+1.261	+1.164	+2.314	+2.566	+2.549	+
D. Eggert's Sons	872	Auxiliary balance, with Airy's supplementary	—0.457	— .890	— .960	+ .940	+ .236	— .936	—1.756	—1.273	—
do.	870	Same	+1.043	+1.217	+ .647	+ .547	+ .093	+2.671	+3.030	+2.620	+
Wm. Bond & Son	466	Ordinary balance	—1.386	—2.033	—4.367	—2.031	+ .879	+ .671	— .026	+ .120	—
Extremes of temperature, per maximum and minimum thermometer			58.2° 44.	54.0° 44.7	46.5° 43.0	56.5° 54.5	71.2° 69.–	86.2° 84.2	93.7° 85.0	86.5° 84.0	7 6
Mean temperatures, per chronometrical thermometer			52.7°	50.9°	45.1°	55.2°	69.9°	85.0°	89.7°	84.9°	7
Relative humidity			.69	.69	.74	.70	.68	.62	.62	.59	.

1	2	3	4	5	6	7	8	9	REMARKS.
			s.	s.	s.	s.	s.	s.	
120	Loseby.	70°	— 1.78	+ 1.64	— 3.42	+ 2.08	+ .44	3.86	This chronometer has Loseby's auxiliary to balance. It had been cleaned six months previous to trial.
		85°	— 1.72	+ 2.15	— 3.87	+ 3.02	+ .87	4.74	
		85°	— .24	+ 2.88	— 3.12	+ 3.46	+ .58	3.70	
		70°	+ 1.08	+ 2.55	— 1.47	+ 3.13	+ .58	2.05	
		55°	+ .85	+ 2.13	— 1.28	+ 2.05	— .08	1.28	
725	Negus.	70°	— .28	+ .31	— .59	+ .08	— .23	.59	This chronometer has the ordinary balance. It had not been cleaned for four years, and the bearings had become dry.
		85°	— 2.10	— 1.60	— .50	— 2.20	— .60	.60	
		85°	— 2.36	— 1.99	— .37	— 2.41	— .42	.42	
		70°	— .92	— 1.32	+ .40	— 1.74	— .42	.82	
		55°	— 1.52	— 1.00	— .52	— 1.20	— .20	.52	Sent to the maker and oiled, without being taken apart.
		73°	— 2.54	— 1.87	— .67	— 2.11	— .34	.67	
		73°	— 2.20	— 1.84	— .36	— 2.07	— .23	.36	
1262	Negus.	70°	+ 1.41	+ 1.81	— .40	+ 1.45	— .36	.40	This chronometer has the ordinary balance. It had been running four years, and the bearings had become dry.
		85°	+ .40	+ .65	— .25	+ .55	— .10	.25	
		85°	— .24	± .00	— .24	— .04	— .04	.24	
		70°	+ .96	+ 1.30	— .34	+ 1.26	— .04	.34	
		55°	+ 1.72	+ 2.13	— .41	+ 2.30	+ .17	.58	Sent to the maker and oiled, without being taken apart.
		73	+ 1.71	+ 2.01	— .30	+ 2.47	+ .46	.76	
		73	+ 1.72	+ 1.91	— .19	+ 2.44	+ .53	.72	
700	Negus.	73°	+ .03	+ 1.05	— 1.02	+ 1.18	+ .13	1.15	This chronometer has the ordinary balance. It had been cleaned nine months previous to trial.
		73°	+ .14	+ .70	— .56	+ .87	+ .17	.73	
919	Negus	73°	- .37	+ .55	- .92	+ .56	+.01	.93	This chronometer has the ordinary balance. It had been cleaned nine months previous to trial.
		73°	- .23	+ .90	- 1.13	+1.00	+.10	1.23	
1347	Negus	73°	-1.47	-1.32	- .15	- 1.06	+.26	.41	(same notation)
		73°	-1.36	-1.42	+ .07	- 1.37	+.05	.07	
596	Negus	73°	+ .59	+ .38	+ .21	+ .57	+.19	.21	This chronometer has the ordinary balance. It had been cleaned two years previous to trial.
		73°	+ .62	+ .54	+ .08	+ .44	-.10	.18	

315 Weekly rates recorded in the 1883-84 chronometer competition at the U.S. Marine Observatory at Annapolis, over 25 weeks, for 25 marine chronometers, with data including mean, maximum and minimum temperatures of the individual weeks plus the relative humidity.
Proceedings, U.S. Naval Institute, 10/171

316 Mean daily rates (every test took place over six days) of an informative experiment on the influence of humidity, at the U.S. Naval Institute, Annapolis, in 1884, with six selected marine chronometers.
(M. Whitney)

…ERVATORY, FROM JANUARY 1 TO JULY 1, 1884, ARRANGED IN THE ORDER OF MERIT.

…o …7	March 17 *to March 24.	March 24 to March 31.	March 31 to April 7.	April 7 to April 14.	April 14 to April 21.	April 21 to April 28.	April 28 to May 5.	May 5 to May 12.	May 12 to May 19.	May 19 to May 26.	May 26 to June 2.	June 2 to June 9.	June 9 to June 16.	June 16 to June 23.	June 23 to June 30.	Temper ture of compens tion.
	s.	s.	s.	s.	s.	s.	s.	s.	s.	s.	s.	s.	s.	s.	s.	s.
·4	− .150,	± .000	+ .143	+ .264	+ .407	+ .471	+ .307	+ .364	+ .657	+ .429	+ .450	+ .386	+ .360	+ .276	+ .306	49.8°
9	− .721	+ .286	+ .214	+ .230	+ .479	+ .436	+ .271	+ .257	+ .336	+ .250	+ .236	+ .279	+ .324	+ .240	+ .163	69.8
1	+ 1.064	+ 2.321	+ 2.179	+ 2.264	+ 2.443	+ 2.400	+ 2.271	+ 2.507	+ 2.443	+ 2.571	+ 2.486	+ 2.600	+ 2.824	+ 2.954	+ 2.841	73.8
i	+ 1.171	+ .643	+ .786	+ .800	+ .693	+ .793	+ .486	+ .614	+ .729	+ .571	+ .593	+ .564	+ .574	+ .597	+ .591	87.0
9	+ .100	+ .036	+ .143	+ .229	+ .371	+ .400	+ .057	+ .221	+ .407	+ .286	+ .021	+ .064	+ .146	+ .169	+ .091	128.0
9	− .900	+ .750	− .607	− .514	+ 1.086	+ 1.043	+ 1.236	+ 1.329	+ 1.407	+ 1.321	+ 1.129	+ 1.279	+ 1.181	+ 1.097	+ 1.056	70.8
3	+ .314	+ .607	+ .500	+ .836	+ .586	+ .757	+ .700	+ .578	+ .514	+ .429	Main spring broke.					60.2
0	+ .029	+ .893	+ .893	+ 1.229	+ 1.229	+ 1.257	+ 1.164	+ 1.436	+ 1.693	+ 1.464	+ 1.307	+ 1.136	+ .824	+ .454	+ .484	71.0
0	+ .314	+ .679	+ .393	+ .336	+ .586	+ 1.186	+ 1.343	+ 1.435	+ 1.229	+ 1.607	+ 1.236	+ 2.029	+ 1.931	+ 2.240	+ 2.492	56.0
6	+ .064	− .714	− .357	− .236	− .879	− .921	− .943	− .814	− .771	− 1.143	− 1.264	− 1.114	− 1.283	− 1.260	− 1.373	92.8
1	+ .850	+ .679	+ 1.107	+ 1.086	+ .693	+ .867	+ .414	+ .686	+ .979	+ .321	+ .736	+ .243	+ .217	+ .026	− .159	58.1
9	+ 2.171	+ 2.286	+ 2.571	+ 2.764	+ 2.621	+ 2.364	+ 1.879	+ 2.400	+ 2.657	+ 2.250	+ 2.593	+ 2.243	+ 2.100	+ 2.026	+ 1.842	51.1
03	+ 2.020	+ 1.286	+ 1.536	+ 1.514	+ 1.479	+ 1.471	+ 1.700	+ 1.721	+ 1.479	+ 2.036	+ 1.736	+ 2.207	+ 2.217	+ 2.490	+ 2.642	66.7
0	− .793	+ .750	+ .821	− .593	− .771	− .421	− .050	− .350	− .414	− .429	− .371	− .257	− .211	− .081	− .159	54.3
36	+ 1.[illegible]	+ 1.000	+ 1.250	+ 1.371	+ .979	+ .971	+ .521	+ .757	+ .443	+ 1.143	+ .986	+ 1.279	+ 1.503	+ 1.954	+ 2.127	65.9
36	− 4.971	− .250	− .571	− .593	+ .264	+ .471	+ .986	+ .507	− .093	+ .429	+ .629	+ 1.207	+ .931	+ .954	+ .841	77.5
o	56.0° 49.0	68.0° 53.0°	65.0° 57.5	64.5° 58.7	69.0° 62.5	70.5° 63.5	76.7° 68.7	76.5° 68.5	73.5° 66.2	80.5° 68.7	76.7° 65.	80.7° 69.0	82.7° 69.5	85.7° 71.0	85.2° 73.0	
o	51.1°	64.5°	61.0°	60.9°	65.3°	65.5°	72.4°	71.3°	69.1°	74.5°	70.2°	75.4°	75.7°	78.3°	78.2°	
	.76	.65	.63	.67	.69	.69	.65	.72	.67	.72	.72	69.	.73	72	.70	

zero, 15 and 30 degrees Celsius, were prescribed. As of 1858 the chronometer tests for the navy were finally turned over exclusively to the naval depot, which carried out all tests from then on and gave out prizes of 1000 to 2500 francs in competitions.

In addition, the first precise chronometer evaluations were introduced at the French clockmaking center of Besançon in 1885. Since 1888 yearly competitions for marine chronometers have taken place there, with medals of enamel, bronze, silver and gold being given. In the time thereafter, the standards used at Besançon became more and more like those used in Switzerland.

The Naval Observatory in Washington, D.C., was founded in 1830 and given the job of providing chronometers, maps and other navigational aids. In 1893 the observatory moved into the Simon Newcomb Laboratories, newly built for it; it was expanded in 1962.

Marvin Whitney reports exhaustively on the American methods of testing chronometers in a series of articles[4] extending over two years: "The activities covered not only the testing of American products, but also took part in their further development." The table in #315 shows the weekly rates from a chronometer test in 1883-84, which was carried out essentially like the tests in Greenwich. In the table of #316 are recorded the results of an interesting study of the influence of dampness in the air on chronometer behavior, which took place in 1884 — likewise at the US Naval Observatory — and was carried out in a manner similar to that used at Greenwich.

The administration and regulation of chronometer use in the U.S. Navy was very strict. Whitney writes of the individual measures taken in the handling of a chronometer:

"When a chronometer was to be delivered to a ship, it was carried there by hand by a messenger. It was already running and set at GMT (Greenwich Mean Time). Records on its fast or slow running during testing on land were either stated, in simple cases, on a certificate giving the expected daily rate (rating certificate), as shown in #317, or in more important cases by a diagram (rate curve) showing the influence of the surrounding temperature, as shown in #318, and made available to the users of the chronometer."

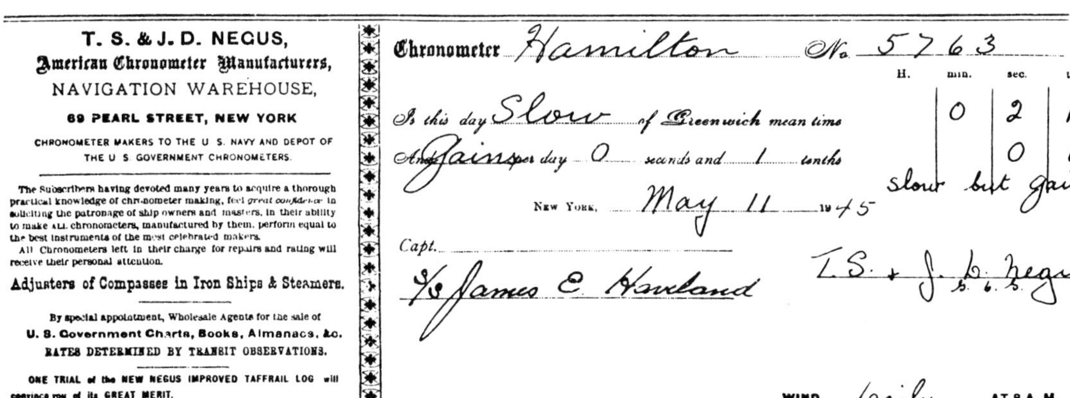
T. S. & J. D. NEGUS,
American Chronometer Manufacturers,
NAVIGATION WAREHOUSE,
69 PEARL STREET, NEW YORK
CHRONOMETER MAKERS TO THE U. S. NAVY AND DEPOT OF THE U. S. GOVERNMENT CHRONOMETERS.

The Subscribers having devoted many years to acquire a thorough practical knowledge of chronometer making, feel great confidence in soliciting the patronage of ship owners and masters, in their ability to make ALL chronometers, manufactured by them, perform equal to the best instruments of the most celebrated makers.
All Chronometers left in their charge for repairs and rating will receive their personal attention.

Adjusters of Compasses in Iron Ships & Steamers.

By special appointment, Wholesale Agents for the sale of
U. S. Government Charts, Books, Almanacs, &c.
RATES DETERMINED BY TRANSIT OBSERVATIONS.

ONE TRIAL of the NEW NEGUS IMPROVED TAFFRAIL LOG will convince you of its GREAT MERIT.

Chronometer Hamilton No. 5763

	H.	min.	sec.	ths.
Is this day Slow of Greenwich mean time		0	2	1
And Gains per day 0 seconds and 1 tenths			0	1

slow but gains

New York, May 11 1945

Capt. Y/S James E. Howland

T. S. & J. D. Negus

Wind. Daily at 8 A. M.

317 Rate difference certificate of the chronometer firm of T. S. & J. D. Negus, New York, May 1945.
(M. Whitney)

NAVSHIPS 701

NAVY DEPARTMENT

BUREAU OF SHIPS

SHIP'S CHRONOMETER RECORD

U. S. Naval Observatory, Washington, D. C., OCT 28 1944

Maker Hamilton No. 4030

Date of last overhaul New Average deviation of daily rate 0.15

Return to the U. S. Naval Observatory for overhaul on or before Oct. 26, 1947

Chronometer Trial Supervisor, U. S. Naval Observatory.

Chronometer Inspection Officer, U. S. Naval Observatory.

RATE CURVE OBTAINED DURING TRIAL

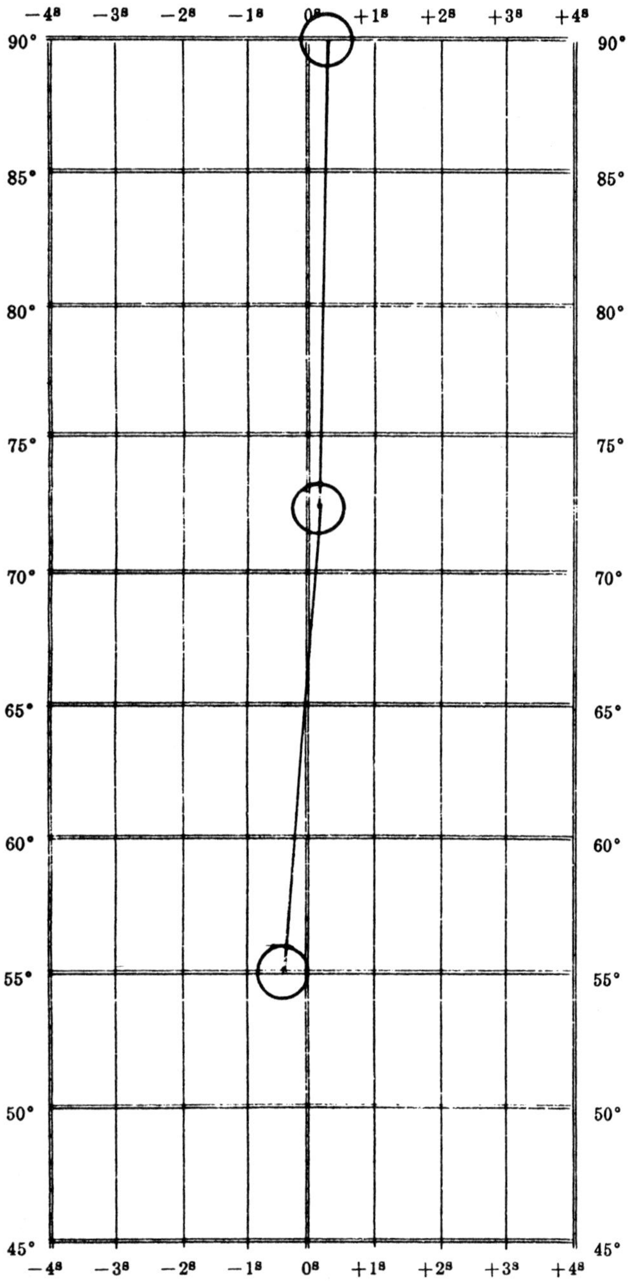

INSTRUCTIONS FOR THE CARE AND HANDLING OF SHIPS' CHRONOMETERS

A complete and regular record of the performance of ships' chronometers shall be kept as outlined in the "Instructions" at the end of this form.

Chronometers shall be wound regularly every 24 hours. A 56-hour instrument usually requires about 6 or 7 half-turns of the winding key. Insure closing of the winding hole by the dust plate on the back of the instrument after each winding.

The bezel of a ship's chronometer shall be removed only during corking of the balance for shipment. Detailed instructions for this operation are given under "Transportation," paragraph 2 (*a*), NAVSHIPS Form 702.

The indication of the hour and minute hands shall never be changed manually, except by an authorized person at a chronometer supply pool, from the time an instrument leaves the Naval Observatory until it is returned to the Naval Observatory for overhaul.

If a chronometer develops an erratic rate or stops through mechanical derangement, it shall be turned in for replacement at the first opportunity. An instrument which stops through lack of winding and is restarted may take an entirely different rate from that previously established and should be observed closely for a change in performance.

A chronometer shall be returned to the Naval Observatory for overhaul 3 years after the date of last overhaul shown above. An instrument which has been in service for 3 years, and which is turned in at a chronometer supply pool for replacement, shall be forwarded without delay to the Naval Observatory for overhaul.

318 Rate curve for a Hamilton marine chronometer in 1944, depending on the temperature (Fahrenheit). *(M. E. Whitney)*

SHIP'S CHRONOMETER RECORD

ERROR, MEAN DAILY RATE, MEAN DEVIATION IN DAILY RATE, TEMPERATURES, AND REMARKS

Maker HAMILTON No. 4030

U. S. S. BON HOMME RICHARD Type of Vessel CV

Entry No.	Place, Longitude	G. C. T. and Greenwich Date	Error on G.C.T. +Fast, −Slow	No. of Days	Mean Daily Rate +Gaining, −Losing	Mean Dev. in Daily Rate	Temperature High	Temperature Low	Remarks: Radio Station; Days Under Way; Heavy Seas; Gun Firing; Bombing; Any Occurrence Likely to Affect Rate
1	BROOKLYN, N.Y 73-58W	1600 12-20-44	(+) 0-00-37.5	—	—	—	—	—	
2	BAYONNE, N.J. 74-04W	1600 12-30-44	(+) 0-00-46.	10	(+) 0.85	—	80	66	
3	GRAVESEND BAY 74-02W	1600 1-9-45	(+) 0-00-55	10	(+) 0.90	0.025	82	60	
4	AT SEA 74-32W	1600 1-19-45	(+) 0-01-04	10	(+) 0.90	—	80	60	
5	PORT OF SPAIN, TRINIDAD 61-31W	1600 1-29-45	(+) 0-01-14	10	(+) 1.00	0.05	85	64	7 DAYS UNDER WAY. – GUN FIRING - 1 DAYS
6	AT SEA 62-14W	1600 2-8-45	(+) 0-01-25	10	(+) 1.10	0.05	87	81	9 " " " GUN FIRING - 7 DAYS
7	PORT OF SPAIN TRINIDAD 61-31W	1600 2-18-45	(+) 0-01-37.5	10	(+) 1.25	0.075	90	81	9 " " " GUN FIRING - 4 DAYS
8	At Sea 74-30W	1600 2-28-45	(+) 0-01-49.5	10	(+) 1.20	0.025	88	63	8 " " "
9	BAYONNE, N.J 74-04W	1600 3-10-45	(+) 0-02-03	10	(+) 1.35	0.075	84	58	2 " " "
10	At Sea 74-15W	1600 3-20-45	(+) 0-02-15	10	(+) 1.20	0.075	83	66	1 " " "
11	AT SEA 112-52W	1600 3-30-45	(+) 0-02-29	10	(+) 1.40	0.10	89	72	9 " " "
12	PEARL HARBOR 157-58W	1600 4-9-45	(+) 0-02-43.5	10	(+) 1.45	0.025	82	63	4 " " "
13	PEARL HARBOR 157-58W	1600 4-19-45	(+) 0-02-57.5	10	(+) 1.40	0.025	88	78	4 " " "
14	AT SEA 166-03W	1600 4-29-45	(+) 0-03-12	10	(+) 1.45	0.025	88	78	8 " " " GUN FIRING - 2 DAYS
15	PEARL HARBOR 157-58W	1600 5-9-46	(+) 0-03-26.7	10	(+) 1.47	0.05	87	77	2 " " "
16	PEARL HARBOR 157-68W	1600 6-19-46	(+) 0-03-41.5	10	(+) 1.47	0.00	87	79	5 " " " Gun Firing - 3 Days
17	AT SEA 159-41E	0400 5-30-45	(+) 0-03-58.5	10.5	(+) 1.47	0.00	90	80	8 " " " Gun Firing - 7 Days
18	At Sea 133-03E	0400 6-9-45	(+) 0-04-15	10	(+) 1.65	0.09	90	84	10 " " " " " 6 Days
19	San Pedro Bay P.I. 125-04E	0400 6-19-45	(+) 0-04-34	10	(+) 1.90	0.125	92	84	4 " " " " " 2 Days
20	San Pedro Bay P.I. 125-04E	0400 6-29-45	(+) 0-04-52	10	(+) 1.80	0.05	91	85	3 " " " " " 3 Days
21	AT SEA 146-43E	0400 7-9-45	(+) 0-05-11	10	(+) 1.90	0.05	91	83	10 " " " " " 3 Days
22	AT SEA 146-37E	0400 7-19-45	(+) 0-05-27	10	(+) 1.60	0.15	87	69	10 " " " " " 1 Day
23	AT SEA 138-15E	0400 7-29-45	(+) 0-05-45	10	(+) 1.80	0.10	88	77	10 " " " " " 2 Days
24	AT SEA 144-51E	0400 8-8-45	(+) 0-06-02.5	10	(+) 1.75	0.025	90	80	10 " " " " " 5 Days
25	AT SEA 144-02.5E	0400 8-18-45	(+) 0-06-19.5	10	(+) 1.70	0.025	89	77	10 " " " " " 3 DAYS
26	AT SEA 141-59E	0400 8-28-45	(+) 0-06-37.5	10	(+) 1.80	0.05	88	81	10 " " " " " 1 DAY
27	At SEA 138-10E	0400 9-7-45	(+) 0-06-54.5	10	(+) 1.70	0.05	89	79	10 " " " " " 3 DAYS
28	Tokyo BAY 139-45E	0400 9-17-45	(+) 0-07-11	10	(+) 1.65	0.025	88	79	9 DAYS UNDERWAY. GUN FIRING, 1 DAY
29	AT SEA 140-04E	0400 9-27-45	(+) 0-07-27	10	(+) 1.60	0.025	85	83	6 DAYS UNDERWAY. GUN FIRING, 2 DAYS
30	AT SEA 144-50E	0400 10-7-45	+ 0-07-44.5	10	(+) 1.75	0.075	90	86	8 DAYS UNDERWAY - MINOR Collision AT GUAM.
31	AT SEA 148-16W	0300 10-17-45	(+) 0-08-01	10	(+) 1.65	0.05	90	71	10 DAYS UNDERWAY
32	ALAMEDA, CAL 122-18W	2000 10-26-45	(+) 0-08-16	10	(+) 1.50	0.075	77	65	4 DAYS UNDERWAY
33	AT SEA 129-11W	2000 10-30-45	(+) 0-08-22.5	4	(+) 1.625	0.0625	75	64	2 DAYS UNDERWAY – HEAVY SEAS
34	PEARL HARBOR 157-58W	2000 11-5-45	(+) 0-08-32	6	(+) 1.58	0.02	85	71	4 DAYS UNDERWAY
35	AT SEA 154-05E	0400 16-16-45	(+) 0-08-50.5	10	(+) 1.85	0.135	90	80	7 DAYS UNDERWAY.
36	AT SEA 175-41W	0400 11-26-45	(+) 0-09-08	10	(+) 1.75	0.05	91	75	8 DAYS UNDERWAY.
37	Hunters Pt., Cal. 122-28W	0300 12-6-45	(+) 0-09-25.5	10	(+) 1.75	0.000	80	65	6 Days underway.
38	At Sea 160-56W	2100 12-16-45	(+) 0-09-41.5	10	(+) 1.60	0.075	69	59	6 Days underway - Heavy seas.
39	At Sea 164-51E	0400 12-21-45	(+) 0-09-41.5	4	0.00	0.800	80	72	4 Days underway – Heavy seas.
40	San Pedro Bay, P.I. 125-03E	0400 12-31-45	(+) 0-10-06.5	10	(+) 2.50	1.250	88	73	6 Days underway.
41	At sea 168-54W	2000 1-9-46	(+) 0-10-18.5	10	(+) 1.20	0.650	88	71	10 Days underway.
42	Hunters Pt., Calif. [illegible]	[illegible]	[illegible]	10	(+) 2.15	0.275	78	64	7 Days underway.
43	At sea 125-05W	0300 1-30-46	(+) 0-10-58	10.5	(+) 1.71	0.220	84	59	2 Days underway.
44	Puget Sound 122-32W	0300 2-9-46	(+) 0-11-16	10	(+) 1.8	0.045	87	58	1 Day underway.
45	Elliott Bay, Wash. 122-23W	0300 2-19-46	(+) 0-11-34	10	(+) 1.8	0.000	89	55	1 Day underway. Earthquake on night of fourteenth.
46	[illegible], WASH. [illegible]	[illegible]	[illegible]	[illegible]	[illegible]	[illegible]	[illegible]	[illegible]	
47	ELLIOTT BAY, WASH. [illegible]	[illegible]	[illegible]	[illegible]	[illegible]	[illegible]	[illegible]	58	
48	ELLIOTT BAY, WASH. [illegible]	[illegible]	[illegible]	[illegible]	[illegible]	[illegible]	[illegible]	[illegible]	
49	Elliott Bay, Wash. 122-23W	0300 3-31-46	(+) 0-12-40	10.5	(+) 1.52	0.040	79	53	
50	Elliott Bay, Wash. 122-23W	0300 4-10-46	(+) 0-12-55	10	(+) 1.50	0.010	78	57	1 Day underway.
51	Elliott Bay, Wash. 122-23W	0300 4-20-46	(+) 0-13-10	10	(+) 1.50	0.000	79	60	
52	Elliott Bay, Wash. 122-23W	0300 4-30-46	(+) 0-13-25	10	(+) 1.50	0.000	77	56	
53	Elliott Bay, Wash. 122-23W	0300 5-10-46	(+) 0-13-39	10	(+) 1.40	0.050	80	58	
54	Elliott Bay, Wash. 122-23W	0300 5-20-46	(+) 0-13-53	10	(+) 1.40	0.00	79	56	Moved standard chronometers.
55	USN Shipyard, Brem. Wash. 122-38W	2400 5-30-46	(+) 0-14-06.5	10	(+) 1.35	0.025	77	61	1 Day underway. Moved standard chronometers.
56	USN Shipyard, Brem. Wash. 122-38W	2400 6-9-46	(+) 0-14-24.5	10	(+) 1.80	0.225	82	62	Moved Standard Chronometers.
57	USN Shipyard Brem. Wash. 122-38W	2400 6-19-46	(+) 0-14-40.5	10	(+) 1.60	0.100	81	62	½ Day Underway
58	USN Shipyard Brem. Wash. 122-38W	2400 6-29-46	(+) 0-14-54	10	(+) 1.35	0.125	92	66	
59	U.S.N. Shipyard Brem. Wash. 122-38W	2400 7-9-46	(+) 0-15-11	10	(+) 1.70	0.175	86	69	
60	USN Shipyard Brem. Wash. 122-38W	2400 7-19-46	(+) 0-15-27.5	10	(+) 1.65	0.025	84	67	
61	ELLIOTT BAY, WASH. 122-23W	2000 8-1-46	(+) 0-15-52	14	(+) 1.75	0.050	77	72	TIME TICK NOT AVAILABLE FOR 14 DAYS
62	ELLIOTT BAY, WASH. 122-23W	1700 9-9-46	(+) 0-16-5[illegible]	39	(+) 1.67		80	70	TIME TICK NOT AVAILABLE FOR 39 DAYS
63	ELLIOTT BAY, WASH. 122-23W	1700 10-8-46	(+) 0-17-52	2[illegible]	(+) 1.83		81	63	TIME TICK NOT AVAILABLE FOR 29 DAYS
64	ELLIOTT BAY, WASH. 122-23W.	1700 10-18-46	(+) 0-18-07	10	(+) 1.50		89	53	
65	ELLIOTT BAY, WASH. 122-23W.	1700 10-28-46	(+) 0-18-21	10	(+) 1.40		89	52	
66									

319 Extract from the chronometer log book (Ship's chronometer Record) of a Hamilton chronometer for the period from December 20, 1944 to October 28, 1946, with notes on routes, longitude, Greenwich mean time and Greenwich date, rate differences found between the individual entries, mean daily rate differences and mean daily deviations derived from them, plus highest and lowest temperatures in the interval and other important notes, such as those on combat activity and presence in various ports.
(M. Whitney)

On board the ships, extraordinarily accurate log books were kept, recording the external conditions to which the chronometers were exposed, which is a real continuation of what had already been required in the orders for Captain Cook's second voyage in 1779; in every single determination of longitude the appropriate local time of the location, the time of the point of departure (today always that of Greenwich), as indicated by the chronometer aboard the ship, was to be recorded in writing including the total error shown since the last determination. An interesting log-book page from a warship of the U.S. Navy from about the end of World War II is shown in #319.

The most primitive calculation was thus the addition or subtraction of the product of the daily rate times the mean rate since the last calculation; it was more precise, though, to arrive at a sum by taking the arithmetic mean of all the individual daily rates, for which the rate difference diagram had to be used. These were the two methods that were used, before the availability of wireless time signals (time ticks), to determine the exact Greenwich time on shipboard through the use of a chronometer. With the simultaneous observation of two or more chronometers of different construction on the same ship, it was possible to calculate somewhat more precise time values by using complex mathematical-statistical methods. Since the introduction of radio time, though, there is no more difficulty in determining the daily rate of a chronometer.

It is obvious, though not universally known, that nothing about a chronometer may be changed during its time on a ship. Whitney stresses this also: "As long as the chronometer is on the ship, it is never reset until it has been brought back to land for overhauling after two or three years. The only correction on the ship is limited to a regular winding at the same time every day; on the ships of the U.S. Navy this is usually done at 11:30 A.M., and its having been done is reported to the officer on duty at 12 o'clock. At the same time — but generally only at intervals of several days — the chronometer and broadcast time indications, notations on environmental conditions, and the calculations that have been made are recorded in the ship's chronometer record."

It is not easy to gain a comparative overview of the individual observatories and test centers that have been active in testing chronometers, and even harder to understand how the evaluations from the various places can be compared with each other. This is even harder because many changes were made in the individual rules over the years. Valuable contact points for such considerations were published by Captain Rottok[5] at the end of World War I.

Guyot remarks in a discussion of the difficulties of comparing the test results at the various test centers: it is definite that the progress in chronometer making is attributable to the effect of the strict rules of the observatories at Geneva, Neuchâtel, Besançon, Greenwich, Kew, Teddington, Washington and Hamburg.

The many test records and log books are, of course, worthless for further development of mechanical computers, since they have been displaced by the modern methods that use electronic-atomic phenomena. But the numerical contents of the old research records are still of general interest because they afford a clearer understanding of very general natural behavior — far beyond the behavior of the chronometer. Sections 5 and 6 of #318 are very informative; they give the daily mean rates in comparison with GMT and the corresponding (mean) variations in daily rate. The complete lack of rules in these records (even for as costly and modern a chronometer as Hamilton No. 4030) shows the limits in precision that cannot be overcome by mechanical means. Even the best mechanical chronometer system will always deviate more or less from the test statistics when used at sea, on account of the varying and constantly changing conditions prevailing in its workplace (to say nothing of the process of aging in the chronometer). This practical inevitability is at the heart of the very large problem area that dominated chronometer making for more than a century. The daily variations in chronometer indications show that one must recognize that the perfectionists' conception that there is absolute precision or reproducability has ceased to exist here.

The completely irregular variations in the mean daily rates of chronometers are a phenomenon that is incompatible with the philosophical-scientific conception of the positivists that has dominated the natural sciences for 200 years. The positivists regard it as a dogma that every natural scientific phenomenon can be explained completely and precisely formulated in mathematical terms.

The first great crack in positivism that had been strengthened by many successes in the 19th Century was made by Werner Heisenberg with his famous theorem of indefiniteness, namely that in the atomic realm there are insuperable limits to the precision of observations, for example, of the speed and position of an electron. In the most recent years, similar limits to the predictability of probability of both play and economic decisions when the number of participants exceeds four have been shown by John von Neumann.[6] Thus the rate variations shown in Section 7 of #318 can be regarded as a gray boundary area in the attainment of striven-for mechanical precision in the behavior of the oscillating system with its escapement. Thus in dealing with mechanical chronometers one must accept an outlook on life such as has recently been shown by experience in news-gathering technology and information theory (Shannon),[7] which is that one must become familiar, on a broad front, with the limits set for mankind.

1. D. de Carle, *Watch & Clock Encyclopedia*, London, 1959, pp.242 ff.
2. Regierungsrat Dr. Repsold, *Chronometer- und Taschenuhrprüfungen auf der Deutschen Seewarte*, Die Uhrmacherwoche, No. 1, 3, 34, 35, 37, 40, 42 of 1931 and 34 of 1932.
3. Edmond Guyot, *Histoire de la détermination de l'heure*, LaChambre Suisse de L'horlogerie, La Chaux-de-Fonds, 1968.
4. Marvin E. Whitney, *The Ships Chronometer*, Hor. Times (USA) 1978, 79, 80.
5. Rottok, *Chronometerprüfungen bei der franzöischen, englischen und amerikanischen Marine*, Annalen der Hydrographie, Vol. 45, 1917, Hamburg.
6. John von Neumann, *Theorie der Spiele*, 19.
7. Claude E. Shannon, *Mathematische Theorie der Nachrichtenübertragungung*, 1948.

VII: Chronometer Development in the 20th Century

Mechanical Perfection via Modern Metallurgy and Technology

While the greater part of the essential problems in the building of chronometers had already become known in the heroic epoch, without the scientific basis and quantitative needs for optimal functioning having been met, the fine mechanical period brought not only the requirements in workmanship to achieve more economical production and durability to withstand the many demands of use, but also the correct insights into the behavior of springs and their suitability for the oscillating system. The dependably good chronometers made by the British builders in the 19th Century are proof of that.

To the extent that the demands for the highest precision in running and greater durability had risen with the achievements of the fine mechanical period, the means of accomplishing this in the following modern, metallurgic-technologically oriented first decades of the 20th Century became more and more specialized and refined, and became less and less understood and visible.[1] Despite that, we shall try here to make these interesting and very admirable achievements clearer to the chronometer admirer and amateur through brief, concise descriptions of the quintessence of the individual steps or concepts. Only such an approach will lead to a correct evaluation of the chronometers, which grew ever simpler and more similar in their external appearance and construction but had reasons beneath the surface for their improved running and reliability.

Through the introduction into chronometer making of new metallurgical discoveries gained just before the turn of the century, these impulses were born and led to a great improvement over the previously attained achievements. Unfortunately, the new electronic principles entering the field after 1950 brought an unexpected end to this interesting development. But understanding the significant achievements of the first five decades of the 20th Century requires knowledge of the level attained before.

The 19th Century had taken over from the previous one the knowledge that it is possible with the wheel-clock principle to attain accurate rates to within seconds per day, even with a clock at sea over long periods of time, and to calculate longitude. To be sure, the costs of the first such instruments were so high that their widespread use was not yet possible.

An important goal during the 19th Century was to lower the price of marine chronometers considerably through simplified construction. But another main problem was how to limit the middle temperature error, a problem to which the chronometer builders of the time devoted the greater part of their constructive imagination. Thus the entire 19th Century is an era of mechanical skill without much theoretical basis. Ed. Phillips in France and several others did, mainly between 1865 and 1880, formulate highly valuable theoretical explanations of the right spring shapes, the bending of the spring's terminal curves and the choice of spring materials, but these were not easy to understand and were thus available to the makers only to a limited extent.

The metallurgical discoveries that became so important for watch and chronometer production in the 20th Century came from C. E. Guillaume. Working on a very extensive — and very independent — new study of the general properties of alloys at the Bureau International des Poids et Mésures at Sèvres in the second half of the last decade of the 19th Century, he first discovered an iron-nickel alloy that showed a similarly favorable thermal expansion characteristic to platinum-iridium over a wide range of temperatures and could also be produced much more inexpensively. That was, in addition to having other important practical uses, the starting point for the progress in chronometers during the 20th Century.

To trace the path of this great technical progress that

became possible in clock and chronometer production through the use of Guillaume's iron-nickel alloy, it is rewarding to trace Guillaume's path to it back to its origins. Guillaume, born at Fleurier in the Swiss Jura in 1861, was the brother, son and grandson of remarkably well-trained watchmakers of that region, received his training at first in Neuchâtel (high school and one year at an academy), then received his diploma in physics at the ETH in Zürich, where he wrote his dissertation on electrolytic condensers. Thanks to a recommendation by the director of the Neuchâtel Observatory, Guillaume obtained a position at the Bureau International in Sèvres. There obtained in 1891 he was assigned the task of finding substitutes for the extremely expensive platinum-iridium that was then the only material of which the comparative measures for instruments of measuring length — could be made. Comparisons of length were then of particular importance, as it was necessary for the determination of the extent of the earth's surface to set extended geodesic bases by laying out ordinary metallic measuring rods, four meters long, with microscopic end-to-end positioning and rigidly precise determination of any thermal expansion. For protection against the sun's influence, measurements were made under tents, and a work crew of sixty men could, under favorable conditions, measure 400 meters in a day. Since a bases was usually ten kilometers long and hundreds of bases had to be made, one can easily imagine the amount of work needed and likewise understand the importance of simplifying this process, which had been used for measuring the earth's surface since the beginning of the 19th Century, by providing measuring rods of greater length and, if possible, without expansion.

In full awareness of the importance of his task, Guillaume at first undertook basic research that properly extended over several years; many fascinating individual results were included in the address which he delivered in 1920, when his work was honored with the Nobel Prize.[2]

The goals sought by Guillaume were not simple: stability of the material in terms of its constant length over long periods of time, absence of oxidation, sufficient mechanical rigidity and as little thermal expansion as possible. These demands severely limited the choice of materials from the start. Since nickel, which was of interest from the start because it expanded much less than the formerly used measuring materials, had to be eliminated because all the nickel producers of the time refused to produce four-meter-long measuring rods of nickel, it was only after long consideration and a stroke of luck that he became aware of the physical properties of iron-nickel alloys and decided to study them.

The unusual properties of iron-nickel alloys that first caught Guillaume's attention were as follows: 1) The observation made by John Hopkinson shortly before, that iron (Fe) - nickel (Ni) alloys with nickel contents of about 25% are non-magnetic and soft after the melting process, but become stiff and strongly magnetic after cooling in dry ice; 2) The fact that M. I-R. Bénoit of the technical section of the French artillery had made known in 1895 that a three-metal alloy of Fe + 22% Ni + 3% chromium (Cr) was not ferromagnetic but possessed the high expansion of brass; 3) That a piece of iron with 30% nickel content, which Guillaume had just obtained, showed an expansion coefficient about one-third below that of platinum but was ferromagnetic.

These contradictions, at first inexplicable, particularly the fact that a 22% nickel-iron alloy was not ferromagnetic but a 30% Ni-Fe alloy was, and major changes of properties that had appeared, led Guillaume to the decisive realization that the anomalies found disagreed with the conceptions of alloys then held, according to which the properties changed proportionally to the proportions of the components, and suggested the existence of generally significant points of great change in the behavior of the various material properties.

From this Guillaume concluded that the old concept of constant changes in the properties of alloys with varying proportions was basically untenable. Proceeding from this, when one of the characteristic properties of the material — such as the ferromagnetism in this case —shows radical changes with certain alterations in the proportions, unexpected changes in other characteristics as well, such as the thermal expansion in that area where the magnetic properties changed so much, can be expected.

Thus Guillaume decided to research the area of changes in nickel-iron alloys thoroughly and to use the magnetic property of them as a suitable characteristic for determining the radical areas. This was closely related to several other researchers' observations that had just been published, which had stated in general terms that magnetic substances, according to their composition, show either thermally reversible or thermally irreversible susceptibility to magnetization, and thus differ in that though all of them lose their magnetism at increasing temperatures, some of them regain it at the same point when temperatures are lowered, while others (which are irreversible) regain it only at much lower temperatures.

Guillaume's consideration of the utility of a thorough investigation of the alloy proportions marked by radical magnetic action found thermal expansion to occur as shown in #320 for Fe-Ni alloys with varying proportions. He was able to determine as a particularly interesting result that in the reversible region there was even a certain unexpected area where the thermal expansion coefficient decreases very strongly, namely at 20 degrees Celsius to about 1/15, while at 300 degrees Celsius it is still about 1/3 of the known coefficient of pure iron, and that these values could be reduced even more by additional mechanical treatment.

The discovery of the unexpectedly small temperature expansion had only been possible because Guillaume had examined the whole range in steps of a few percentage points of nickel content carefully for any sign of disturbance, and thus could present a previously unknown picture of the behavior of coefficients and of expansion equalization, which was also to be of basic significance for many other uses of alloys.

The physicist described the linear expansion in the following equation:

$$l = l_0 (1 + \alpha\theta + \beta\theta^2)$$

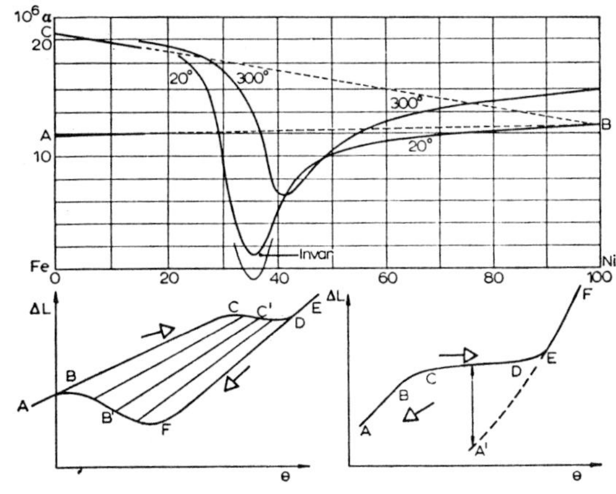

320a) Expansion coefficients and α_{300} for Fe-Ni alloys at 20 and 300 degrees Celsius for various mixture conditions (the percentages of nickel are noted on the abscissa). The two straight lines AB and CB enclose the alloy proportions with irreversible expansion behavior.
b) Length changes for an alloy with irreversible expansion behavior.
c) Length changes for an alloy with irreversible expansion coefficient behavior.
From Guillaume's Nobel Prize address, 1920

For 20% nickel content in iron, α equals 19 . 10-6 or 18 . 10-6 at 20 degrees Celsius or 300 degrees Celsius. Guillaume recorded the results of his experiments in 93 scientific reports, which are listed along with his aforementioned address when he received the Nobel Prize.

For the iron-nickel alloys found by him that had extremely low expansion coefficients, the Swiss Professor Marc Thury suggested the name INVAR. With the discovery of Invar, Guillaume had made an important contribution to the realm of alloyed and special steels, which was to expand and become increasingly important in the following decades. Today this has become another area of metallurgy that is described in numerous detailed handbooks[3] which presently describe the many mixtures of two and more elements with a basis of iron and their properties, often stressing phase diagrams and transitional considerations derived from them. This approach scarcely means anything to anyone who is not a specialist, even if he has an idea that different microscopic grid patterns (such as two- or three-dimensional arrangements) have an essential influence on the physical properties of the substance in question. The practically complete lack of references to the historical process of development, and especially to the beginning of the metallurgical method of observation in itself, in these inclusive books weakens the basic information for those with a growing interest in scientific and technological history today. But around 1890-1900 — the time when Guillaume worked at and accomplished these developments that were so important for chronometry — the present-day perspective to the formation of solid bodies and crystals was practically non-existent.

For those who are interested in the origins and occurrences of technological innovations, the initial stages of every new development and the chances that often lead unexpectedly to success are of importance, just as are the brilliant handling of them by certain personalities. The authors of specialized books should therefore always give at least some information that indicates the historical origins of their subjects, in order to let their readers trace the details backward in time. This will be attempted here.

Without one's own involvement in the new metallurgical phenomena entering into the building of chronometers, one can scarcely evaluate most of the determining factors of chronometer development in the 20th Century.

Here Guillaume's already mentioned Nobel Prize address forms an important bridge with the past, and the fact that the prize was granted to Guillaume proves to be a just recognition of a highly important achievement in the beginnings of a realm that later expanded to global importance.

The development of Invar first resulted in the possibility of replacing the 4-meter oak rods used in basic geodetic measurements with longer measuring lines. On this basis, Ed. Jägerlein introduced a measuring process using 24-meter Invar measuring tapes, with which in one day a crew of only 1/5 the previous size could cover five kilometers. This measuring procedure first proved itself splendidly during the Swedish-Russian Spitzbergen expedition of 1898-99.

The success of the Spitzbergen expedition and the role that the new Invar measuring tapes played in it were quickly made public through the media.

But before that, Guillaume had already thought of using his newfound alloy in pendulum clocks, and made his ideas public in Paris[4] and Geneva.[5] In this respect, Ditisheim[6] recalls that P. Berner, in the Journal Suisse d'Horlogerie of 1897, had already been the first to discuss the possibility of using Invar for temperature compensation in clocks. Thanks to these suggestions — especially in the Journal Suisse d'Horlogerie of 1897 — the chronometer adjuster P. Perret of La Chaux-de-Fonds had been inspired to examine the interesting possibilities of the new alloy and thus to turn directly to Guillaume in Sêvres and request a test of the material for the making of hairsprings for pocket watches.

Perret immediately received the requested material from Guillaume and, with much skill had made a hairspring and installed it in a pocket watch. Unlike the then-customary behavior of pocket watches with steel springs, this pocket watch with its Invar hairspring worked remarkably well in high temperatures. Until the end of the 19th Century, pocket watches made in the usual way, with a brass balance and steel hairspring, possessed a temperature-dependent residual error of about 11 Secs./degree C per day, so that in a daily temperature range of from zero to thirty degrees they lost up to six minutes per day, with the hairspring itself about 90% responsible. Only marine chronometers with bimetallic compensated balances had an error of only 2 to 3 seconds per day (in the realm between zero and 30 degrees C), while above and below those boundaries a quickly growing error occurred. This effect, is known as the middle temperature error. The partial elimination of this error was a primary task of the chronometer makers all through the 19th Century.

P. Perret was not able to progress further with the unexpected behavior of his test watch, which differed so much from what usually occurred. He traveled to Sèvres at once to ask Guillaume to find an explanation and, if possible, advise and help him to find a way to utilize what he had found. This trip resulted directly in valuable suggestions and also awakened Guillaume's lasting interest in the improvement of hairspring materials.

In 1897 and 1899 Perret had already obtained two patents on the use of Fe-Ni hairsprings, and had thus obtained noteworthy accuracy in pocket watches with hairsprings made of such alloys. Whereas rate changes of 12 to 14 seconds per degree Celsius had previously been expected with normal steel hairsprings, the new hairsprings showed a difference of about one second per day per degree, which under average conditions meant a daily rate of half a minute, as opposed to four to six minutes per day previously. Numerous other Swiss watch manufacturers soon followed Perret's direction.

P. Perret died in 1904; afterward an evaluation of the new hairspring concepts was continued by the Société des Fabriques des Spiraux Réunies SA in further collaboration with Guillaume. During the following years, many pocket watches were equipped with Fe-Ni hairsprings, which went well because of their precision being so much greater than before.

The following works by Guillaume and his collaborators led in the next decades to further, even more impressive progress in the realm of hairspring materials; they are discussed in the section on self-compensating balance springs.

The simplifications and improvements in clock construction in general and chronometers in particular through the use of special alloys in hairsprings were subjects of particular interest and much practical work during the lifetime of P. Ditisheim. Shortly before his death (in 1945), he published another thorough monograph on this subject;[7] Described in depth is the metallurgical developments of Guillaume and his colleagues, his own work in the field of clock technology and those of other, similarly equipped watchmakers and chronometer makers, as well as many series of tests at numerous test centers and observatories and the attained results. This monograph is probably the best thorough, technically inclusive, scientific documentation of the last phase of development of the mechanical chronometer.

Guillaume's Integral Balances and Other Balance Improvements

Parallel to the introduction of iron-nickel steel for hairsprings and balances as led by P. Perret, a second advance in chronometer making resulted directly from Guillaume's experiments with alloys. At that time Guillaume also set out to develop a simple, temperature-

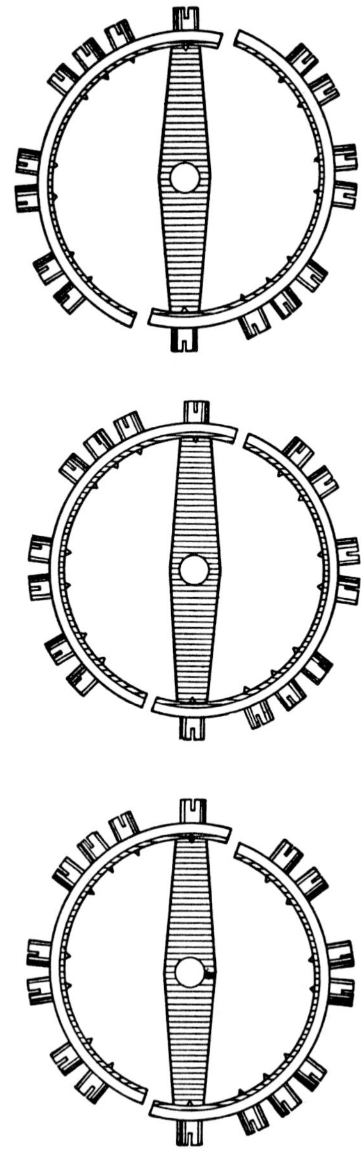

321 The mechanical play of a temperature-compensated balance of classical design at various temperatures: a) state at manufacturing temperature, b) at higher temperatures than a), the long bimetallic strips approach the center and thus compensate to some extent for the influence on heat on the steel hairspring, c) at lower temperatures, the ends of the bimetallic strips move outward and thus compensate to some extent for the influence of cold of the steel hairspring.

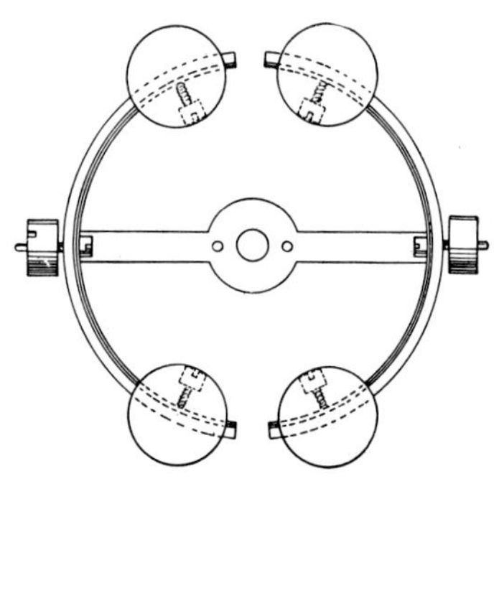

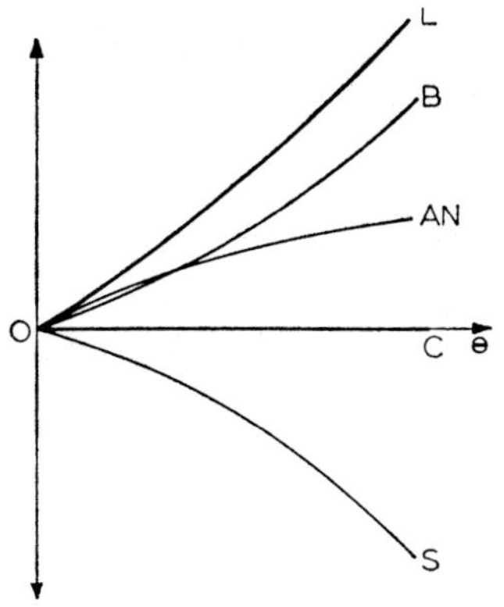

322 Guillaume's diagram to explain the elimination of the middle temperature error in chronometers with bimetallic balances according to the integral principle.

To the ordinate axis, there are added both the influence of temperature on the expansion coefficient Δl of the balance components L, AN, and also of the elasticity module E of the hairspring; temperature Θ is subtracted on the abscissa.
L = brass
AN = iron-nickel alloy (Annibal)
S = steel hairspring
B = bimetallic system of L and AN
The temperature range of the elasticity coefficient E of the hairspring S includes a component increasing with Θ^2. The expansion coefficient of the Fe-Ni alloy used — in direct combination with the brass — as a bimetallic strip tends toward an end value and also includes a quadratic component. The bimetallic strip made of AN and L thus takes on a tendency opposed to but equal to the elasticity coefficient of S. Through the opposing effects of these two tendencies, the result when temperatures change is thus a reaction of the balance assembly resulting from B and S, a straight line C falling from the T axis independent of temperature.
From Guillaume's Nobel Prize address, 1920

323 Balance design after Guillaume's 1899 "integral balance"; symmetrical construction of the bimetallic segments with the adjusting weights opposite the crossarm. This arrangement gives the classic compensated balance, a design that goes back to Pierre Le Roy, greater stability because of the shorter segments.

compensated balance for chronometers of high value to eliminate the still-disturbing middle temperature error. Whether this line of development was inspired by Perret's visit cannot be ascertained. In his eulogy to Ditisheim, Torrens[8] made the brief remark: "On a spring evening in 1899, C. E. Guillaume got the idea that the middle temperature error — the nightmare of all chronometer makers up to that time — could perhaps be eliminated by using a suitable nickel-iron alloy." Guillaume's idea was to replace the iron in the bimetallic strips of chronometers, which since Arnold and Earnshaw, almost exclusively used balances with cut rings of bimetallic strips (brass-iron combinations), with a particular iron-nickel alloy. Guillaume had published this in Switzerland soon thereafter.[9]

The effect of the classic temperature-compensated balances, typical of the whole fine mechanical epoch, is shown in #321, where a sharp bending of the balance arms causes a change in the moment of inertia; adjustments to the hairspring were made so that at two temperature values practically the same rate resulted, but at all other temperatures, as shown by #75 in Chapter 5, rate deviations occured that introduced the middle temperature error.

Guillaume's experiments on the bending behavior of the usual bimetallic strips made him aware that a 45% Ni-Fe alloy in place of the usual iron strip could practically eliminate the temperature dependence of the hairspring that was so important for chronometers, as shown in the diagram of #322.

Two Swiss chronometer makers, Paul Ditisheim of La Chaux-de-Fonds and André Nardin of Le Locle, quickly picked up Guillaume's new idea, and at the Paris World's Fair of 1900 they showed two marine chronometers with balances constructed according to the new ideas. For the first time in the history of chronometers, they could show an almost complete elimination of the middle temperature error without additional attachments. How this striking parallelism in all details of the works of Ditisheim and Nardin came about could not be documented. The design of the balance made with the new bimetallic compensation, called "integral balance" by Guillaume himself but generally known as the Guillaume balance, is shown in #323. The symmetrical design of the balance weights to the right and left of the arms of the wheel is striking; it decreases the length of the free carrying arm to about half that of the classic design, thus considerably increasing the stability of the compensated balance. Through this, as well as the elimination of other insecure features, the assembly time for chronometers could be reduced to a fraction of what it was before.

Thanks to their simpler manufacture and being less prone to asymmetry, it is easy to understand why numerous continental European chronometer makers quickly switched to using the integral balance in their chronometers. Of the chronometers from the first half of this century that are shown in this book, fully half are equipped with such Guillaume integral balances, while the conservative British chronometer makers stayed with the classic design.

Along with eliminating the middle temperature error and simplifying the work of adjustment, Guillaume's

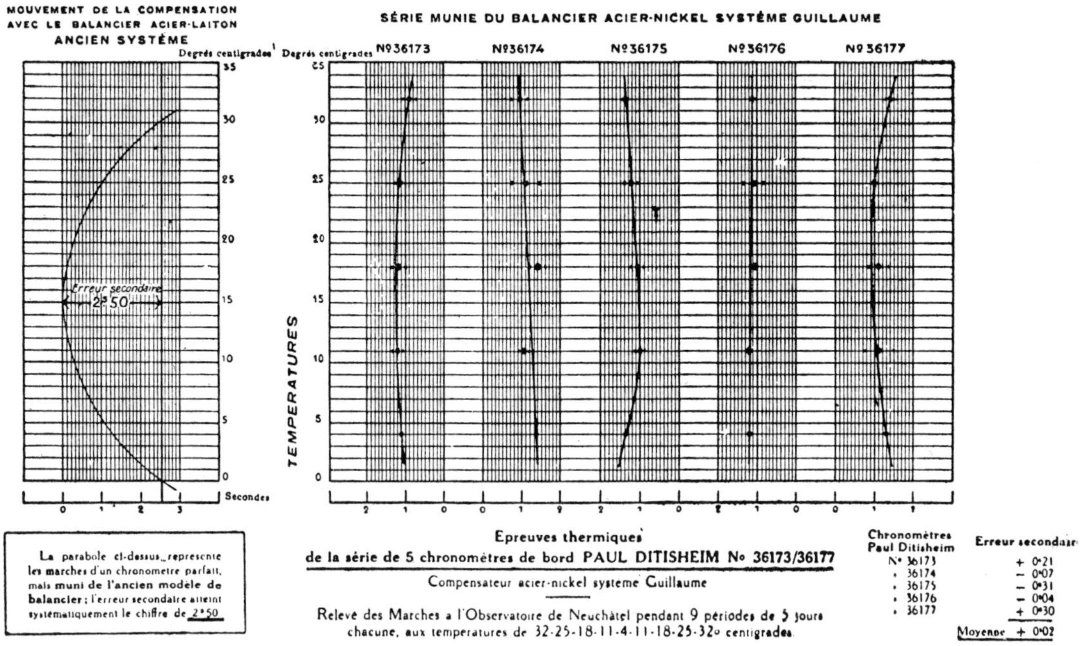

324 Compensation properties of marine chronometers with Guillaume balances in comparison to an identical system with a steel hairspring. The classic system gives a middle temperature error of 2.50 seconds, while the five tested chronometers (P. Ditisheim No. 36,173 to 36,177), in tests extending over 45 days at nine different temperatures, gave a mean middle temperature error of only + 0.02 seconds.
Journal Suisse d'Horlogerie No. 11-12, 1942

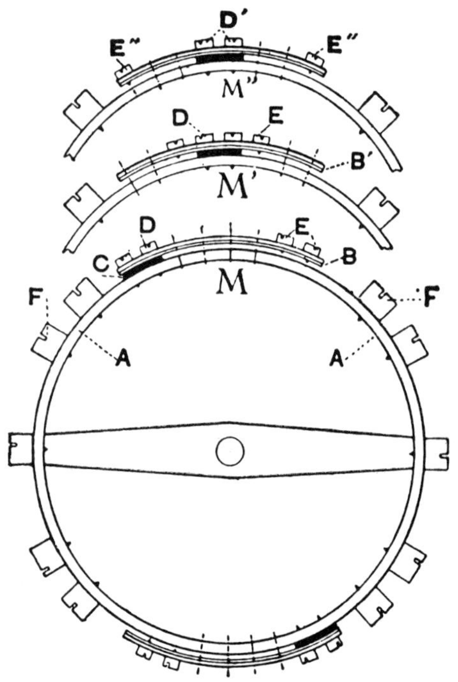

a

b

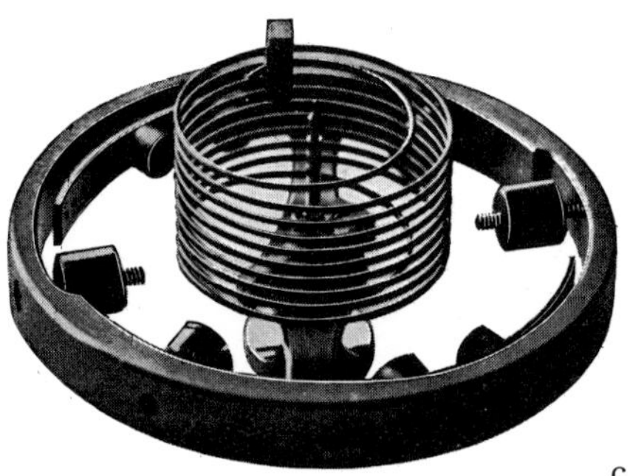

c

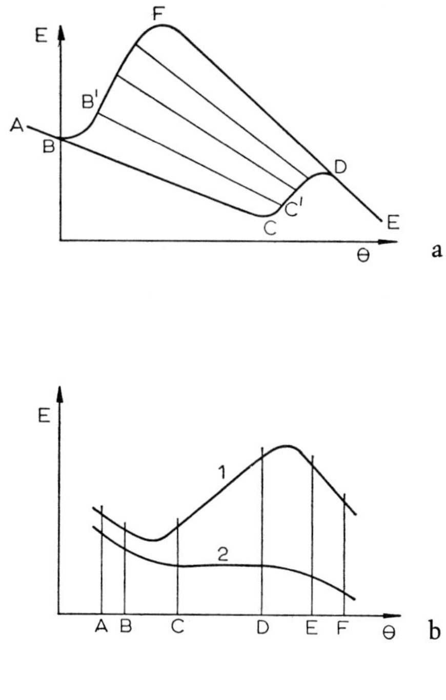

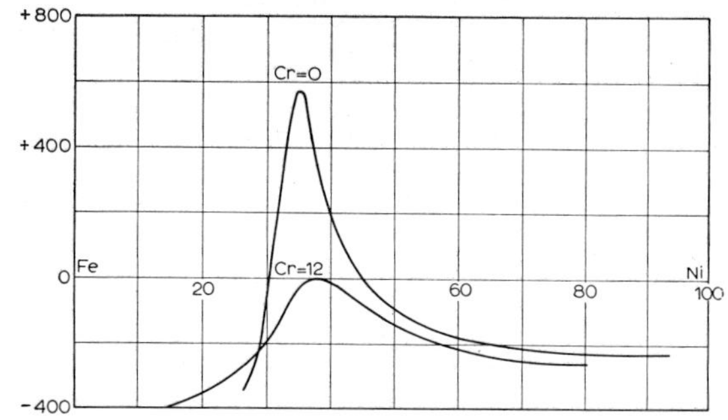

327 Influence of the behavior of the thermoelastic coefficients (dE/d0) of Fe-Ni alloys by the addition of 12% chromium to alloys of varying nickel content compared to the (dE/d0) changes of Fe-Ni alloys without added chromium at 20 degrees Celsius. (dE/d0) gives the change from E per degree; (0) . . . temperature in degrees Celsius.

(From Guillaume's address, Stockholm 1920)

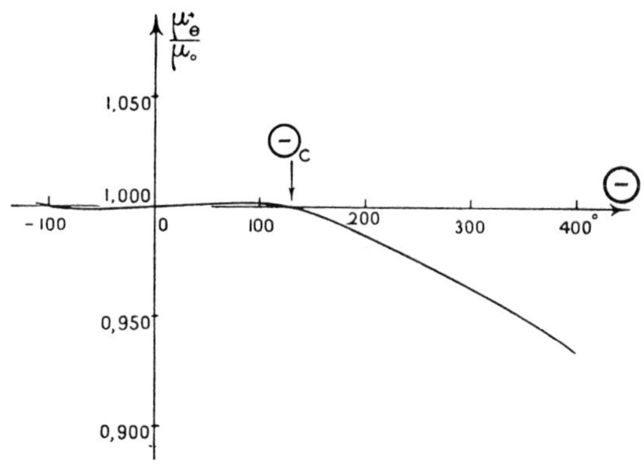

328 Thermoelastic behavior of an iron-nickel alloy with 34% Ni and 10% Cr, defined by the change in elasticity constant as opposed to in a chronometrically interesting area, where the difference is less than +/-1.0 degree/00, (u=1/E).

("Le Spiral Reglant", p. 290

integral balance improved the precision of high-value chronometers approximately fivefold. In addition, the shortening of the balance arms reduced the residual influence of the centrifugal error to one-sixth.

A centrifugal error in bimetallic compensated balances occurs when the main spring runs down and the original amplitude of the balance decreases from about 400 degrees (+/-200 degrees), and the weights on the ends of the balance arms are pushed a little less toward the outside because of the decrease in velocity, the moment of inertia thus becomes a little smaller, so that a tendency toward quicker beats sets in.

Thus the integral balance proved to be a great step forward in chronometer making, above all in terms of simplification and economy. The gain in the direction of eliminating the middle temperature error is shown by the diagrams in #324. In the same direction, there has also been a tendency, such as has also been done successfully in costly pocket watches since the twenties, to use uncut balances and self-compensating hairsprings in chronometers. To deal with the last residual error that still remains despite the spring's self-compensation, P. Ditisheim developed his uncut balances with added compensation[10] (Affix compensateur), as shown in #325, shortly after 1920. He had placed the added element to the outside in pocket watches; he mounted them inside on marine chronometers, with the intention of thus keeping air resistance as little as possible. In his 1945 report on this subject there are extensive test results of movements equipped with such balances, all of which were fitted with the Elinvar hairsprings described in the next section.

With such designs, the following daily errors were found: a) in Biel (1926), for watches of Class I, 0.57 — 3.5 seconds per day; b) at the Neuchâtel Observatory (1927-1929), for pocket chronometers, 0.16 — 0.45 seconds per day; c) at the Neuchâtel Observatory (1928-1931), for marine chronometers, 0.14 — 0.32 seconds per day.

The remarkably good results with Ditisheim balances with uncut rings and additional compensation, and the gain in time-saving thanks to the simple construction and overhauling, finally moved other chronometer builders to take similar steps, including Glashütte and Hamilton.

Self-Compensating Balance Springs
(spiraux autocompensateurs)

It was known for a long time in the 19th Century that the temperature dependence of the period of the balance and hairspring is very much (+/-80%) caused by the hairspring itself. While at first only its changes in length at different temperatures was seen as a cause, it was seen much later that it was above all the thermal sensitivity of the spring's elasticity; thus the ring proved to be the main cause for the temperature-related rate changes in watches and chronometers. The turning moment M, that affects the axial turning of a hairspring when wound around its axis, is proportional to the angle of turning and the module of elasticity E (also called Young's modulus) with the same geometrical configuration Gk:

$$M = Gk \cdot E \cdot \alpha$$

It was again Guillaume who dealt with the influence on the thermal elasticity behavior of the spring material, and who could finally report on it in 1920;[11] that the addition of other materials to iron-nickel alloys brought about a favorable influence in the sense of independence from temperature of the elastic behavior of the hairspring. The modulus of elasticity coefficient was determined by the percentages of the portions in the additional mixture. Illustration #326 shows that the behavior of the modulus of elasticity, in a similar manner to the expansion coefficient, also shows partly an irreversible (a) and partly a reversible (b) behavior. The cause of this very complex phenomenon is the variation in mixing conditions between the individual grades of iron crystallization and the nickel, which results in strong changes of the thermoelastic coefficients. For its use in chronometer hairsprings, the horizontal area between C and D of line 2 in #326b is of special significance. Its behavior in the middle of the temperature range for Fe-Ni alloys between 0 and 12% Cr, brought about by the favorable behavior of line 2, is shown in #327. In other temperature ranges, though, displaces and changes itself, which is the cause of the complex changes in diagrams a and b of #326.

The new Fe-Ni alloy with added chromium was named ELINVAR by Guillaume because of its almost independent temperature coefficient of elasticity — in the chronometrically interesting temperature range between 0 and 150 degrees C. this alloy contains some 39% Ni and 10% Cr. The great constancy of temperature of less than 1% for a similar alloy is shown in #328.

The introduction of Invar hairsprings in pocket watches, already mentioned in reference to Perret, has resulted in similar designs by other watchmakers only to a limited extent, because the production of hairsprings from iron-nickel spirals was only possible with special processes originated by Perret — for Invar was soft as a raw material but difficult to harden. In addition, balances equipped with binary Fe-Ni hairsprings showed a middle temperature error that could not be disregarded.

The development and production of Elinvar has largely eliminated the qualities of Invar that hindered its extended use. The first tests of Elinvar were turned over by Guillaume to Ditisheim, who installed a hairspring made of it in his lever pocket watch NE 44,811 of 43 mm diameter, with an uncut ring balance. This watch was tested for over four and a half months in 1920, partially at the National Laboratory at Teddington, under varying conditions, and showed extraordinarily good results.[12] Over a period of 25 days, for example, in which the mean daily temperature changed between 5.4 and 23.6 decrees Celsius, the entire rate difference was only some 2.5 seconds. Subsequent further tests at the Neuchâtel Observatory finally showed that this watch restricted the daily error the 0.17 — 0.19 seconds per day and thus attained a true chronometer precision.

With Elinvar, increased use in general watchmaking for the purpose of improving running precision was possible; because of the lower limit of elasticity, though, the hairsprings made of Elinvar had to be installed very carefully. But for chronometers and costly pocket watches, the good self-compensating qualities of Elinvar could be utilized to a greater degree, and the resistance to rust was an advantage not to be overlooked. By 1923 some fifty million watches had been built this way.

It took the better part of two more decades, to 1936, when it could be reported at a conference that a successful solution to the problem of how to combine the high elasticity limits of steel with the favorable self-compensating behavior of Elinvar had been found — a problem that had long been regarded as insoluble.

Professor Chevenard, for a long time Guillaume's first assistant, was able to do it. Shortly before that, he had been able to adapt the new process of superhardening (hypertremping), which had made the production of duralumin possible, for complex iron-nickel alloys and attain structural hardening (durcissement structural) there too, gaining the sought-for exceptional constancy of specific physical and thermal characteristics in a large range of temperature.[13] In this regard, three new alloys were introduced: Metelinvar (Fe-Ni + molybdenum and vanadium); Durinval (Fe-Ni + aluminum and titanium); and Nivarox; the first two are especially suitable for the hairsprings of watches and chronometers. The results of several series of tests,[14] including two at Neuchâtel in 1944, show the progress made in the reduction of the mean values of rate deviations and the middle temperature in broad temperature ranges, as follows:

Alloy	Balance Type	Thermal Coefficient	Middle Temp. Error
Metelinvar	monometallic	-0.07	-0.6
	monometallic	+0.09	-0.4
Durinval	monometallic	+0.10	-0.5
	monometallic	+0.15	-0.8

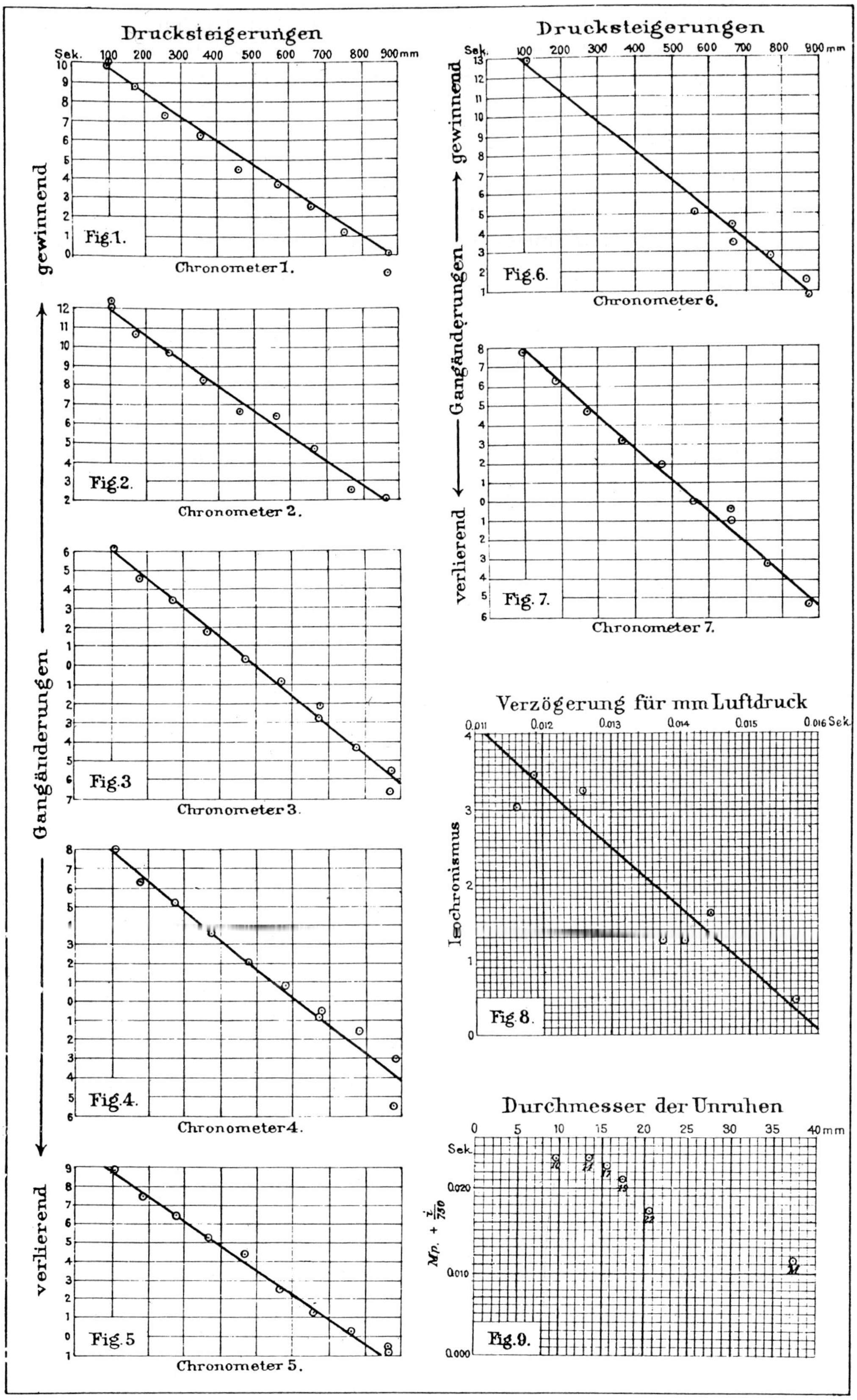

329 The influence of barometric pressure on chronometer rates, shown by seven rate change graphs for pocket chronometers, made in 1904 by P. Ditisheim in his laboratory in the large pressure area of 100 ./. 900 mm mercury column. Along with the striking slowing of rate at rising pressure, the graphs show the tendency, changing from case to case, toward additional influence on isochronism depending on the diameter of the balance.

The thermal coefficient is the mean value of the daily rate difference at the stated temperature limits at one degree Celsius, while the secondary error is the mean value between the rate deviation at 19 degrees and the mean of the other extreme values.

The improvements achieved in going behavior are striking; the particularly important range from 4 to 32 degrees Celsius makes the use of hairsprings made of this material very suitable for watches with uncut balances. In the temperature range in question, the weak positive expansion coefficient still present in the alloy has a favorable effect, because it is just sufficient to counteract the thermal increase of the balance's moment of inertia.

Another important factor determined by Chevenard that is of general interest is that eliminating the thermal influences could not be united with the suppression of magnetic reverse effects; materials such as those cited in the table above will always be prone to take on a lightly magnetic polarization.

Consideration of Influences

Until after World War I, the basic expectations that were applied to chronometers were roughly the same as in the 19th Century; that they be used to take the time of the point of departure along for orientation on the earth's surface, mainly on ocean voyages, but also on expeditions to isolated regions of the earth, in order to be used along with the astronomical determination of the local time to calculate the geographical longitude. In the process, the temperatures that the chronometers were intended to encounter varied from a few degrees below zero Celsius to about 40 degrees in the tropics. For the rare expeditions to the polar regions, appropriate measures were taken ways to limit or completely rule out the well-known effect of cold on the chronometers such as the expedition members wearing them on their bodies.

The rise of regular air travel and the probes into the upper levels of the earth's atmosphere originally relied on the previous methods of using chronometers for whatever orientation was done. In the process, though, considerable modifications were made in view of the environmental conditions met by the chronometers, above and beyond what had been customary before, and particularly regarding temperature and air pressure, since they had far-reaching effects on all moving and lubricated parts. Both problem areas soon became the objects of extensive scientific and practical experimentation, intended to allow chronometers to be used reliably in new, and unusual working conditions, and to provide important new experience for the solution and feasibility of other complexes of problems that had been acquired from the new realms in the human sphere of action. These problems will be dealt with, or at least noted, in the following text.

Barometric Influence on Chronometer Indication

Very early in his career as a chronometer maker, P. Ditisheim had determined that his chronometers, adjusted to run precisely at Neuchâtel (498 meters above sea level), always ran a little slowly during the tests in Paris (69 meters above sea level).[15] He was not satisfied with the general explanation that barometric pressure affected the chronometer's rate, but sought a quantitative explanation for this phenomenon — which had already been mentioned by Urban Jürgensen in 1826, and to which the Neuchâtel astronomer Hilfiker had referred in 1888, noting that rising barometric pressure always caused the slowing of clocks.

After making contact with E. Guillaume, Ditisheim decided to build a testing apparatus in his laboratory at La Chaux-de-Fonds (mean barometric pressure 675 mm), with a device to achieve any desired air pressure between 675 + 200 and 675 - 600 millimeters of a mercury column — where the chronometer movements could work and where at the same time the temperature and moisture levels could be observed.

The very first series of tests with eight pocket chronometers showed clearly that the chronometers all ran faster as barometric pressure dropped. Unexpectedly, though, the amounts of change in the rates of the fully identical chronometers (46.6 mm movement diameter, 20.2 mm balance diameter) for the same changes in barometric pressure were very diverse, as #329 shows.

As a result of thorough consideration and extensive statistical work, Ditisheim was able to set up a basic equation to show both the varying rating and the influence of various barometric pressures on the rate of a given chronometer:

$$M_p + \frac{i}{750} = K$$

Here Mp is the daily rate change in seconds for one millimeter of barometric pressure as opposed to a total vacuum, where undisturbed isochronism should be present.

i equals the affect on isochronism for the individual chronometer in question, that is, the rate difference of the first 12 hours minus that of the second 12 hours at 750 mm of barometric pressure, which amounts to a few seconds. K is a number characteristic of the chronometer type, essentially determined by the diameter of the balance, for example, 0.0102 for 37 mm diameter, 0.0162 for 20 mm diameter, or 0.0223 for 10 mm diameter. For a barometric pressure of p = 750 mm, the mean daily rate G is derived for a pocket chronometer with 20 mm balance diameter, where i may equal 1.9 seconds:

$$G = M_p \cdot p = K \cdot p - \frac{i}{750} \cdot p =$$

$$0{,}0162 \cdot 750 - \frac{1{,}9}{750} \cdot 750 = 10{,}25$$

seconds per day

compared to working in a total vacuum; but if the chronometer is adjusted for a rate of 0 at 750 mm of barometric pressure, then the chronometer will run 12.25 seconds fast per day.

In conclusion, one can sum up the influence of barometric changes as follows: 1. When the barometer rises, all chronometers run more slowly. 2. The extent of rate change is determined by the change in pressure and the inherent isochronism of the chronometer. 3. The intensity of the rate change caused by the pressure change is inversely proportional to the diameter of the balance.

Interference with Isochronism according to Ditisheim

(Defaut or ècart d'Isochronisme)

In the course of his thorough investigation of the influence of air pressure on the rating of chronometers in 1903, P. Ditisheim also came upon a property of chronometers that had not been recognized before; the oscillating system with balance and hairspring does not work isochronically, with constant amplitude, during the entire running of the movement from being fully wound to stopping, but that the amplitude of the balance assembly, and thus also the length of the period of the individual oscillations, vary somewhat according to how fully the spring is wound.

To understand this effect in chronometers with a going duration of one day, Ditisheim used the rate difference between the first twelve hours after a full winding and the second twelve hours before the next winding. This value, designated the isochronism error (ècart d'isochronisme) by him, was first observed by him in a series of tests with eight pocket chronometers (chronomètre à bord) of the same type, with 49.6 mm movement diameter and 20.7 mm balance diameter, at the same surrounding temperature and barometric pressure. He determined that the individual chronometers probably show the same isochronal error day after day, but that it varies from piece to piece, and the extent of deviation was 0.4 to 3.4 seconds per 24 hours. In all cases it amounted to a faster rate in the second 12 hours with smaller amplitudes, which Ditisheim could explain as a result of the lesser loss to air resistance at the smaller amplitudes, since his experiments on chronometer behavior at different barometer pressures had shown that in a vacuum the interference with isochronism in all the tested chronometers had disappeared. When the balance moves in an airless atmosphere, it carries with it an air bulge that is several square millimeters in cross section and can cause energy-destroying turbulence; its presence also makes it understandable that the screws and weights attached to the balance rings show relatively little influence on its rate, as Ditisheim was able to prove on another occasion.[16]

Lubricants, Long-Term Behavior of Chronometers, and Extreme Working Conditions

Mechanical systems with moving parts (gears, arbors with pivots, discs with fillets, etc.) need lubrication between their parts that make contact, to reduce friction wherever metal surfaces slide over other metal surfaces; without lubrication much more power is needed to carry out the motions and considerable loss of material takes place because of the greater abrasion.

For many years various oils, tallow, fats, graphite, and other materials have been used as lubricants, with varying success.

When mechanical clocks were first used, plant oils were used rather randomly to lubricate clockwork, but they soon turned rancid, and the acids that were created attacked the metal parts and eventually gummed up the bearings and pivots. In the 18th Century, the better clockmakers turned to special animal oils such as neat's-foot oil, which naturally varied from one city and manufacturer to another. On the continent of Europe it was usually the oils extracted from the bones of hooved animals. In America the oil from the head of an aquatic mammal was used. The porpoise was used with more or less success, but sometimes with catastrophic results, until the beginning of the 20th Century. P. M. Chamberlain[17] mentions the following misfortunes in his important book, which deals with many chronometer designs and makers: corrosion of bearings, tendency to creep (leaving the bearing), boiling away, freezing at low temperatures, too strong for balance pivots and not strong enough for bearings with high pressure on them — all difficulties that were encountered by many generations of clockmakers and users before the development of new oil technology around 1925.

The first job of any lubricant is to form a thin filmlike layer on each of the metal parts that contact each other, covering any rough spots in the metal and thus creating a situation in which the transition from the movement of one part to that of the other takes place gradually within the two films of lubricant without the metal surfaces coming into contact at all.

The individual molecules of lubricant move toward each other at a growing distance from the metal surfaces, so that an inner friction takes place in the lubricant, in which the resulting energy losses depend essentially on the constitution of the lubricant. This is called viscosity, and its intensity is described by a (temperature-dependent) viscosity coefficient n η , which makes it possible to compare various lubricants and other substances, including gases. The viscosity coefficient thus constitutes an important characteristic of lubricants, in that the viscosity should correspond to the intended use. Both the safety of the lubrication and the degree of effectiveness on the system depend on the right viscosity, but numerical values for viscosity are rarely stated.

The basic concepts in these considerations are easy to understand at first: to determine a viscosity coefficient, a thin flat plate is drawn lengthwise through the medium at a speed u, with the necessary pulling power Z being measured simultaneously. The viscosity coefficient n is then computed by using this equation:

$$Z = k \cdot u = \eta \cdot \frac{F}{x} \cdot u$$

Here x equals the distance of the plate surfaces from the walls of the container, which are parallel to the moving plate, or the thickness of a bordering layer of the medium where it is no longer influenced by the plate's motions.

The concept of the bordering layer is a very important one; the quantitative determination of it through formulas, though, leads only to rough estimates, while the numerical values of n η of the various substances are very useful for comparison purposes. In the following table, the viscosity values for various well-known substances extending over numerous sizes are to be seen:

Sustance	Temperature	Viscosity Constant* in Grossdynsec/m²=Poise (1 Grossdyn=0.102 Kilopond)
Air	20°	$1.7 \cdot 10^{-5}$
Liquid CO2	20°	$7 \cdot 10^{-5}$
Benzol	20°	$6.4 \cdot 10^{-4}$
Water	0°	$1.8 \cdot 10^{-3}$
	20°	$1.0 \cdot 10^{-3}$
	98°	$0.3 \cdot 10^{-3}$
	-21.4°	$1.9 \cdot 10^{-3}$
Mercury	0°	$1.6 \cdot 10^{-3}$
	100°	$1.2 \cdot 10^{-3}$
	300°	$1.0 \cdot 10^{-3}$
Glycerine	0°	$1.0 \cdot 10^{-3}$
	20°	$8.5 \cdot 10^{-1}$
Pitch	20°	10^{7}

From: R.W. Pohl, *Mechanik, Akkustik und Wärm*

* For lubricating oils, rather than the viscosity constant, the viscosity = viscosity constant: thickness of the substance is given, which is listed in stokes (deci-, centistokes); lubricating oils have viscosities in the range of 50 to 700 centistokes.

Along with the viscosity, especially for very smooth, finely worked surfaces, the adhesion counteracts the beneficial effect of a lubricant under certain conditions; quantitative evaluations of this are almost impossible.

The distribution of the lubricant over the surface of the bearing is a factor that must not be neglected; it is necessary that the lubricant stay where it is needed and form a "complete" covering.

Most important, though, is the behavior of the lubricant over time; it is a question of whether, and how long, the effective lubricating qualities present on application last or remain over a practically sufficient time, as well as if and when disadvantages such as solidification, gum formation, etc. occur.

The problems involved in lubricating balances were trade secrets for many years, and scarcely any information on them was made public.

Despite all the recurring difficulties during the whole fine mechanical era, there were no systematic attempts to explain and improve the lubricating situation. It is surprising today that, in the testing of chronometers, there is not the slightest reference made to the essential lubrication of the individual objects, and only the lack of any method of evaluation, or of any reference to the supply and production of the oils that were used, can be regarded as explanations of this lack. Thus today one is compelled to wonder whether many mechanically well-made chronometers performed poorly in the tests on account of unsuitable lubricants. The involvement of numerous effective components, each individually dependent on the composition as well as the origin of the lubricant, surely helped to create the complex lack of clarity in the problem of lubrication.

As far as the author has been able to determine, the decisive steps that brought light to the mystery-shrouded realm of lubrication problems in clocks and chronometers were begun by P. Ditisheim in collaboration with P. Woog, later the chief chemist of the Compagnie Française de Raffinage in Paris around 1920. As significant results, the following facts from three reports on the subject[18 19 20] may be cited:

1. All lubricants — to differing degrees — are influenced by aromatic substances and change (decrease) their lubricating ability more or less quickly in the process. The aromatic substances include not only the perfumes of the various flowering plants (particularly roses and carnations), but also the emanations of woods (such as nut oil) and electric electrical insulation materials, plus the fumes of benzol, toluol and the like. The viscosity can be decreased very much by such effects.

2. The nature of the metals to which the lubricants are applied can, under certain conditions, result in a reaction with the lubricant through hardening and lead to the adhesion of the moving parts to each other. Metals that contain sulfur or sulfur compounds are especially affected.

3. According to the construction nature of the parts to be lubricated and the type of movements, various lubricants are to be recommended. The suggestion often made previously that a universal oil for clocks be produced by mixing various lubricants has been disproved. Ditisheim and Woog recommend that three different groups of lubricants be considered:

a) Mixtures of animal oils with mineral oils, whose tendencies to oxidize are suppressed by special treatment and to which light-absorbing pigments have been added — CHRONAX oils, which are suitable for small and medium-sized clocks at temperatures to -15 degrees Celsius.

b) Pure mineral oils with added active substances to increase their viscosity. The use of pure mineral oils to lubricate clocks was impossible as long as their tendency to spread out and leave the actual area to be lubricated could not be controlled. In 1925 P. Woog invented the so-called Epilame process, in which the lubricated area is coated with a very thin film that does not let the film of mineral oil go beyond it. When lubricated points are so treated, pure mineral lubricants can also be used at very low temperatures (-75 degrees C), for which different grades of mineral oils with different viscosities are available depending on the low temperatures to be expected. This development has become very important not only for use in chronometers, but in all instruments with moving parts, such as those in high-altitude aircraft, satellites, meteorological instruments, etc.

c) Synthetic oils on a silicon basis, which had first been developed by P. Woog.

4. The viscosity of the lubricating oils decreases sharply with low temperatures. This influence of temperature over the wide range between -30 and +50 degrees Celsius for one and the same lubricant is shown in #330.

The developments in clock lubrication that derived from the ground-breaking work described above were influenced considerably by the new insights already noted. At various chronometric conferences, particularly the International Congresses of Chronometry, the problem area of lubrication and wear has always had several meetings devoted to it, in which additional valuable advances and new insights have been reported.

In practice, the collected oil specifications used in the clock supply trade are of value, including those on synthetic products: 1. Escapement oils for pocket and wristwatches; 2) Oils for jewel bearings, cylinder escape wheels and adjusting screws; 3) Oils for pivot holes without jewels and main springs in pocket watches, wristwatches and alarm clocks; 4) Escapement oils for large clocks without alarms, chronographs, measuring devices; 5) Oils for pivot holes without jewels and drive springs in large clocks; 6) Greases for winding, barrel arbors, anchor rods, main springs of all sizes, tower and outdoor clocks, rotor bearings, etc., as well as various other special lubricants, some of them including graphite or even molybdenum sulfide additives.

For those who know and love chronometers, there is thus a possibility of preparing their chronometers best at their critical areas and thus of maintaining their accuracy under favorable conditions.

Chronometer Testing en route

It used to be common to hear in chronometer evaluations: "As soon as a piece is outside the workshop, laboratory or observatory, its behavior is considerably (sometimes

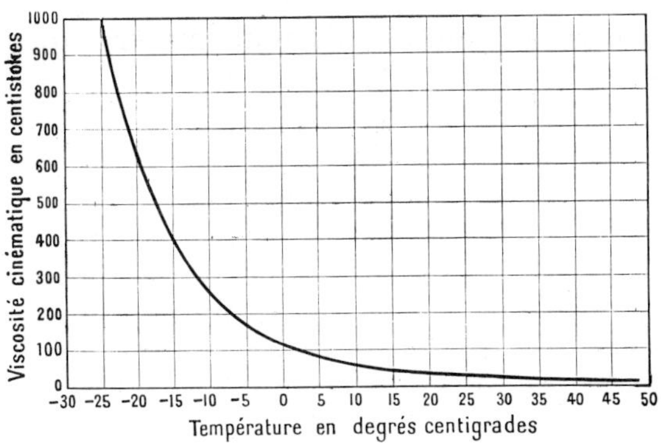

330 Influence of cinematic viscosity of a lubricating oil by its temperature; between +4 and -30 degrees Celsius, the cinematic viscosity increases tenfold and, correspondingly, so does the inner friction.
From Ditisheim, "Lubrication des Montres", p. 14

up to three times) worse." There is a kernel of truth in such a statement, for when a clock is on the move, uncontrolled and unexpected surrounding temperature changes and mechanical influences, such as jolting and shaking take place as never before. In order to get to know his chronometers under such unusual conditions, which had much to do with actual use, P. Ditisheim carried out various programs that were a continuation of Huygens' travels at the end of the 17th Century — the chronometer transportation of Sully over the bumpy Paris pavement, the renowned ocean voyages of Harrison's H4 or of Fleurieu and other expeditions. Here the most important of these programs will be described in a few words that characterize the general situations. Details of the tested chronometers, as well as on their remarkably good performance, will be found in the special reports on those programs.

1. The determination[21] of the geographical difference in longitude between Neuchâtel and Paris in 1902, using five chronometers (with lever escapement, 22 lignes = 49.6 mm diameter). Transported by rail in padded boxes hung by rubber bands A mean difference of 18 minutes, 28.80 seconds of arc occurred, while previously, networks with numerous measuring points had been established by the European observatories, giving seconds of arc that differed slightly only beyond the decimal point, namely 28.86 seconds of arc in 1886 and 28.77 seconds in 1893.

2. The first explanation of chronometer behavior at various altitudes was the subject of the observations of the same instrument in Paris (67 meters), Neuchâtel (489 meters) and La chaux-de-Fonds (1017 meters), as well as at the summit of the Chasseral (1586 meters), by Ditisheim along with Messrs. Arndt, Berner, Bigourdan and M. H. Wehrli — as reported in an Academy report of 1903.

3. The reaction of chronometers to rougher transport:

a) Placing the chronometers in the baggage net of a third-class railroad passenger car, in a compartment over the axle, running from Neuchâtel to St. Maurice and back (some 252 km).

b) Packing the chronometers in the trunk, over the rear axle of a car, for trips of 311 and 230 kilometers in two days.

These tests were mentioned in retrospect by Defossez[23] in 1945 and probably took place during the twenties.

4. Explaining chronometer behavior in large-scale changes of altitude.[23]

As early as 1904, Ditisheim had drawn up a chart, in units of 500 kilometers, of the predicted mean barometer pressures for the great range between -500 and +9000 kilometers, in which the points of contact were cited as:

km of Altitude	Mean Barometric Pressure (mm hg)	Mean Temp. Celsius Summer	Winter
-500	809		
0	760	13°	1.9°
+1000	670.3	11.8°	-2.40°
3000	522.02	2.1°	-6.0°
6000	358.65	-8.3°	-17.0°
9000	246.55	-38.0°	-46°

and in which the corresponding mean summer and winter temperatures for Paris were also cited.In 1923, in connection with the annual convention of the Swiss Natural Science Society (Soc. helvetique des Sciences Naturelles) in Zermatt, several marine and pocket chronometers, along with several other small clocks, were tested at several intermediate stations between Viege (654 meters) and Gornergrat (3136 meters). The results included a satisfactory proof of the barometric formula stated by Ditisheim in 1904.

Magnetic Influence on Chronometers and its Prevention

The occurrence of the longitude error has shown clearly that the effect of additional forces and moments on the balance assembly causes rate changes. Magnetic influences can also cause disturbances in the rate of a chronometer or a clock in the same manner, if they occur newly as opposed to the conditions at manufacture. It is the magnetizable parts of the balance and the balance spring that are affected. The possibility of magnetic disturbances was basically known in the 19th Century, and the possible use of non-magnetic palladium for balance springs was the first step taken to avoid disturbing magnetic influences.

As long as it was just a matter of the earth's magnetic field and its occasional variations, such as magnetic storms, counteracting magnetic influences was only of secondary importance. In 1918 Dr. Möller of Hamburg made a thorough study[24] of the problem of protecting chronometers from strong magnetic fields. At the close of his study, though, the essentially theoretical nature of this problem area was stressed when he noted that such strong magnetic fields were of no importance in reality.

The increasing use of large electromagnets in laboratories and test centers in the years after World War I, and especially the equipping of warships in World War II with very large demagnetizing equipment for the premature detonation of magnetic mines, made protecting chronometers from magnetic influences much more important than before.

In 1928 Ditisheim reported that the Elinvar hairsprings available since 1920 were advantageous for use in precision clocks and chronometers because they could not be magnetized, while a simple one-piece balance made of brass would make the system insensitive to magnetic influences.

Finally, in 1945 Ditisheim wrote briefly[26] of the magnetic shielding of the movements of precision clocks and chronometers with additional mantles of Permalloy, a method that is of special importance for pocket chronometers that may come near large magnets in laboratories at times. The presence of a pocket chronometer in a strong magnetic field for a time will naturally cause a definite spatial polarization of the mainspring, which will remain in the spring even after the chronometer is removed from the magnetic field but will slowly be displaced when the clock runs in a different position from the original one when the barrel or the spring continues to turn. These changes nevertheless can cause disturbing reverse effects when a high degree of precision is required, through eddy currents, even on unmagnetizable parts of the balance assembly.

A capsule of Permalloy over the entire movement prevents much of the disturbing magnetism of the mainspring, thanks to the weak magnetic qualities of this material.

All in all, though, the provision of special defenses against magnetic disturbances does not appear to be a generally appropriate practice for chronometers, but one rather to be used only in special cases.

Chronometer Development and Chronometer Production

Until the end of World War I, chronometer production continued as it had at the end of the 19th Century. There were a number of chronometer makers in various countries, though the number of firms decreased while some of them expanded. In the 1917-1918 winter competition at the Deutsche Seewarte,[27] for example, 76 marine chronometers were still submitted by six firms: Chronometerwerke Hamburg; L. Kurtz of Münster in Westphalia; A. Lange & Sons of Glashütte, Saxony; Fr. Lidecke of Geestemünde; Union-Glashütte of Saxony; and G. Wiegand of Peine. In the years that followed, though, chronometer production in Germany decreased sharply and most of these firms disappeared from the scene, with the exception of Glashütte. Here, under the direction of Studienrat Alfred Helwig, chronometers and tourbillons were created whose workmanship and achievements can be regarded as pinnacles of chronometry.

In Switzerland, though, the production of precision clocks and chronometers continued to advance, and the pioneering achievements of Ditisheim regarding the increasing use of Guillaume's new Fe-Ni alloys proved to be very important.

In the years before and during World War II, the need for marine chronometers increased again. In Germany at this time the firm of Wempe took over and expanded the old Chronometerwerke Hamburg facilities.

All in all, German chronometer work has been based on a continuation of the fine mechanical efforts with the best possible use of the achievements of the previous era; it has been founded on the outstanding technical and practical work of Moritz Grossman and Professor Ludwig Strasser in Glashütte.

A very important problem area in the realm of fine technological work concerns the correct raising of the balance. Important progress has been made here through improved understanding of the dynamic behavior of rotating masses, as has been expressed in the Handbook for Clocks.[28] Thus the original cutting method has been replaced first by the 2- or 3-point method and finally by an electronically registered, practically automatic solution that saves the tester much time and trouble. But since this knowledge was very important in terms of economic success, it was kept secret for a long time.

In Switzerland the quick familiarization with the new special alloys, beginning with Invar and proceeding via Elinvar to the new special alloys such as Metelinvar, Durinval and Nivarox, has been a very important stimulant for the development of chronometer making. Thus in the first half of the 20th Century the last high point of mechanical chronometer construction has been attained through the use of the most modern metallurgy and technology in an epoch particularly marked by them. Ulysse Nardin has stood out in producing outstandingly precise and reliable marine chronometers. P. Ditisheim himself has also produced some noteworthy technical progress of his own; above all, he has achieved vitally important success through sharing and spreading new scientific knowledge in numerous comprehensive writings.

Decades of close cooperation with Guillaume in the realm of metallurgical developments to make the chronometer independent of temperature changes may be cited as most important of all, then his involvement in studying the influence of the altitude of the workplace on chronometers, and also his involvement and promotion of work on the problems of clock lubrication, particularly at low temperatures, and finally his thorough study of magnetic disturbances in clocks and chronometers and how to combat them.

The provision of electric contacts for transmitting time signals and the preparation for the possibility of electric winding and power should not be forgotten at this point; they indicate the ongoing change to the utilization of new physical phenomena in place of classical mechanics.

Through his publications, Ditisheim has done much to promote the making of chronometers in the 20th Century; without his publications it would scarcely have been possible to work out an overview of the course of chronometer making in its last phase.

A further effect of the inclusion of new ideas and materials in the ongoing developments in chronometer making in Switzerland has resulted in a considerable development of special uses, particularly in sporting events, a field in which Dr.Ing.e.h. Reinhard Straumann of Waldenburg has stood out particularly.[29]

In many minds the concept has arisen that the prominent technician must make all the important steps in the development of his products or inventions personally. To a limited extent, this has been the case, as applied to such technical pioneers as James Watt, Werner von Siemens or Ghega, though even these were supported by important assistance from capable colleagues. In the complex present-day structure of all technical apparatus, facilities and process, teamwork, above all has been vital for success. In this respect the greatness of Paul Ditisheim in sharing important new technical and scientific discoveries with those directly involved and making them easily accessible, has been all the more important.

Through the developments noted above, Switzerland has taken the lead in producing precision clocks during the first half of the 20th Century.

Good, prompt information, correct understanding of problems and quick reaction to making it practical have certainly been very important here as in many other cases, and were instrumental in leading the Swiss clock industry to its leading position in the world.

In the 1940-1950 era, the chronometry of the 20th Century blossomed one last time, and interestingly so, in the developments of HAMILTON in the USA.[30] Before World War II there was no native mariner chronometer production in the USA; only to a small extent were imported parts assembled. When the war broke out, the Axis Powers prevented the delivery of Swiss chronometers to anyone but themselves. Thus it happened that, as reported interestingly and exhaustively by Marvin E. Whitney, the USA developed its own production in a remarkably short time. Hamilton and the Elgin Watch Company took up the manufacture of marine chronometers, and in 1942 Hamilton produced the first significant chronometer; this was the start of a successful production. At first, in 1942 and 1943, twenty to thirty chronometers were delivered per month. Then production increased radically, so that in October 1944 546 chronometers could be provided. The basis for Hamilton was the Nardin chronometer, yet the use of a monometallic balance with an Elinvar hairspring and a design including 24 equally spaced adjusting screws made for very significant progress. They made assembly much easier and also increased the average attainable accuracy. Through the use of a monometallic balance, the centrifugal error could be eliminated entirely and the weight of 10.2 grams for the Nardin balance, not unimportant in terms of required power, could be reduced to 5.1 grams, all means to the end of providing very good lasting results at sea, with errors of only about one second per day.

In the fifties, though, electrical systems — first the electric tuning-fork watch and soon afterward the electronic quartz watches — brought a very quick end to the role of the mechanical chronometer in sea and air traffic, a role that ended about 1955. This abrupt end suddenly became obvious to clock collectors when, in England, a great number of brand-new "government surplus" pocket and marine chronometers were put up for sale.

The only firm that still produces chronometers in England is the old house of Thomas Mercer in St. Albans, where tradition lives on and beautiful marine chronometers — such as for the Queen's Silver Jubilee — and traveling chronometers are still produced.

The dying out of the tradition-rich British chronometer history and the fate of the firms involved in it have recently been described very clearly and instructively by Davies in relation to the fate of production areas.[31]

1. H. Bertele, *Chronometrie des 20. Jahrhunderts und Paul Ditisheim*, Report of the 10th Congress for Chronometry, Geneva, 1979.
2. *L'Invar et L'Elinvar*, Nobel address by Charles-Edouard Guillaume, 1920, Soc. Suisse de Chronométrie, Liestal, 1974.
3. a) Gustav Tamman, *Lehrbuch der Metallkunde*, Springer, 1932, or 1st to 3rd editions under the title "Lehrbuch der Metallographie", 1914 ff.; b) Franz Rapatz, *Die Edelstähle*, 1951; c) E. Houdremont, "Handbuch der Sonderstähle", Heidelberg, 1943.
4. Charles-Edouard Guillaume, *Les Aciers au Nickel et leurs applications à la chronometrie*, comptes rendus Ac. Sciencs.; Paris, 1897.
5. Ch.-E. Guillaume, *Calcul d'un pendul composé en acier-nickel*, Journal Suisse d'Horlogerie, Geneva, 1897.
6. Paul Ditisheim, *Galilée et le Pendule*, Journal Suisse d'Horlogerie 1942, p. 27.
7. Paul Ditisheim, *Le Spiral Réglant et le Balancier depuis Huygens jusqu'a nos jours*, Editions Journ. Suisse d'Horlogerie, 1945.
8. D. S. Torrens, *Paul Ditisheim, A Tribute to a great Horologist*, Brit. Hor. Journal, London, December 1945.
9. C.-E. Guillaume, *Etudes sur le balancier compensateur*, Journal Suisse d'Horlogerie, Geneva, 1899.
10. Paul Ditisheim, *Le Balancier á Affixe compensateur*, from" Le Spiral Réglent et le Balancier", Lausanne, 1945, and *Demonstration of (a) A New Balance for Compensating the Temperature Error of Watches and Chronometers, and . . .*, Proc. Phys. Soc. London, Vol. 35, 1923.
11. Ch. E. Guillaume, *Dilatabilité et Thermoélasticité des Aciers au Nickel*, Comptes rendues, Ac. Science, Paris, 1920.
12. P. Ditisheim, *Le Spiral reglant . . .*, p. 45.
13. P. Chevenard, *Traitement thermique des ferronickels complexes à deux constituants*, Comptes rendues de l'Academie des Sciences, 189, p. 846, Paris, 1936.
14. P. Ditisheim, *Le Spiral reglant . . .*, p. 48.
15. Paul Ditisheim, *La Variation des Chronometres avec la Pression Atmosphérique. Correction de cette Variation*, Journal Suisse d'Horlogerie, Geneva, 1904.
16. Paul Ditisheim, *Influence de la Pression Barométrique surla Marche des Chronometres*, Journal Suisse d'Horlogerie, 1945, pp. 255-265.
17. Paul M. Chamberlain, *It's About Time*, New York, 1944.
18. Paul Woog, *Contribution à l'étude de graissage; onctuosité et Influence moléculaire*, Paris, 1926.
19. Paul Ditisheim, *Lubrication des Montres et d' autres Appareils, fonctionnant aux basses Temperatures*, Millot frères, Besançon, 1935.
20. Paul Ditisheim, *Les Parfums et les Montres*, Journal Suisse d'Horlogerie, No. 11-12, 1937.
21. Paul Ditisheim, *Les Chronomètres de Petit Format*, Essai d'une Détermination de la Difference de Longitude Paris- Neuchâtel, Journal Suisse d'Horlogerie, Geneva, 1905.
22. L. Defossez, *Paul Ditisheim, chronométrier*, Journal Suisse d'Horlogerie, Lausanne, 1946.
23. P. Ditisheim, *La Variation des Chronometres avec la Pression Atmosphérique*, Journal Suisse d'Horlogerie, Geneva, 1904.
24. Franz Möller, *Uber den Schutz von Chronometern vor starken magnetischen Feldern durch eiserne Gehäuse*, Annalen der Hydrographie, Vol. 46, Berlin, 1918.
25. Paul Ditisheim, *Correction de l'effet du champ magnétiquesur la marche des montres*, Comptes rendus, Ac. Sciene, T.187, 5. XI, Paris, 1928.
26. Paul Ditisheim, *Le Spiral reglent . . .*, pp. 50-51.
27. (Amtlich) *Bericht über die vierzigste auf der deutschen Seewarte abgehaltene Wettbewerb-Prüfung von Marine-Chronometern*, (Winter 1916 17), Ann. d. Hydr. etc., Vol. XXXXV(1917), No. VIII.
28. Prof. Dr. D. Glaser, Vol. II, Cf. 3. 2. 9. (2).
29. *Fortschritte in der Uhrentechnik durch Forschung*, Festschrift 1952, Steinkopf, Stuttgart (5).
30. Described in detail by Marvon E. Whitney, *The Ship's Chronometer*, in Horological Times (USA), January 1978- February 1980.
31. Alan C. Davies, *The Rise and Decline of Chronometer Manufacturing*, A. H. XII, 1980.

331a Marine chronometer in wooden case with gimbal mounting, signed "Konrad Wiegand, Peine No. 33", ca. 1920-25, with auxiliary dials for second and winding indication.
Babelsberg Observatory

331b Movement of Wiegand's chronometer, 56-hour movement, integral balance with blued helical hairspring.

332a Marine chronometer, signed "Paul Ditisheim No. 112, La Chaux-de-Fonds", ca. 1912, for siderial time; dial diameter 9.5 cm, with central second and up-down dial, in mahogany case with gimbal mounting.
Asprey, London

332b Movement of Ditisheim's chronometer, gilded two-day movement without fusee, direct barrel drive, chronometer spring detent escapement, balance assembly and escapement in a small box with protective glass lid, replaceable by a similar system with lever escapement, Guillaume balance, extra wheel on the arbor of the fourth wheel for addition of an electric contact.

333a Marine chronometer, signed "Johs Raabe, Glashütte No. 102", ca. 1916-17, in wooden case with gimbal mounting, 56-hour movement.
Privately owned

333b Movement of Raabe's chronometer, integral balance, chronometer spring detent escapement, helical hairspring, chain and reversed fusee.

333c Dial of Raabe's chronometer no. 102, 56-hour movement.

333d Rear of the movement of Raabe's chronometer no. 102.

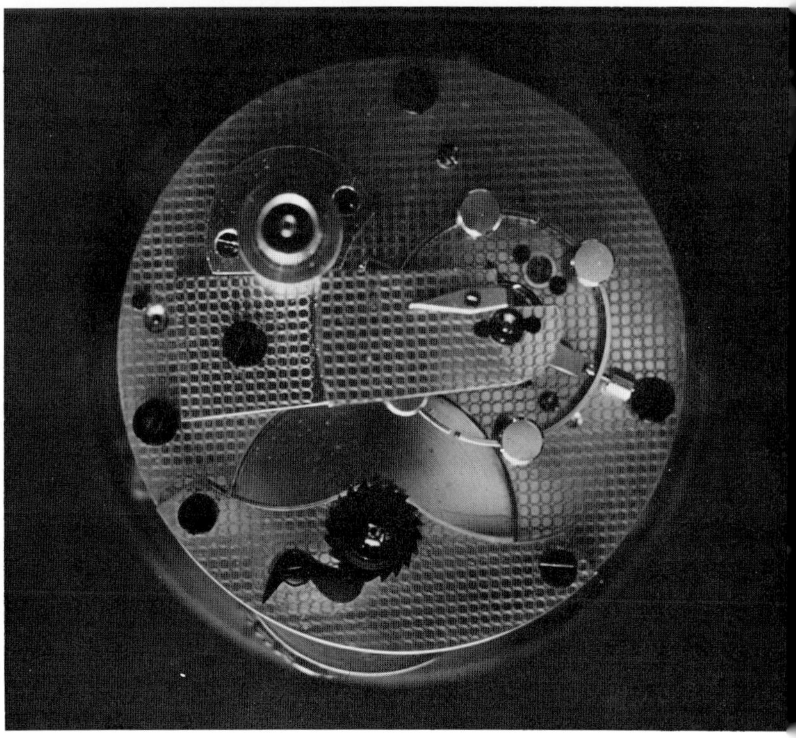

334a Dial of a marine chronometer signed "HORO No. 50819", ca. 1940; dial diameter 104 mm, winding indication and second dial.
French Navy

334b Movement of the HORO chronometer, with chain and fusee, integral balance, helical hairspring, chronometer spring detent escapement.

335a Marine chronometer in mahogany case with gimbal mounting and additional carrying case, signed "G. Gerstenberger, Glashütte, SA No. 56", ca. 1920.
Privately owned, Germany

335b Side view of the movement of Gerstenberger's chronometer, two-day movement, chain and reversed fusée, integral balance and helical hairspring, chronometer spring detent escapement.

334 △

334b △

335a ▽

335b

336a Marine chronometer in mahogany case with gimbal mounting, signed "P. Ditisheim/Solvil Genève, No. 1031", ca. ?, dial with winding indicator by the XII and second dial between V and VII.
Atwood, Time Museum, Rockford, Illinois

336b Back of the movement of Ditisheim's chronometer, with two-armed compensated balance, Breguet hairspring, lever escapement.

337a Marine chronometer in wooden case with gimbal mounting, signed "Paul Ditisheim, No. 2, La Chaux-de-Fonds", ca. 1925
Privately owned

337b Movement of P. Ditisheim's chronometer varying completely from typical chronometer con struction, with interesting electric auxiliary device whose working program and control cannot b determined — apparently a direct electromagneti power, divided two-armed balance with numerou adjusting screws.

337d Diagonal view of the movement of P. Di sheim's chronometer, with a different view of t contact system.

337c Side view of the movement of P. Ditisheim's chronometer.

342a Marine chronometer in wooden case with gimbal mounting, signed "Franz Lidecke, Geestemünde, No. 463", ca. 1930-40; winding indicator by the XII, second dial between V and VII.
Babelsberg Observatory

342b Movement of Lidecke's chronometer, with chain and reversed fusee, simple two-armed compensated balance with blued hairspring, chronometer spring detent escapement.

343a Marine chronometer movement, signed "L. Leroy & Cie No. 1633, Paris", ca. 1936, 92 mm dial diameter, winding indicator.
French Navy

343b Movement of Leroy's marine chronometer, 49-hour movement, Guillaume balance with helical hairspring, chain and reversed fusee, chain broken, chronometer spring detent escapement.

338a Marine chronometer, signed "Paul Ditisheim, La Chaux-de-Fonds, No. 1072", dial diameter 10 cm, central second, 52-hour movement, ca. 1920.
Smithsonian Institution, Washington D.C.

338b Movement of Ditisheim's chronometer, movement diameter 8 cm, lever escapement, Breguet hairspring, two-armed compensated balance, fully shielded, direct barrel drive.

340a Marine chronometer in wooden case with gimbal mounting in second carrying case with leather strap, signed "Paul Ditisheim, La Chaux-de-Fonds No. 2010", ca. 1925; 50-hour movement.
Th. Beyer Museum of Timekeeping, Zürich

340b Movement of Ditisheim's marine chronometer with escapement and balance component removed.

340c Movement of Ditisheim's chronometer, two-day movement, removable balance assembly with lever escapement, Breguet hairspring, two-armed compensated balance.

339a Marine chronometer in wooden case with gimbal mounting, signed "F. N. Tietz, Kiel, No. 2107", ca. 1920; dial diameter 97 mm, up-down dial by the XII, second dial between V and VII.
French Navy

339b Movement of F. N. Tietz's chronometer, with chain and reversed fusee, integral (Griesbach) balance, helical hairspring, chronometer spring detent escapement.

341 Chronometer movement by A. Lange & Sons, Glashütte, Saxony, with integral balance, plate diameter 85 mm, ca. 1925, chain and reversed fusee, chronometer spring detent escapement.
German Museum, Munich

344a Marine chronometer in mahogany case with gimbal mounting, signed "Ulysse Nardin, Locle Suisse, No. 4233", 1940; dial diameter 100 mm, winding indicator, second dial by the VI.
French Navy

344b Back of the movement of Nardin's chronometer, 56-hour movement, chain and fusee, Guillaume balance, helical hairspring, chronometer spring detent escapement.

345a Marine chronometer in wooden case with gimbal mounting, signed "Ulysse Nardin, Locle, Suisse, No. 7680", ca. 1940; dial diameter 7 cm, steel hands, winding indicator and second dial.
Private collection, Bremen

345b Movement of Nardin's chronometer, integral balance with Breguet steel hairspring and regulation, chain and fusee.

344c Side view of the movement of Nardin's marine chronometer, with chain and fusee, 56-hour movement.

6a △ 316b ▽

347a △ 347b ▽

Previous page:

346a Nautical chronometer by A. Lange & Sons, No. 676, Glashütte, 1934, showing not the clock time but a scale of degrees; the "hour hand" turns from 0 to 360 degrees once in 24 hours, the "minute hand" from 0 to 60 degrees once in four hours, and the "central second" from 0 to 1 degree (= 60 degrees of arc) once in four minutes, small up-and-down indicator by the zero.
Privately owned

346b Movement of the nautical chronometer by Lange & Sons, chain and reversed fusee, integral balance with helical Nivarox hairspring, chronometer spring detent escapement, hand setting by a wheel on the case rim.

347a Marine chronometer, unsigned (Japanese?), ca. 1920, No. V-334, in mahogany case with gimbal mounting; dial diameter 10.2 cm, 56-hour movement.
Smithsonian Institution, Washington D.C.

347b Movement of the Japanese chronometer, Guillaume balance, chronometer spring detent escapement, chain and fusee, presumably by U. Nardin, See. #344.

348a Marine chronometer in wooden case with gimbal mounting and additional carrying case with strap, signed "Wempe, Hamburg, No. 2042", ca. 1940.

348b Movement of Wempe's chronometer, with chain and reversed fusee, 56-hour movement, integral balance with helical hairspring.

349 Movement of a marine chronometer by Wempe, No. 5366, ca. 1940, chain and reversed fusee, integral balance and helical hairspring, chronometer spring detent escapement.
Privately owned

a

350b

350c

350a Airtight marine chronometer from the Wempe Chronometerwerke, Hamburg, post-1940, developed especially for use in submarines to counteract the effect of strong deviations in air pressure that occur inside the boats, small second by the 6, up-and-down indicator by the 9, two-day movement.

350b Under-the-dial view of Wempe's airtight chronometer, the two strong wheels at the left are for winding, which is done as usual from the rear via this system.

350c Side view of the Wempe chronometer's movement; the large, wide pinion with the cone operates the up-and-down indication.

350d Movement of the airtight chronometer by Wempe; geared barrel, lever escapement, integral balance with helical hairspring.

350d

352a Dial of a marine chronometer in mahogany case with gimbal mounting, signed "Hamilton, Lancaster PA U.S.A. No. 6979, 1941"; steel hour and minute hands from the center, two separate dials for winding and second indication.
French Navy

352b Movement of the Hamilton chronometer, 56-hour movement, chain and fusee, chronometer spring detent escapement, undivided two-armed balance and helical hairspring.

353a Deck watch, signed "Hamilton, Lancaster PA U.S.A. 1942"; aluminum case, two separate crowns for winding and setting.
National Maritime Museum, Greenwich

353b Back of the movement of the Hamilton deck watch (Model 22) with notation: Temperature adjusted and regulated in 6 positions; completely closed balance ring, Breguet hairspring with regulator, fine setting via curved disc, lever escapement.

351a Marine chronometer in mahogany case with gimbal mounting, signed "Wempe, Hamburg, No. 9027" (with electric winding), ca. 1940, dial diameter 10.5 cm.
Private collection, Bremen

351b Movement of the Wempe chronometer; integral balance with helical hairspring, electric winding.

354a Marine chronometer in wooden case with gimbal mounting, signed "A. Lange & Sons, Glashütte bei Dresden, No. 5100", ca. 1940; hour and minute hands from the center, auxiliary dials for second and winding indication.

355a Marine chronometer in wooden case with gimbal mounting, signed "A. Lange & Sons, Glashütte bei Dresden, No. 1223", ca. 1939-40; auxiliary dials for second and winding indication.

354b Movement of Lange's chronometer, 56-hour movement, between barrel and fusee is a flat steel band instead of a chain, integral balance and helical hairspring, chronometer spring detent escapement.

355b Movement of Lange's chronometer, 35-hour movement, direct barrel drive without fusee, lever escapement, integral balance, helical hairspring, setting by the case rim.

356 Movement of a marine chronometer made for the Russian Navy by G.U.B., Glashütte, after the end of World War II. Two-day-movement, chain and reversed fusee, integral balance with helical hairspring, chronometer spring detent escapement. *Photo by Tony Mercer*

355c Side view of the movement of Lange's chronometer; the small wheel in the foreground is for hand setting.

357a Traveling chronometer with leather case, signed "Thomas Mercer, No. 1173", ca. 1970, with winding indication.
Mercer, St. Albans

357b Side view of movement of Mercer's chronometer, eight-day movement with chain and fusee, rear winding, two-part compensated balance with helical hairspring, chronometer spring detent escapement.

358 Chronometer movement by Thomas Mercer, ca. 1970; two-day movement with chain and fusee, two-part compensated balance and helical hair-spring, chronometer spring detent escapement.
Mercer, St. Albans

359 Chronometer movement by Thomas Mercer, ca. 1970, eight-day movement with chain and fusee, two-stage structure with iron protector for the two-armed compensated balance on the lower level, simple cock bored out to save weight.
Mercer, St. Albans

VIII: Tourbillons & Carousels

Tourbillon and carousel clocks form a small family of their own, called turning-frame clocks, for which many people have a very special esteem. They are clocks that have apparatus that does not allow a certain type of disturbance, the positional error, to have an effect.

In the heroic epoch, during the exhaustive testing of Harrison's No. 4 at Greenwich, it had been realized that the position in which a clock was placed had an influence on its accuracy, that is, that even in cases of great movement precision, differences appear according to whether the clock is standing, lying or in an angled position during the observation period. At first there were only unclear ideas about the origin of such positional errors, but soon it was seen clearly that it was gravitational errors in the most sensitive part of the balance that were responsible for the position error. Illustration 360 shows in schematic fashion how a gravitational error in the balance, at different positions and through the resulting different angles between the center of gravity and the moving parts of the hairspring, makes for different power conditions and thus also somewhat different oscillating conditions.

It had been recognized very early that a good poise (as precise an identity of the balance's center of gravity with its turning axis as possible), is of vital importance to its running, and that the poise of the balance is an important phase of the setting or regulating of the clock. But this precision of poise has its limits, and in addition, it is unavoidable in all combined systems, particularly in balances with bimetallic arms, that over the course of time changes will occur in the overall center of gravity. On the one hand, minimal gradual displacements in the construction can be the cause, but on the other hand, it must not be forgotten that metal volumes can move over long periods of time, especially when alloys are involved. Thus it can occur that very precisely round wheels warp and become elliptic over years and decades, because alloys can change more in one direction than in another under certain conditions.

To eliminate the position error completely, A. L. Breguet had the idea as early as 1785 to make the critical part, the balance assembly which is fixed in a definite position to the body of the movement by the hairspring attachment to the movement plate, rotate around a horizontal axis as opposed to the body of the watch by means of a special auxiliary apparatus and thus eliminate the gravitational error in the balance. It is particularly observation watches, such as deck watches and pocket chronometers that are worn on the body, that were predestined for Breguet's manner of eliminating the position error, since in them the plane of the balance is generally vertical, but when held in the hand the XII can be in various positions; in marine chronometers, where the plane of the balance remains basically horizontal thanks to their usual housing of the movement in a case with gimbal suspension, gravitational errors can scarcely occur, so that marine chronometers hardly come into the picture as far as the use of turning frames goes.

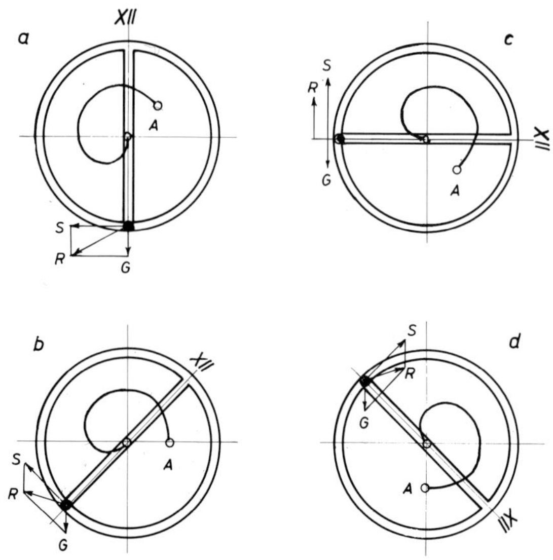

360 The position error: In the balance rim, even with careful retention of the circular form, a somewhat one-sided division of mass can occur. The center of gravity of the disturbance in the even distribution of weight is raised by one strong point, it possesses the weight G, so that a corresponding power affects it in the direction of the lead.

If the axis of the balance is horizontal, then when the watch takes different positions in the pocket — depending on whether the XII on the dial is upright or to the side (various possibilities are indicated by a through d) — the resulting R from G and the hairspring tension S occur, thus the oscillating behavior from point to point changes, and thus the beat of the balance and the going rates of the watch in different positions deviate from each other.

361 Breguet's tourbillon, original heavy construction, cross-section:
d Fixed bottom plate
k Cage
f Pinion for cage
m Drive from movement (3rd wheel)
g Escape wheel
t Pinion for escape wheel
s Fixed crown wheel, concentric to axis of system
u balance
(after Pelleton)

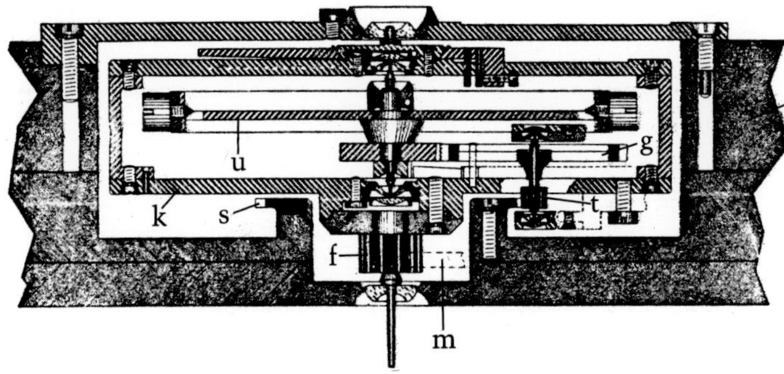

362 Breguet's tourbillon, modern, lighter construction after A. Helwig:
c Front plate, fixed
d Bottom plate, fixed
k Cage
f Pinion for cage
e Drive from movement (3rd wheel)
g Escape wheel
m Pinion for escape wheel
s Fixed crown wheel, concentric to axis of system
u Balance
(after Helwig)

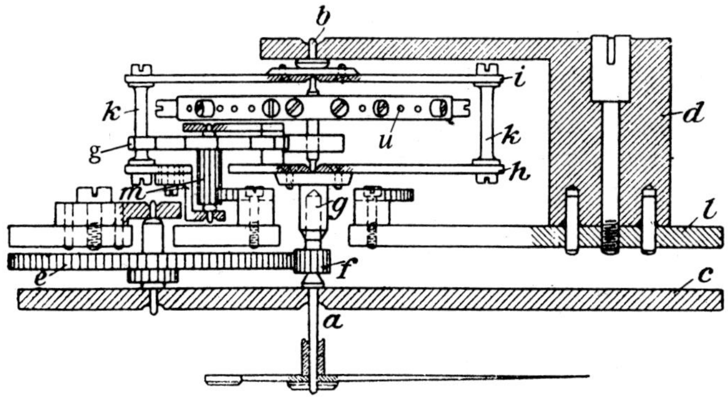

Breguet had received a specific patent in 1791, at the time of the French Revolution, for changes in the balance position, in which further advantages of his new idea were the avoidance of one-sided bearing wear and the even lubrication of the balance pivots.

As simply as the turning-frame principle can be explained as it has been above, its practical execution is very complex. Illustration 361 shows the design of Breguet's original tou the design is that not only the balance but also the escapement has to revolve around the balance axis. A tourbillon needs three types of motion: (a) the rotation of the cage, which contains the balance assembly and escapement, around the balance axis, (b) the oscillating of the balance, and (c) the turning of the escape wheel. In tourbillons made after A. L. Breguet, as seen in #361, the third wheel of the movement is used to turn the cage, and at the same time the pinion of the escape wheel is turned by a fixed wheel concentric to the cage axis (this corresponds to the fourth or second wheel in other clocks) and thus creates the turning motion for the escape wheel.

A. L. Breguet fortunately used a rotating time of one minute in his tourbillons; thus he was able to attach the second hand to the axle of the cage. When 15 teeth are used for the escape wheel and one second includes 2.5 full or 5 half swings, then the cage, in rotating once a minute, gives out 300 impulses per minute or five per second (providing a lever escapement is used). This high number of impulses includes a great inherent difficulty in the elimination of the position error. Since the cage holds the balance, the escape wheel with its pinion and the block needed for it, plus the entire escapement system, the mass that is exposed to the impulse at every move of the escape wheel is much greater than in normal pocket watches, where there is only a second hand in addition to the usual wheels. The succession of the four greater individual impulses is, on the one hand, a greater output of energy, and secondly, it causes greater wear on the moving parts, approximately in proportion to the increased mass.

The beginning and end of tourbillon construction is thus the need to make the cage as light as possible, but at the same time to make it sufficiently firm and stable. This is a great challenge to fine mechanics. Elementary, simple rules for an optimal solution do not exist; here the fine mechanical expertise must rise to the level of a fine mechanical art, for the quality of the functioning tourbillon is the practical proof.

Design, Production and Special Installation of a Tourbillon

Even in production, these factors make greater demands than those of normal pocket-watch construction. This makes it understandable that the materialistically minded merchant, thinking only of production runs and profits, has little interest in tourbillons. A. L. Breguet was also an outstanding businessman in his day, but he also built a good number — something over thirty — of tourbillons in order to document his great ability. Since then, many important watch manufacturers have taken great pride in producing good and reliable tourbillons. Discriminating customers have always appeared too, who were ready to pay the considerably higher prices, which were unavoidable in view of the construction. The fact that the high expenditure also pays for itself in terms of precision is seen in, among others, the fact that

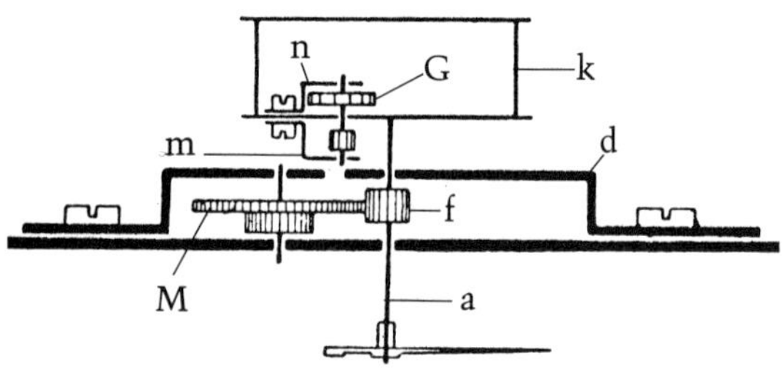

363 Basic design for a flying tourbillon, using the design of the German Watchmaking School at Glashütte, Saxony:
a Pivot for second hand, carries cage
k Cage
d Fixed housing
M (3rd) intermediate wheel
f Second pinion
G Escape wheel
n Upper cock for escape wheel
m Lower cock
T Pinion for escape wheel
(after Helwig)

364a Five-minute tourbillon from the German Watchmaking School, Glashütte:
a)Back of the movement, two-barrel drive;
b) Cage with built-in escapement and balance, the latter shown only in part.
(after Helwig)

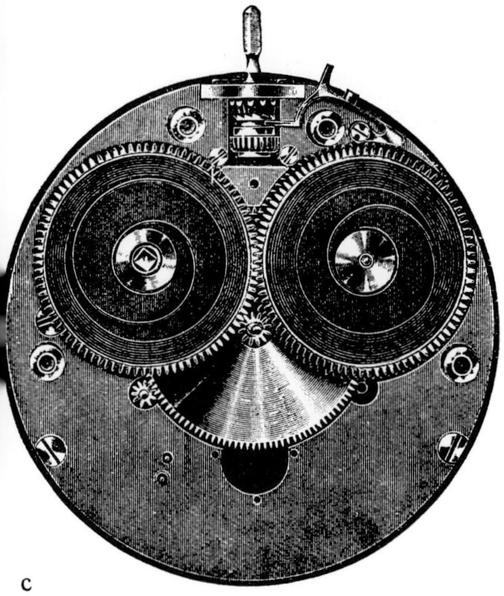

c

d

365

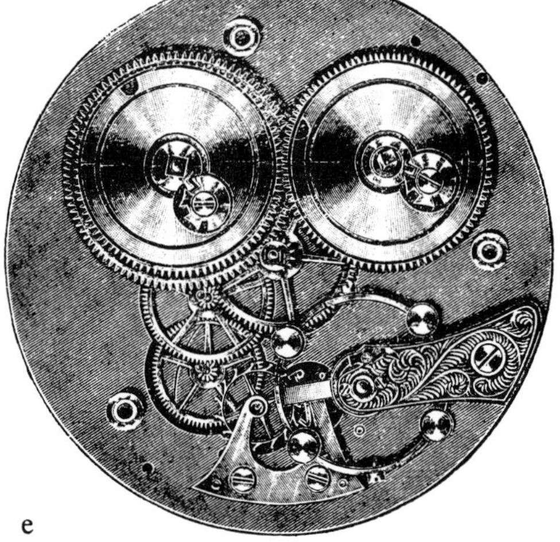

e

c) Movement with top plate and barrel cover removed, winding by the arbor of the left barrel, the teeth of the two barrels mesh, transmission to the movement via the arbor of the right barrel through the large wheel seen in d, which is fixed to the arbor of the right barrel.

d) Movement with top plate and both barrels removed, at right is the contact of the first wheel with the pinion of the minute wheel (arbor in middle of movement with pinion and large disc); minute wheel turns pinion at lower left at ratio of 1:12.

365 Rear of a typical pocket watch with two barrels working in sequence; the Maltese cross positions to limit spring running are seen above. *(after Helwig)*

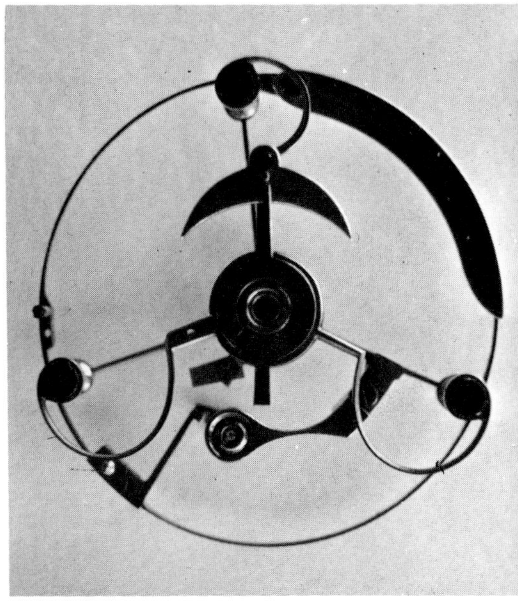

366 △

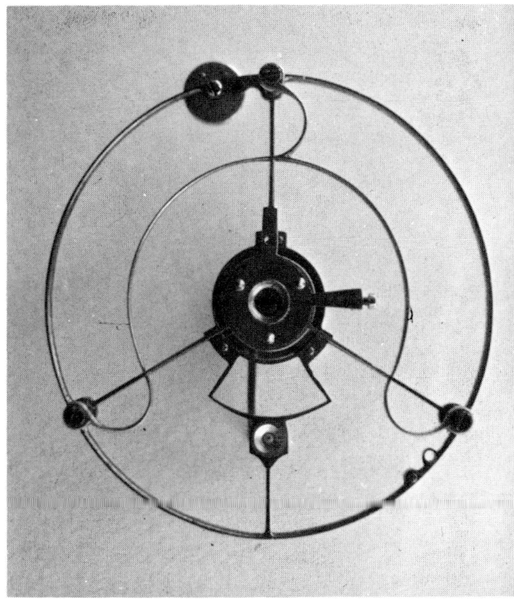

367 △

368 ▽

366 Turning frame of a flying tourbillon, a relatively heavy version.

367 Turning frame, an improved version after A. Helwig. A balancing counterweight, at left near the upper arrow, is still necessary.

368 New design for a floating-bearing turning frame with skeletal steel tower. The outer wheels are gone and the two pillars (above) simultaneously form the counterweights for balancing. *(R. Miklosch)*

369 Pocket chronometer by Girard-Perregaux, with tourbillon after Breguet, chronometer pivoted detent escapement.

370 Gold pocket chronometer by N. Nielsen, built for Frodsham ca. 1920. Classic tourbillon with lever escapement. This watch received 83.1 points of a possible 100 in the 1922 chronometer test.

371 Pocket chronometer with flying tourbillon, made by Alfred Hübner at the Watchmaking School in Glashütte in 1931. Chronometer spring detent escapement, two sequential barrels, up-and-down indication.

369 △

370 △

371 ▽

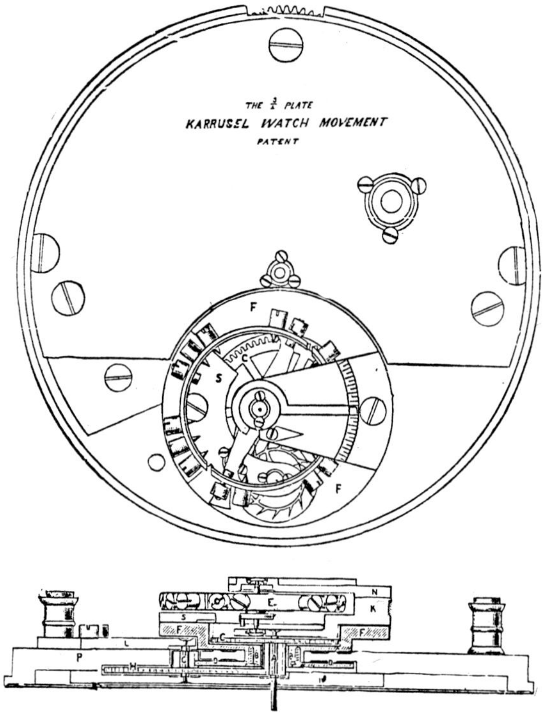

372 Carousel watch after B. Bonniksen, produced as of 1894 in three sizes: 14, 16 and 18 lines:
a) View of the bottom plate
b) Cross-section of movement and plate
p, p Bases
H, G 3rd wheel with pinion
M Cover plate under dial
L Cock for 3rd wheel
D, D Drive wheel for carousel plate
G Pinion for D, linked with H
E Balance
F, F, B Carousel system, also carries escapement
K Cock for 4th wheel with bearing for balance pivot
S Cock for escapement (not shown on b)
N Balance cock
(after B. Bonniksen, 1905 catalog)

since the introduction of the tourbillon, tourbillon watches have won the highest and best honors at public tests, with running results generally twice as good as those of the best normal chronometers.

In Breguet's design, a double bearing with a bridge or cock over the cage is used (#361-362); Studienrat A. Helwig of the German Watchmaking School at Glashütte, Saxony, developed cages with flying bearings (#363) at the beginning of this century, for particular lightening and reduction of the cage mass. Very interesting work has been done in this area at Glashütte, and Helwig has described it in detail.[1] In #363 and 364b the principal layout and the turning frame are shown, while #364a, b and c help to explain the two-spring design that the German Watchmaking School at Glashütte likes to use in its tourbillons.

A. L. Breguet had his tourbillons and the greater part of his special movements and their components made in the Swiss Jura, so it is not surprising that some of these Swiss masters produced tourbillons on their own after Breguet's death. Above all, J. F. Houriet (1768-1830), one of Breguet's most noteworthy collaborators, did so, and also Frédéric Louis Favre Bulle (1770-1849), Louis-Jean Richard, Guinand, Grether, Roulet, Albert Pellaton-

Favre, Ch. F. Perret, James C. Pellaton, Georges Pellaton, Marc Inäbnit, André Bornard, Marcel Wuillemier, and others later in the next century.

Many of these masters were also notable teachers at the Swiss watchmaking schools. Independently of these schools, Studienrat A. Helwig of the German Watchmaking School at Glashütte later devoted great personal interest in tourbillons and greatly developed this art.

In the course of the development of turning-frame watches at the German Watchmaking School in Glashütte, the desire to produce enough energy to drive them led to the use of two springs working together (#364, 365). Helwig adopted a remarkable means of using two springs working sequentially, unlike that of A. L. Breguet in which two springs worked simultaneously. In the 18th Century, Thomas Mudge first used two springs working in series in his first chronometer (#23). The cooperation of two or more springs is an idea that has been tried again and again in practice for many years, but the principles have never been analyzed basically in mechanical terms as to their advantages and the peculiarities of their functioning. At the beginning of the 20th Century it was also used occasionally in gramophones to turn records for long periods, where at times more than two springs were linked together. In the serial linkage used at Glashütte, special locks in Maltese cross form were used, similar to those already used by Breguet, to make use of relatively short springs; this certainly afforded a very constant force but, on the other hand, reduced the total available energy so much that the final decision between a gain in constant force and greater expenditure for spring volumes had to be made.

A. Helwig's investigations have also concerned slowly turning tourbillons, a variation that Breguet had also tried. Here the cage turning speed is reduced; if one changes from a one-minute to a five-minute (half-tourbillon) type, as in the movement shown in #364a-b), then this gives the same effect in use, basically speaking, as the one-minute tourbillon — for in such short times different positions of the watch are not to be expected. But such a reduction of the cage turning speed reduces the strength of the impulses to the system by one-fifth, so that the five-minute design can be built somewhat more simply, though such turning-frame watches are also heavier.

Despite the significant running achievements of tourbillons and their inherent power technology, and despite the undying interest of those who admire them, the literature on tourbillons is sparse. A. Helwig has written a thorough book on them, and J. Pellaton, later the director of the school in Neuchâtel and also the son of the very renowned tourbillon builder Albert Pellaton, has published his experiences with tourbillons in a brochure.[3]

The Carousel Principle

A further step in the same direction as the five-minute tourbillons consists of the carousels, whose basic idea goes back to Bonniksen (#372), who built a number of these watches —likewise classified as turning-frame watches — in England at the end of the 19th Century. They are essentially turning-frame systems with 52.5-minute rotating time (there are also those that turn in 34 minutes), in which the carriage can be built considerably more heavily and the drive can be simplified. In the carousel system the two turning motions are worked into the rotating system separately of each other. Bonniksen[4] published a catalog in 1904, showing the types built by him, describing the carousel principle, and including several test certificates. The detached lever escapement, as shown in #372, is generally used.

The Cinematics of the Tourbillon Watches

Common to all turning-frame watches is the necessity of having two different speeds affecting the balance assembly: that which drives the escapement and that which turns the entire system to eliminate the position error. Thus it is necessary to use special gears, specifically planetary gears,[5] such as epicyclic and differential gears. On the one hand, these allow two different speeds to be combined into one, as was often done in the 16th and 17th Centuries in astronomical clocks[6] for astrolabes and other calendar indications, a technique that was even a requirement for a master's certificate in some cities (but that is largely forgotten today); on the other hand, they are used to divide one turning motion into two different rotating speeds, such as are required for the turning-frame system.

Special Requirements of Tourbillons

These are described by A. Helwig as follows:
The escapement springs (called spring detent by Helwig) in turning-frame watches make special demands; they must be extraordinarily strong and yet lie in their resting position with moderate tension; then the locking jewel can be set with very little pressure. One must not overlook that this pressure is something very different in a chronometer escapement from a lever escapement. The detent is kept at rest by the strength of the spring, and the firmness of the locking jewel is already assured in that only a very small angle of pressure needs to be given, quite unlike the lever escapement, where it is only this pressure that holds the lever at its limit. A large pressure angle in a tourbillon may not look right, as one recognizes the escape wheel's moving back to the turning frame all too clearly by an unpleasant twitching. The unlocking of the teeth of the escape wheel must take place remarkably quickly, since the tips of the teeth in all chronometer running wheels, especially those of the tourbillons, are strongly exposed to abrasion. Sometimes steel wheels are used in turning-frame watches. This is done to decrease the wear on the tooth tips, and the slight inconvenience of having to oil the mechanism is accepted gladly; changing the oil is insignificant compared to having the escape wheels worn down too fast."

The problems described here have impelled the watchmaker who thinks in terms of improving the tourbillon or carousel principle to various further developments.

Naturally these systems are always of great practical

significance when the watch is carried in a pocket, so that the watch's vertical position can be turned slightly to the left or right. Such observation watches (deck watches) and pocket chronometers are gladly used at sea as precision watches for carrying time from the ship's chronometer to the place of observation, since their high level of precision and reliability are of great significance for this task.

Variants and Further Development

The refined concept of the turning frame, of having the particularly sensitive element — the balance — that is so decisive to the watch's precision rotate, has often moved watchmakers with technical imagination to think out, and sometimes also to work out, variants and further developments of the basic concept. Such steps are, of course, considerably rarer to find and harder to understand than the many variations and further developments of, for example, the lever escapement, automatic self-winding or calendar movements, but because of the fascinating idea of the turning frame, they are of particular interest to all who take a deep engineering interest in clocks and chronometers.

For further development of the turning-frame concept there are several possibilities of using other axles in the movement than that of the second hand — as was done in Breguet's one-minute tourbillons — as carriers of the turning-frame apparatus, the cage. Sometimes the third wheel was used as the carrier, both by Breguet and at the German Watchmaking School in Glashütte, while in Bonniksen's carousel the second wheel was used, which increased the rotating time of the turning frame to about one hour.

A particularly refined arrangement is the Benoit turning frame, which amounts to a carousel rotation with the speed of the escape wheel but requires absolutely no additional masses and thus involves no additional weight impulses. A. H. Benoit was the director of the watchmaking school in Cluses in his day (circa 1865) and worked out the principle illustrated in #373a-b-c. The essence of this process is that the dynamics of the balance alone, without additional coupling elements, accomplishes the unlocking and locking of the escape wheel via a two-armed detent with hooks. In Benoit's design, the balance is on the same arbor as the escape wheel but can turn against the escape wheel thanks to a special intermediate bearing, but is also held fast to the escape wheel itself by the lower end of the balance spring. Thus at first the balance is pulled along by the escape wheel, but when the wheel (L) is stopped and held by a tooth, it can swing farther out in a clockwise direction. Benoit designed the process so that the outward swinging of the balance simultaneously takes control of the escape wheel with the two pallets, L and L', seen in #373b, which are mounted on a lever system turning around pivot A. The escape wheel is thus held by hook L as long as the balance swings around to the right. Thanks to the intermediate balance spring, while the escape wheel stands still, the balance can also swing back (counterclockwise). In this swing back, a moment will finally occur in which, resulting from reversed pressure on the balance spring, which is now pushed backward, the escape wheel turns back a little bit and tooth d of the resting hook L behind it projects as far as the limiting pin P_1 allows. The spring R, working with weak pressure, lies lightly against the lever system and holds it fast, so that the escape wheel and the balance can now swing forward. In the process, the balance finally receives the necessary force to make the escape wheel move forward one tooth, until tooth d, via release hook L', again pulls the resting hook L into the escape wheel, this time as far as the second limiting pin P_2 allows it to. Now the process of self-controlling repeats itself as described above.

The Benoit system can, as Helwig stresses, obtain good locking and even sufficient pressure without having the slightest influence on the ease of the unlocking, for the escape wheel moves backward itself out of its generous locking position, since it is pushed backward by the balance spring which is tensed backward during unlocking.

As far as we know, the German Watchmaking School in Glashütte made one model of the Benoit turning frame for its collection of instructional models, but whether Benoit himself ever made a watch according to his principles is not known. In very recent years, though, a turning-frame watch according to Benoit's system and with a Breguet hairspring was made in Switzerland, and is said to function well.

The Biaxial or "Torkel" Tourbillon

Thinking out Breguet's idea farther logically leads to rotating the balance assembly and the escapement on two axes, because in this way all possible gravitational effects caused by moving a carried watch out of the vertical dial plane are eliminated. In 1979 this path was taken by the unusually talented Anthony Randall of Birmingham. The basic arrangement is shown in #374. The fact that all external magnetic influences on the balance assembly can be eliminated by such a double axis is an advantage of this design that is not to be underestimated.

Karlstein Masterpieces

The production and adjustment of turning-frame systems, as has often been noted, make considerable demands on the watchmaker, but the attainment of the goal of making a precisely working tourbillon gives particular fulfillment and strengthens the self-reliance of the student who wants to become a master, as has often been stressed by Master Becherer, Senior Instructor at the Austrian National Trade and School for Watchmakers at Karlstein on the Thaya, in Lower Austria. Thus this school has also taken up the designing and making of turning-frame systems in its work program, to the extent that talented students make a tourbillon as their masterpiece during their last year at the school. Illustrations 375 and 376 show two fine achievements of recent years along with their movement drawings.

1. Alfred Helwig, *Drehganguhren*, Verlag der Deutschen Uhrmacherzeitung, Berlin, 1927.
2. Richard Miklosch, *Die Entwicklung des fliegenden Drehgestells in Glashütte/Sachsen*, A. U. 2/80.
3. James C. Pellaton, *Le Tourbillon*, Annales Françaises de Chronometrie, 1951.
4. B. Bonniksen, *The Karussel Watch*, Reprint by Malcolm Gardner, Kent, 1905.
5. Fr. Assmus, *Technische Laufwerke einschliesslich Uhren*, Springer-Verlag, Berlin-Göttingen-Heidelberg, 1958.
6. H. Bertele, *The Origin of the Differential Gear and its Connection with Equation Clocks*, Transactions Newcomen Society XXX, 1955-56, 1956-57.

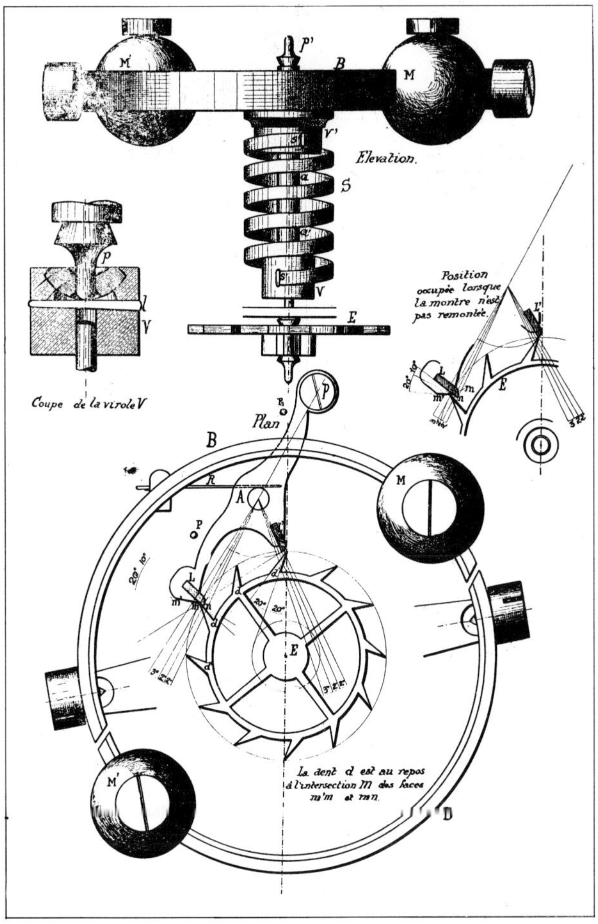

373 Highly refined turning-frame system after A. H. Benoit, 1862:
a) Side view:
E Escape wheel
V Balance arbor
B Bimetallic compensated balance
M, M Balance weights
S Hairspring
Z Intermediate piece, both mount for lower balance pivot and holder of lower end of hairspring, linked centrally with escape wheel.
b) Drawing of Benoit's design:
W Control detent
A Bearing pivot
L' Discharge hook
L Resting hook (stone)
P1, P2 Curb pins
R Braking spring for detent
A and R are fastened to the body of the movement;
c) Intermediate piece Z (locking piece), enlarged, with bearing and cap stones. *(after Helwig)*

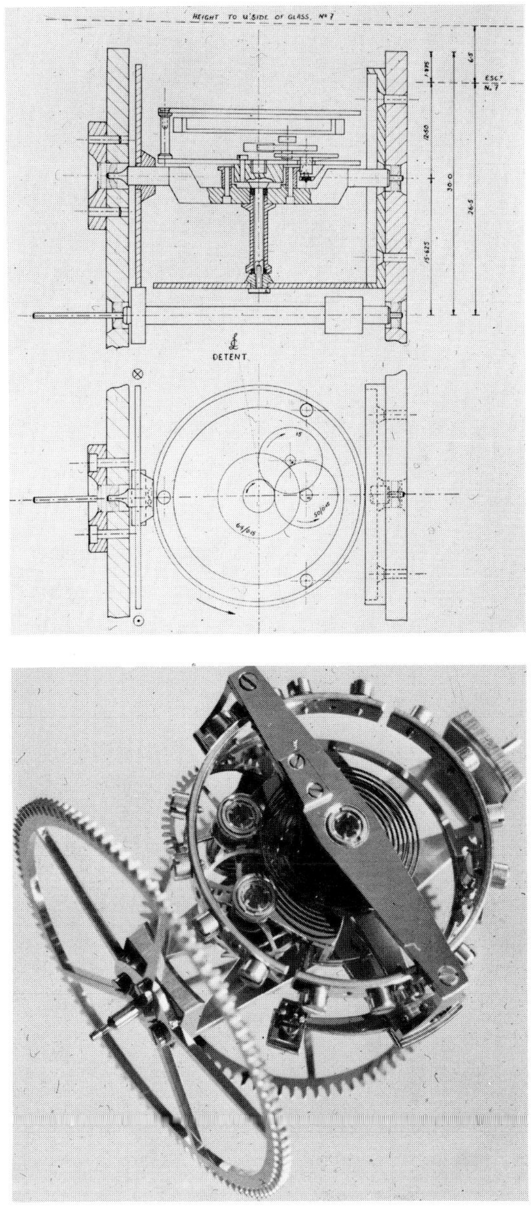

374 Basic design of a biaxial (Torkel) tourbillon: the two axes of rotation WX and YZ are perpendicular to each other. Power is transmitted in 1 by wheels 2 and 3, causing turning around axis WX; crown wheel 4 is centrally attached to axis WX on the housing and, when axis WX turns, transmits the motion via wheel 5 to turn axis YZ, by which the escapement and balance are turned. *A. G. Randall (British patent application, 1978)*

375 Factory drawing of a tourbillon:
a) Top view
b) Cross-section
1 Second arbor
2 Second pinion
3 Second wheel
4 Cup
5 Front plate
6 Brass cocks
7 Bottom plate
8 Escape wheel
9 Lever
10 Balance
11 Hairspring
12 Balance cock
13 Cap jewel
14 Turning
15 Lower cock
16 Upper cock
17 Curb
18 Anchor lever
19 Pillar (short)
20 Novostone bearing
21 Regulator indicator
22 Balance bearing jewel
23 Small cock
24 Escape wheel pinion
25 Pillar (long)
26 Pillar for cock
27 Third wheel
28 Minute wheel
29 Screw for second pinion
30 Screw for cup
31 Bearing jewel for second pivot
Karlstein Watchmaking School

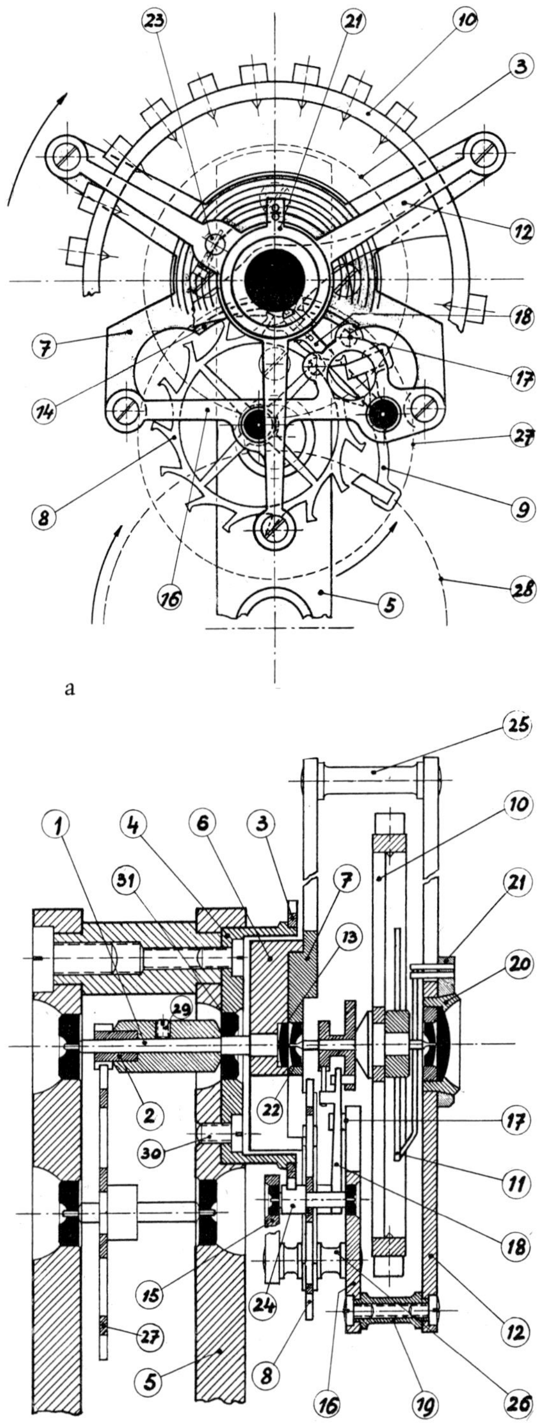

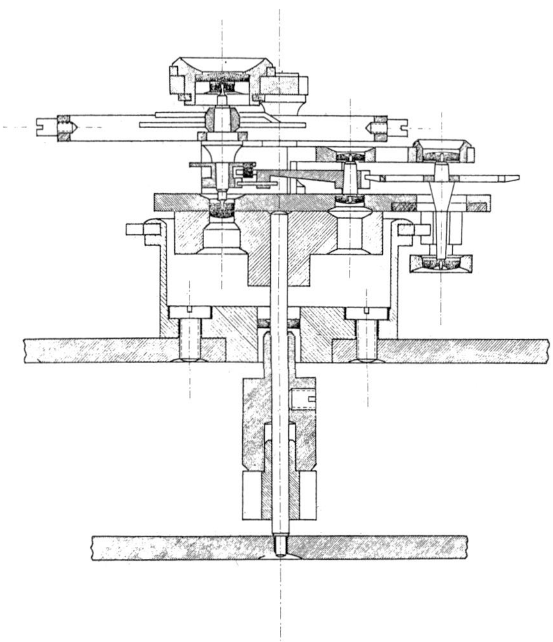

376 Factory drawing of a tourbillon (Masterpiece of 1966-67). *Karlstein Watchmaking School*

377a Movement of a pocket chronometer with four-minute tourbillon, signed "Breguet et fils, No. 2555", Paris, ca. 1809; chain and fusee, chronometer spring detent escapement (Peto's cross detent), three-armed bimetallic balance, Breguet hairspring. *Time Museum, Rockford, Illinois*

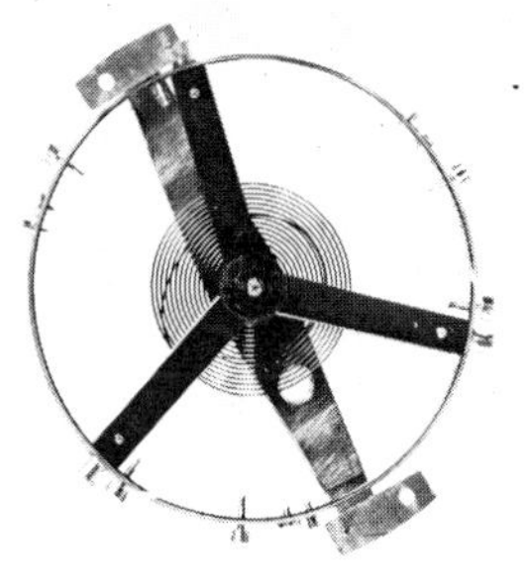

377b Movement with tourbillon removed. The tourbillon is driven via a second large pinion (upper left in #377a) to the second arbor and the teeth of the outside wheel on the bottom plate of the turning frame. Concentric to the lower turning-frame bearing is the fixed crown wheel, on which the pinion of the gear in the turning frame turns.

377c Turning frame with balance removed. The hairspring (underneath) with the opposite discharge spring can be seen clearly.

378a Pocket chronometer with one-minute tourbillon by J. F. Houriet (1743-1830), Switzerland, ca. 1820. Gold case, white enameled dial with vendor's signature: "Du Bois et Comp."; small dials for seconds (right) and hours (left), minute from the center, thermometer with Réaumur scale below.

378b Movement of the Houriet pocket chronometer, drive via chain and fusee, one-minute tourbillon with chronometer spring detent escapement, heavy chronometer balance with segment-shaped weights, spherical hairspring (invented by Houriet).

378c Under-the-dial view, bimetallic strip of the thermometer at right, maker's stamped FH 58 at lower left.
Th. Beyer Museum of Timekeeping, Zürich

378d Enlarged view of the Houriet tourbillon; one revolution per minute, the spherical hairspring is easy to see, on the balance rim are fine markings for balanced positioning of the segment-shaped balance weights.

379a-d Pocket chronometer with one-minute tourbillon by Louis Jürgensen (1806-1867), signed on the dial: "Urben Jürgensen Sönner." This watch exactly resembles Houriet's watch (#378) in case style, hand arrangement and all construction details.
Watch Museum, La Chaux-de-Fonds

380a Chronometer movement with one-minute tourbillon, signed: "Ecole d'Horlogerie La Chaux-de-Fonds", diameter 70 mm, height 31 mm.
Watch Museum, La Chaux-de-Fonds

381 Pocket chronometer with one-minute tourbillon, signed: "Albert Pellaton-Favre, Le Locle, No. 71." Diameter 53 mm, height 17 mm. Chronometer spring detent escapement, Breguet hairspring and regulation, geared barrel, crown winding, winding wheels with "wolf's teeth."
Watch Museum, La Chaux-de-Fonds

380b Enlarged view of the tourbillon from the watchmaking school at La Chaux-de-Fonds; chronometer pivoted detent escapement, bimetallic balance, Breguet hairspring and regulation, geared barrel, key winding.

383 Pocket chronometer with one-minute tourbillon by Sidney Better, London, under contract to the "Northern Goldsmiths Co. Newcastle upon Tyne", made ca. 1920. Swiss lever escapement, free-sprung balance, geared barrel, crown winding.

384 Pocket chronometer with "flying" one-minute tourbillon by Hans Apel, Glashütte, 1926; silver case; chronometer spring detent escapement, Breguet hairspring and regulation, driven by two sequential barrels, differential drive for up-and-down indication, crown winding.
Th. Beyer Museum of Timekeeping, Zürich

385 Chronometer movement with one-minute tourbillon, as demonstration piece by James Pellaton, Le Locle, made for Ernest Liniger, Brussels, in 1946. Diameter ca. 100 mm; chronometer pivoted detent escapement, Breguet hairspring with regulation, chain and fusee.
Privately owned.

382 Pocket chronometer movement (Savonnette version) with one-minute tourbillon in typical Pellaton style, finished ca. 1950 from a raw movement of ca. 1910 by James Pellaton, the son of Albert Pellaton (1832-1915). Chronometer spring detent escapement, Breguet hairspring and regulation.
Th. Beyer Museum of Timekeeping, Zürich

386a Movement of a deck watch with 34-minute carousel (after B. Bonniksen's patent), signed "S. Better, 4 Sandford Buildings, London E2", central second, diameter 62 mm.
Privately owned

386b Diagonal view of the deck watch movement by S. Better, 34-minute carousel, chronometer spring detent escapement (See #383, tourbillon with lever escapement (!) by S. Better), bimetallic balance, free-sprung balance Breguet hairspring.

386c Wheel train of the carousel watch by S. Better; drive by geared barrel, turning frame driven via a second large pinion on the third wheel arbor, second wheel in center, from which the wheel concentric to the turning frame is driven.

387 Pocket-watch movement with special "carousel" design which, unlike Bonniksen's "flying" design (see #386), has the upper bearing of the carousel in a cock, signed "Henry Hughes & Son Ltd. 59 Fenchurch St. London No. 1792/1534." English ("pointed-tooth") lever escapement, bimetallic balance, free-sprung Breguet hairspring, geared barrel, crown winding.
Time Museum, Rockford, Illinois

388 Movement of a pocket watch with "carousel" with upper bearing in a bridge, signed "Carley & Clémence Ltd. 30 Ely Place, London E. C. No. 57618." English lever escapement, bimetallic hairspring, free-sprung Breguet hairspring, geared barrel, crown winding, case with London hallmark of 1903.
National Maritime Museum, Greenwich

389 Pocket chronometer movement with "flying" six-minute tourbillon, made by A. G. Randall in 1961, on the basis of a raw movement by J. White. Turning frame driven via a large pinion (right front) on the second arbor and the toothed rim of the bottom plate of the turning frame (See #377). Swiss lever escapement, bimetallic balance, free-sprung Breguet hairspring, geared barrel, crown winding.

390 One-minute tourbillon of "flying" type with Swiss lever escapement, cage made of very rigid aluminum alloy weighs 0.64 grams; monometallic balance, self-compensating hairspring and regulation. Made by the master class of the Austrian Trade School at Karlstein in 1956.

391 One-minute tourbillon of "flying" type with Swiss lever escapement in "straight line", mono-metallic balance, self-compensating hairspring and regulation. Made by the master class of the Austrian Trade School at Karlstein in 1966-67.

Bibliography

Abbreviation: AH = The Antique Horological Society, London

Abeler, Jürgen, Meister der Uhrmacherkunst, Wuppertal, 1977.

Airy, George Bidell, "On the Disturbances of Pendulums and Balances and on the Theory of Escapements", in: *Transactions of the Cambridge Philospohical Society*, 1827, Vol. III, pp. 105-129.

Aked, C. A., "Longitude and the Marine Chronometer", in: *AH*, June & September, 1974.

–, "Early Timekeepers at Sea", in: *Antiquarian Horology*, 1969.

Allix, Charles and Brusa, Giuseppe, "Julien and Pierre Le Roy", in: *AH*, March, 1971.

Amis, P., "The Bounty Timekeeper", in: *Precision Timekeeping*; Monograph 3 of the AH Society, London.

Anders, Conrad, "Wer war eigentlich James Cook? Seefahrer und Entdecker aus Passion", in: *Die Presse-Spektrum*, Vienna 17-18, II, 1979.

Andrewes, W., "John Harrison, A Study of his early Work", in: *Hor. Dialogues* I/1979, New York.

Anonymous, "Ein sehr interessanter Zeitmesser mit Spindelhemmung", in: *Neue Uhrmacherzeitung* XIII, 1959, pp. 6-8.

Anonymous, "Outline of Principal Inventions", in: *Journal of Natural Philosophy, Chemistry and Arts*, London, 1806.

Appianus, *Cosmographia*, Landshut, 1524.

Appleby, L., "A Chronometer by Breguet, Neveu et Cie No. 5053", in: *Antiquarian Horology*, March, 1865

Assmus, Fr., *Technische Laufwerke einschliesslich Uhren*, Berlin-Göttingen-Heidelberg, 1958.

Auricoste, J., "Au sujet d'une restauration de l'horloge marine de Piere LE ROY", in *Revue Chronometrique*, Paris, 1939.

Baillie, G. H., *Clocks & Watches*, London, 1947.

–, "The 1730 Harrison M. S.", (Notes on this Historic Find), in: *Hor. Journ.*, July, 1950.

–, *Watchmakers and Clockmakers of the World*, London, 1947.

Bateman, D. A., "Vibration Theory and Clocks", in *Hor. Journal*, July 1977, Jan., 1979.

Baugh, A. Daniel, "The Sea-Trial of John Harrison's Chronometer 1736", in: *The Mariner's Mirror*, August 1978, National Maritime Museum, London.

Berner, G. A., *Dictionaire Professionelle illustré de l'Horlogerie, Chambre d'Horlogerie*, La Chaux-de-Fonds, 1961.

Bertele, H., "Precision Timekeeping in the pre-Huygens Era:, in: *Hor. Journal*, London, 1953.

–, "The origin of the Differential Gear and its Connection with Equation Clocks", in: *Transactions Newcomen Society* XXX, 1955-56, 1956-57.

–, "Entwicklung der Schiffschronometer im Zusammenhang mit der Lösung des Longitudinalproblems", in: *Jahrbuch der Deutschen Gesellschaft für Chronometrie*, Vol. 21/1, 1970.

–, *Chronometrie des 20. Jhs. u. Paul Ditisheim; Berichte des 10. Kongresses f. Chronometrie*, Geneva, 1979.

Berthoud, Ferdinand, "Traité des horloges marines contentant la théorie, la construction, la main d'oeuvre de ces machines at la manière de les éprouver, pour parvenir par leur moyen, à la rectification des cartes marines et à détermination des longitudes en mer", Paris, 1773.

–, "Eclairissements sur l'invention, la théorie, la construction et les épreuves des nouvelles machines proposées en France, pour la détermination des longitudes en mer par la mesure du temps et réponse à un écrit qui a pour titre: Précis des recherches faites en France pour la mesure artificielle du temps", Paris, 1773.

–, "Les longitudes par la mesure du temps ou méthode pour déterminer les longitudes en mer avec le secours des horloges marines, suivie de recueil des tables nécessaires au pilote pour réduire les observations relatives à la longitude et à la latitude", Paris, 1775, 2nd ed. Paris, 1785.

–, "La mesure du temps apliquée à la navigation ou principe des horloges à longitude", Paris, 1782.

–, "Découverte des longitudes en mer par le secours des horloges marines: l'usage marines, l'usage de ces machines pour la conduite du vaisseau et la rectification des cartes", Paris, 1782.

–, "De la mesure du temps, ou supplément au Traité des horloges marines et à l'Essai sur l'horlogerie, contenant les principes d'exécution, de contruction et d'épreuves des petites horloges à longitude et l'application des mêmes principes de contruction aux pontres de poche, ainsi que plusieurs constructions d'horloges astronomiques", Paris, 1787.

–, "Traité des montres à longitudes", Paris, 1792.

–, "Suite au traité des montres à longitudes, contenant: 1. la construction des montres verticales portatives, 2. la description et les épreuves des petites horloges horizontales plus simples et plus portatives pour servir dans les plus longues traversées", Paris an V-1796.

–, "Mémoire instructif sur le travail des horloges et des montres à longitudes inventées par F. Berthoud", Paris, 1792.

–, "Histoire de la mesure du temps par les horloges", Paris an X-1802.

–, "Supplément au traité des montres à longitudes avec appendice contenant la notice ou indication des principales recherches ou des travaux faits par F. B. sur diverses parties des machines qui mesurent le temps depuis 1752 à 1807 accompagné d'observations et d'éclairissements", Paris 1807, 2nd ed., Paris, 1838.

Berthoud, Louis, *Entretiens sue l'horlogerie à l'usage de la marine*, Paris, 1812.

–, *Etude sur les lames bimétalliques des balanciers compensateurs et sur les divers systèmes de compensation supplémentaire qui ont été employés dans les chronomètres,* (Congrès de Chronométrie de 1900), Paris, 1902.

Bonniksen, B., *The Karrussel Watch*, Kent, 1905.

Burgess, M., "Grasshopper Escapement", in *AH*, Dec. 1971.

Campani, Matteo, *Proposizione D'Orioli Giustissimi Utile a Naviganti*, Rome, 1673.

Carle, D. de, *Watch & Clock Encyclopedia*, London, 1959.

Caspari, (C., ed.), *Mémoire sur l'isochronisme des spiraux de Chronomètres (Comptes rendus Ac. Sciences)*, Paris, 1875.

–, *Etude de l'influence de l'inertie du spiral sur l'isochronisme (Comptes rendus Ac. Sciences)*, Paris, 1875.

–, *Etudes sur le balancier compensateur de M. Winnerl (Comptes rendus Ac. Sciences)*, Paris, 1876.

–, *Influence de l'inclination sur la marche des chronomètres (Recherches chronométriques, XVI. Cahier)*, Paris, 1876.

–, *Isochronisme du spiral cylindrique sans courbes terminales (Comptes rendus Ac. Sciences)*, Paris, 1876.

–, *Etudes sur le mécanisme et la marche des chronomètres*, Paris, 1876.

–, *Recherches sur les chronomètres et les instruments nautiques (Cahier XI des recherches chronomètriques)*, Paris, 1876.

–, *Compensateur de Winnerl (C. R. Ac. Sciences)*, Paris, 1876.

–, *Formule pour le calcul des longitudes par les chronomètres (C. R. Ac. Sciences)*, Paris, 1888.

–, *Cours d'astronomie pratique, application à la géographie et à la navigation coordonnées vraies et apparentes. Théorie des instrumentes,* Paris, 1888-89.

–, *Rapport sur le réglage et les perturbations des chronomètres,* Paris, 1889.

–, *Mémoires sur isochronisme du spiral réglant cylindrique (Congrés de chronométrie de 1889)*, Paris, 1890.

–, *Note complémentaire sur l;isochronisme (Congrés int. de chronométrie de 1889)*, Paris, 1890.

–, *Rapport sur les perturbations et le réglage des chronomètres (Congrés int. de chronométrie de 1889)*, Paris, 1890.

–, *Les chronomètres de marine: fonctionnement et emploi des chronomètres de précision, description du chronomètre et théorie de son mécanisme, perturbations qui altèrent sa marche (Bibliographie pagès 195-198)*, Paris, 1894.

–, *Sur les chronomètres de la marine française (Congrés de chronométrie de 1900)*, Paris, 1902.

–, *Mémoires sur l'isochronisme du spiral cylindrique (Congrés de chronométrie de 1900)*, Paris, 1902.

Chamberlain, Paul M., *It's About Time,* New York, 1941.

Chapiro, A., *Was Henri Motel a Pupil of Breguet?* London, 1973.

Chevenard, P., "Traitement thermique des ferronickels complexes à deux constituants", in: *Comptes rendus de l'Academie des Sciences*, 189, p. 846, Paris, 1936.

Christmas, E. L., "A Chronometer Deck Watch by Brockbanks", *AH*, 1963.

Clutton, Cecil, "Julien Le Roy's Lever Revived", in: *AH* IV Sept. 1963, plus letters in March & September, 1964.

–, "Le Roy and the Lever Escapement – Some new evidence", in: *AH* March, 1966.

–, "Mudge and the Lever Escapement", in: *AH* Sept., 1968.

Cuss, T. P. C., "The Huber-Mudge Timepiece with Constant Force Escapement", in: *Pioneers of Precision Timekeeping*, AH Monograph No. 3.

Daniels, George, "Mudge's Lever Escapement", *AH* June, 1968.

–, "The Art of Breguet", London & New York, 1975.

Davies, Alan C., "The Rise and Decline of Chronometer Manufacturing", in: *AH* XII 1980.

Defossez, L., *Les Savants du XVII Siecle et le Mesure du Temps*, Lausanne, 1946.

–, *Theorie Générale de L'Horlogerie, Chambre Suisse de L'Horlogerie*, La Chaux-de-Fonds, 1952.

–, Paul Ditisheim, "Chronometrier", in: *Journal Suisse d'Horlogerie*, Lausanne, 1946.

Delapierre, R., *Les Jürgensen, Une Dynastie de Grands Horologers*, Le Locle, 1974.

British Patent No. 9302 of 1842, by Ed. J. Dent, On the Errors of Chronometers, and Explanation of a New Chronometer-Compensation's-Balance.

Derham, William, F.R.S., *The Artificial Clockmaker, A Treatise ...*, London, 1696.

Ditisheim, Paul, *Correction de l'erreur secondaire de compensation des chronomètres par l'emploi du balancier Guillaume, 2nd ed.*, La Chaux-de-Fonds, 1902.

–, *Sur la relation entre la pression at la marche des chronomètres (Comptes rendus Ac. Sciences*, Paris, 1903.

–, *Sur un enregistreur chronomètrique portatif imprimant les observations en mesurant le 1/100e de seconde*, Paris, 1904.

–, *Essai d'une détermination de différence de longitude par transport de l'heure (Comptes rendus Ac. Sciences)*, Paris, 1904.

–, *Notes sur les montres et chronomètres Paul Ditisheim*, La Chaux-de-Fonds.

–, *La variation des chronomètres avec la pression atmosphérique et la correction de cette variation*, Geneva, 1904.

–, *Les chronomètres de petits format. Essai d'une détermination de la différence de longitude entre Paris et Neuchâtel*, Geneva, 1905.

–, *Description des chronomètres et des montres d'art construits par les ateliers Paul Ditisheim à La Chaux-de-Fonds*. Exposition nationale suisse, La Chaux-de-Fonds, 1914.

–, *Détermination de la différence de longitude Greenwich-Paris pour le transport de l'heure en avion (Comptes rendus Ac. Sciences)*, Paris, 1920.

–, *A new balance for compensating the temperature error of watches and chronometers*, London, 1922.

–, *Demonstration of a new balance for compensating the temperature error of watches and chronometers and a centre seconds marine chronometer with electric contacts*, Proc. Phys. Soc., Vol. 85, Part 5, London, 1923.

–, *Chronomètres observés aux hautes altitudes et dans le gaz hydrogène, Compte rendue Soc. Suisse Geophysique ...*, Geneva, 1923.

–, *Le problème de l'heure, les dernières publications du bureau international de l'heure et du bureau des longitudes*, 1924.

–, *Les progrès du réglage des chronomètres et des montres (Science moderne)*, Paris, 1925.

–, *Correction de l'effet du champ magnétique sur la marche des montres (Ac. Sciences)*, Paris, 1928.

–, *Etat actuel de la question du graissage en horlogerie. Compagnie française de raffinage. Le laboratoire central (Annales françaises de chronométrie)*, Besançon, 1931.

–, *Lubrication des montres et autres appareils fonctionnant aux basses températures, Annales françaises de Chronométrie*, Besançon, 1935.

–, *Le Probleme de l'Heure à l'Exposition de l'Union astronomique international, Annales Françaises de Chronométrie*, Paris, 1936.

–, *Les parfums et les montres, Journal Suisse d'Horlogerie*, Neuchâtel, 1937.

–, *Sur l'emploi des huiles d'horlogerie naturelles et artificielles (Journées internatinales de chronométrie)*, Paris, 1939.

Ditisheim, P., Lallier, R., Reverchon, L., Vivielle, Pierre Le Roy at La Chronométrie, Edition Tardy, Paris, 1940.

Ditisheim, P., *Galilee et le Pendule*, p. 27, Journal Suisse d'Horlogerie, 1942.

–, *Le spiral réglant et le balancier depuis Huygens jusqu'a nos jours, Journal Suisse d'Horlogerie*, Lausanne, 1945, includes among others Le Balancier à Affixe compensateur.

–, "Influence de la Pression Barometrique sur la Marche des Chronomètres", in: *Journal Suisse d'Horlogerie 1945*, pp. 255-265.

Dreier, A. F., "Die Bestimmung der nautischen Breite", in: *Katalog Winkelmessinstrumente*, Staatliche Museen, Berlin 1979.

Earnshaw, Thomas, Explanation of time keepers constructed by Mr. Thomas Earnshaw: for which a reward of three thousand pounds was awarded by the commissioners of longitude (From the communications made by him to the commissioners), London, 1805.

–, Statement as to the reward of three thousand pounds voted to him by the commissioners of Longitude, London, 1806.

–, Longitude: an appeal to the public stating Mr. Thomas Earnshaw's claim to the original invention of the improvements in his timekeepers, their superior going in numerous voyages and also as tried by the astronomer royal, by orders of the commissioners of longitude and his consequent right to national reward, London, 1808.

Egli, Kurt H., "Inspirationen aus alten Uhren für Neukonstruktionen", in: *CIC Konferenzberichte*, Vol. 3, 1979.

Evans, D. W., "The First Hundred Years of the Marine Chronometer", in: *Horological Journal*, London, April 1953.

Farmer, J. J., *An Account of the Improvements in Chronometers, watches and Clocks by E. T. Loseby*, Coventry 1852, London, 1914.

Ferrari d'Occhieppo, Konradin, *Kunde vom Weltall*, Vienna, 1952.

Filchner, W., Prybyllok, W. & Hagel, L., *Route Mapping in Unexplored Regions*, Basel, 1957.

Fleurieu, M. de, *Le Voyage fait par ordre du Roy en 1768-9*, Paris, 1773.

Friesleben, H. C., *Neuzeitliche Navigation, nautisches Jahrbuch und Zifferblatt der Chronometer, Der Seewart*, Hamburg, Vol. 8, 1937.

Frisius, Gemma, *De principis astronomiae cosmonomiae et cosmographiae deque usu globi cosmographici*, Louvain & Antwerp 1530, Paris 1545.

Gazely, W. G., *Clock & Watch Escapements*, London, 1956.

–, "The Chronometer Escapement", *AH* III, Dec. 1961.

Georgi, Johannes, *Im Eis vergraben*, Vol. IV Meteorologie, F. A. Brockhaus, Leipzig, 1935.

Glaser, G., "Die Berechnung, Dimensionierung und Anpassung von Triebfedern für Uhren", in: *Die Uhr*, 2, 3 & 4, 1968.

–, Hdb. *Uhren*, Vol. II, Zf. 3.2.9.

Glasgow, David, *Watch and Clock Making*, London, 1885.

Gohlke, E., *Untersuchungen über Chronometer und nautische Instrumente*, Bautzen, 1893.

Good, R., *The first Lever Watch made by Thomas Mudge, Pioneers of Precision Timekeeping*, AH Soc., Monograph No. 3, London, 1968.

–, *The Mudge Marine Timekeeper, Pioneers of Precision Timekeeping*, AH Soc., Monograph No. 3, London, 1968.

Gould, R. T., *The Marine Chronometer, Its History and Development*, London, 1923.

–, "Early Marine Chronometers", in: *The Horological Journal*, London, Dec. 1923.

Gros, Ch., *Les Echapements des horloges et montres*, Paris, 1913.

Grossman, Jul. & Herm., *Horlogerie Theorique*, Vol. I, pp. 68ff, Bienne, 1912.

Guillaume, Charles-Edouard, *L'Invar et L'Elinvar, Conference Nobel 1920, Soc. Suisse de Chronométrie*, Liestal, 1974.

–, *Les Aciers au Nickel et leurs applications à la Chronométrie, Comptes rendus*, Ac. Sciences, Paris, 1897.

–, "Calcul d'un pendule composé en acier-nickel", in: *Journal Suisse d'Horlogerie*, Geneva, 1899.

–, "Etudes sur le balancier compensateur", in: *Journal Suisse d'Horlogerie*, Geneva, 1899.

–, *Dilatibilité et Thermoelasticité des aciers au nickel, Comptes rendues*, Ac. Sciences, Paris, 1920.

–, *La compensation des horloges et des montres, procédés nouveaux fondés sur l'emploi des aciers an nickel*, Neuchâtel & Geneva, 1923.

Guyot, Ed., *Histoire de la Determination de Longitudes*, La Chaux- de-Fonds, 1955.

–, *Histoire de la determination de l'heure*, La Chaux-de-Fonds, 1968.

Harrison, John, The principles of Mr. Harrison's Timekeeper . . . Published by order of the commissioners of longitude, Richardson & Clarke, London, 1967.

–, Remarks on a pamphlet lately published by the Rev. Dr. Maskelyne, under the authority of the Board of Longitude, W. Sandley, London, 1767.

–, *Extrait de la réponse de Mr. Jean Harrison aux remarques et objections de Mr. Maskelyne*, Avignon, 1768.

–, An account of the going of Mr. John Harrison's watch at the Royal Observatory from May 6th 1766 to March 4th 1767, with appendices: I. the original observations of the going on the watch from day to day. II. Calculations of the going of Mr. Harrison's watch from day to day. III. Appendix containing observations of equal altitudes of the sun taken at Portsmouth, Jamaica and Barbadoes, according to the time of Mr. Harrison's watch, London, 1775.

–, A description concerning such mechanism as will afford a nice or true mensuration of time, T. Jones, London, 1775.

–, Harrison's inventions (de l'année 1713 à 1774 Recueil de pièces figurant à la bibliothèque de la Clockmakers Cy.)

–, Voyage in the Deptford man of war, to Madiera, to Jamaica accuracy of the timekeeper; and the return with a very tempestuous passage, in the Merlin sloop. (British Museum).

Helwig, Alfred, Drehganguhren, Berlin, 1927.

Herkner, Kurt, *Glashütte und seine Uhren*, Dormagen, 1978.

Heyerdahl, Thor, *Kon-Tiki*, Berlin, 1949.

Holder, L., "Nine Modern Developments in Navigation", in: *W. E. May, A History of Marine Navigation*, Henley-on-Thames, 1973.

Horrins, Johan (Pseudonym for John Harrison, grandson of John Harrison), Memoirs of a Trait in the Character of George III, London 1835.

–, Memorial concerning Mr. Harrison's invention for measuring the time at sea, 1762.

–, Proposal for examining Mr. Harrison's Time-keeper at Sea, 1762.

–, A calculation showing the Result of an Experiment made with Mr. Harrison's Time-keeper, in Consequence of the Instructions and Directions of the Commissioners of the Longitude, in a voyage from Portsmouth to Port Royal in Jamaica, in Order to Determine by that Time-keeper, the difference of Longitude between these two Places; and to find out how much the Longitude found by the Time-keeper differs from the Longitude formerly determined by Astronomical Observations.

Houdremont, E., *Handbuch der Sonderstähle*, Berlin, 1943.

Howse, Derek, *Captain Cook's Marine Timekeepers; Captain Cook's minor clocks and watches; Captain Cook's pendulum clocks; Margetts Watch; Regulator clocks of the Royal Observatory Greenwich V.*

Howse, Derek & Hutchinson, Beresford, *The saga of the Shelton clocks*, VI.

Huber, Martin, *Die Uhren von A. Lange & Söhne – Glashütté Sachsen*, Munich, 1973.

Hugenii a Zuglichem, Christiani, *Const. F. Horlogium, Hagae Comitum, Ex officine Adriani Vlacq*, 1658.

Huygens, Christian, *Oeuvres Complètes de Société Hollandaise des Sciences*, Den Haag, 1888, new editions to 1932.

Jagger, Cedric, "Paul Philip Barraud, 1750-1929"; *AH*, London, 1968 – "Paul Philip Barraud", – The Supplement, *AH*, London, 1979.

Jürgensen, Urban, Horloger, *Prinzipes généraux de l'exacte mésure du temps pour les horloges*, Copenhagen, 1805.

–, *Memoires due l'Horlogerie exacte*, Paris, 1842.

–, *Allgemeine Grundsätze der genauen Zeitmessung durch Uhren*, Leipzig, 1840.

Laycock, W., *The Lost Science of John "Longitude" Harrison*, Ashford, 1976.

Le Roy, Pierre, *Un nouvel échappement à repos dans lequel la roue pose et est retenue à chaque demi vibration sue une pièce fixée à la platine et entièrement étrangère au balancier* (Mem. Ac. Sciences), Paris, 1748.

–, *Recherches des longitudes en mer: Exposé succinct des travaux de MM. Harrison et Le Roy dans la recherche des longitudes en mer et des épreuves faites de leurs ouvrages*, Paris, 1768.

–, *Mémoire sur la meilleure manière de mesurer le temps en mer*, Paris, 1770.

–, *Précis des recherches faites en France depuis l'année 1730 pour la détermination des longitudes en mer par la mesure artificielle du temps*, Amsterdam, 1773.

–, *Suite au Précis sur les montres marines de France avec un supplément au mémoire sur la meilleure manière de mesurer le temps en mer*, Leyde, 1774.

Løvenørn, Poul de, *Beretning om en Reise foretaget . . . i Aarene 1782 og 1783 med Fregatten Prøven, for at undersøge de i Dannemark forfaerdigende Søe-Laengde Uhrer . . .*, Copenhagen, 1786.

Maskelyne, N., *An Answer to a Narrative of Facts . . .*, London, 1792.

May, W. E., *A History of Marine Navigation*, Henley-on-Thames, 1973.

Mercer, Tony, *Mercer Chronometers, Radical Tom Mercer and the House he Founded*, London, 1978.

Mercer, T., "The two-day Marine Chronometer and The Detent in the 19th and 20th Cents.", *Hor. Journal*, 1978.

Mercer, Vaudrey, "John Arnold & Son, Chronometer Makers 1762-1843", *AH*, London, 1977.

–, "Edward John Dent and his Successors", *AH*, London, 1977.

Miklosch, Richard. "Die Entwicklung des fliegenden Drehgestells" in *Glashütte Sachsen*, Alte Uhren 2/80, Munich.

Miles, H. A., "Did Julien Le Roy Make the First Detached Lever Escapement?", *AH* III, Sept. 1961.

–, "Tour of Denmark – Summer 1969", (Part 3, Watches and Chronometers seen on the tour), *AH*, March, 1970.

Möller, Fr., *Uber den Schutz von Chronometern vor starken magnetischen Feldern durch eiserne Gehäuse, Annalen der Hydrographie*, Vol. 46, Hamburg, 1918.

Moore, E. K., *Competitive Trial of Chronometers at the US Naval Observatory*, Proc. US Naval Inst. 10/171, 1885.

Mudge, Thomas, *Thoughts on the means of improving watches and more particularly those for use at sea*, London, 1765.

–, *Register of the going of Mr. Mudge's first timekeeper for April 1780 to May 1781*, London, 1781.

–, Report of the select committee of the house of Commons . . . to make enquiry into the principles on which Mr. Mudge's timekeepers have been constructed, London, 1793.

Mudge, Thomas, Jr., A narrative of facts relating to some timekeepers constructed by T. M. for the discovery of the longitude at sea; together with observations, upon the conduct of the Astronomer Royal respecting them, London, 1792.

–, A reply to the answer of the Rev. Dr. Maskelyne, Astronomer Royal, to a narrative of facts, relating to some time keepers constructed by Mr. Thomas Mudge, for the discovery of the longitude at sea, by Th. Mudge jun. of Lincoln's Inn, to which is added a short explanation of the most proper methods of calculating mean daily rate . . . by his Excellency the Count de Bruhl, London, 1792.

–, Memoirs of the life and mechanical labours of the late Mr. Thomas Mudge, with a striking likeness (Universal magazine), portrait, 1795.

–, A description with plates of the time keeper invented by the late Mr. Thomas Mudge to which is prefixed a narrative by Thomas Mudge, his son, portrait, London, 1799.

Neumann, John von & Morgenstern, O., *Theory of Games and Economic Behavior*, New York, 1944.

Nockold, Susanne, "Early Timekeepers at Sea", *AH*, Sept., 1963.

Osterholt, O., "Routenaufnahmen in West-Szetschuan", in: *Petermanns Geogr. Mitt.*, H 255, Gotha, 1955.

Passi-Drewes, La Chronometrie et ses developpements dans le domaine hydrographique au XVIIIe Siecle, Dissertation theme at the University of Paris, the Sorbonne, under the direction of M. Pierre Chaunu, projected for 1981.

Payer, Julias, Die Ost.-Ung. Nordpolexpedition 1876, Vienna. The second North Pole expedition of 1869-70 and the German polar expedition of

1871 are also treated here.

Pellaton, James C., *Le Tourbillon, Annales Françaises de Chronometrie*, Paris, 1951.

Phillips, Ed., *De la Compensation des Temperatures dans les Chronometres*, Paris, 1880.

–, Memoire sur le Spiral-Reglant des Chronometres et des Montres; Annales des Mines XX, 1861 (with various changes and additions, it has also appeared in other important French periodicals).

Pohl, R. W., R. W., *Mechanik, Akustik und Wärmelehre*, Berlin.

Poulle, Em., *Les Instruments astronomiques de l'Occident latin aux XIe et XIIe siecles. Cahiers de civilisation medievale*, University of Poitiers, 1972.

–, *Le Traite d'Astrolabe de Raymond de Marseille. Studie medievali, Centro Italiano die Studii sull'Alto Medioevo*, Spoleto, 1964.

–, *Un Instrument astronomique dans l'Occident Latin, La Saphea, Centro Italiano die Studii sull'Alto Medioevo*, Spole o, 1970.

–, *Bernard de Verdun et le Turquet*, 1964.

–, *A Propos des Tables astronomiques de Pierre d'Aragon, Junta de Investigações de Ultramar*, Coimbra, 1966.

–, *Les Conditions de la Navigation astronomique au XVe sièclem Junta de Investigações de Ultramar*, Coimbra, 1969.

Querens, Henri Motel, "Horloger de la Marine", in *L'Horloger*, 1922.

Quill, Humphrey, *John Harrison, The Man Who Found Longitude*, London, 1966.

–, "The Grasshopper Escapement", *AH* Sept., 1971.

Randall, Anthony G., *Eine Hemmung mit konstanter Kraft, Alte Uhren I*, Munich, 1978.

Randier, Jean, *Nautische Instrumente*, Munich, 1979.

Rapatz, Franz, *Die Edelstähle*, Berlin, 1951.

Rees, Abraham, *Cyclopaedia or Universal Dictionnary of Arts, Sciences and Literature, London 1820 (Reprint. Rees's Clocks, Watches and Chronometers*, Newton Abbot, 1970.

Repsold, A., *Chronometer- und Taschenuhrprüfungen auf der Deutschen Seewarte, Die Uhrmacherwoche* No. 1, 3, 34, 35, 37, 40, 42 of 1931 and 34 of 1932.

–, *Chronometer und Nautisches Jahrbuch, Annalen d. Hydrographie*, Hamburg, May, 1937.

Resal, Henri, *Mémoire sur le frottement des engrenages coniques et de la vis sans fin* (Journ. Ecole polytechnique), Paris, 1850.

–, *Recherches sur la loi des oscillations du pendule à suspension à lames des chronomètres fixes*, Paris, 1856.

–, *De l'influence de la suspension à lames sur le mouvement du pendule conique (Annales des mines)*, Paris, 1860.

–, *De l'influence de la suspension à lames sur les oscillations du pendule conique (Mém. C. R. Ac. Sciences)*, Paris, 1860.

–, *Les échappements (Annales des Mines)*, Paris, 1867.

–, *Des applications de la mécanique à l'horlogerie*, Paris, 1868.

–, *Des engrenages*, Paris, 1869.

–, *Note sur la pendule à oscillations ecliptiques* (C. R. Ac. Sciences), Paris, 1869.

–, *Théorie des effets observés par Savart, sur l'influence mutuelle de deux pendules* (C. R. Ac. Sciences), Paris, 1873.

–, *Note sur le mouvement du pendule conique, en ayant égard . . .*

Rottok, *Chronometerprüfungen bei der französischen, englischen und amerikanischen Marine, Annalen der Hydrographie*, Vol, 45, Hamburg, 1917.

Sabrier, I. C., *Henri Motel et la Chronometrie Française de la première moitié du XIXe, Ancaha*, no. 25, 1979.

Sacro Bosco, Joannes de, Sphaera. *Handschriften in zahlreichen Staats- und Universitätsbibliotheken; verschiedentlich mit Kommentaren gedruckt: z. B. Vhristophori Clavii' Bambergensis ex Societai Iesu in Sphaeram Joannis de Sacro Bosco Commentarius. Venetiis, Apud Bernardum Balam sub signo Solis*, 1606. Editions over several centuries.

Saunier, Cl., *Trippelin & Rigg, A treatise on modern horology in Theory and Practice,* London, 1862.

Schenk, Bobby, *Astronavigation*, Bielefeld, 1977.

Selmer, Carl, *Navigatio Sancti Bredani Abbotis*, University of Notre Dame Press, 1959.

Severin, Tim, *The Brendan Voyage*, New York, 1978.

Shannon, C. E., *Mathematische Theorie der Nachrichtenübertragung*, 1948.

Short, James, F.R.S., presumed author of the anonymous work: An Account of the Proceedings, in Order of the Discovery of the Longitude; in a letter to the Right Honourable . . . Member of Parliament, London, T. & J. W. Pasham.

Spencer-Jones, Harold, "The Royal Greenwich Observatory and Horology", in: *Horological Journal*, London, Dec. 1955.

Sully, Henry, *Horloge inventée et executée pour une plus juste mesure du temps en mer* . . . (Mem. Ac. Royale Sciences), Paris, 1724.

–, *Description abregée d'une horloge de nouvelle invention pour* . . ., Paris, 1727.

Tamann, Gustav, *Lehrbuch der Metallkunde*, Berlin 1932 (first to third editions: "Lehrbuch der Metallographie", 1914ff).

Tardy, *La Montre (Les Echappements: à Recul, à Demi-Recul, à Repos, Libres à Ancre)*, Paris.

–, *Bibliographie Générale de la Mesure du Temps*, Paris, 1943.

–, *Dictionnaire des Horlogers Français*, Paris, 1972.

Thielemann, O., *VEB Uhrenwerk Glashütte, Uhren und Schmuck*, Vol. 17, East Berlin, 1980.

Torrens, D. S., *Paul Ditisheim, A Tribute to a great Horologist*, Horological Journal, December, 1945, London.

Villarceau-Yvon, A. J., *Notice sur les travaux de Yvon-Villarceau*, Paris, 1855.

–, *Recherches sur le mouvement et la compensation des chronomètres*, (Ann. de l'Observatoire de Paris), Paris, 1862.

–, *Réimprimés dans Recherches chronométriques*, Paris, 1876.

–, *Recherches sur le mouvement et la compensation des chronomètres* (C. R. Ac. Sciences), Paris, 1863.

–, Note dur une lettre de M. H. de Mauguay (C. R. Ac. Sciences), Paris 1869. –, Note sur les conditions des petites oscillations d'un corps solide de forme quelcontre et la théorie des équations différencielles linéaires (C. R. Ac. Sciences), Paris, 1870.

–, Remarques relatives à la division décimale des angles et du temps (Comptes rendus Ac. Sciences), Paris, 1870.

–, Reprint, Paris, 1883.

Vivielle, Com., *Les sabliers, horloges du mer. Bull.* officiel du Yacht Club de France, Paris 1934.

–, *Sandglasses, Clocks of the Sea*, NAWCC, February, 1977.

Weaver, I. D., "Treatise on the Balance Spring" (Transpation of the memoir on the subject by E. Phillips), Monograph no. 15, *AH*, London, 1978.

Wissenschaftliche Ergebnisse der deutschen Grönlandexpedition Alfred Wegener's, 1929 und 1930-31. Im Auftrag der Notgemeinschaft der deutschen Wissenschaften von Kurt Wegener herausgegeben. Leipzig, 1940.

Whitney, Marvin, "The Ship's Chronometer", in: *Horological Times* (USA), 1978-79-80.

Wood, Chr., "Earnshaw Chromometer No. 550", in *AH*, Sept., 1972.

Woog, Paul, *Contribution a l'étude de graissage: Onctuosité et influence moleculaire*, Paris, 1926.

Photo Credits

The photos were made available to us by the sources noted in the captions – museums, companies and private owners. The photographs listed here were generally commissioned by the publishers for use in this volume.

Abeler, Wuppertal, 212a, b
Ahrens, Lucerne, 125a, b, c, d, e
Alpenland, Vienna, 95a, b
Bodenheimer, Munich, 122

Christies, A. C. Cooper Ltd, London, 61a, b, c, d, e, f
Fleury Claude, Bienne, 337
Fortt Studio Ltd, Surbiton, 94b, 131a, b, 143a, b
Gessing, Offenbach, 228a, c, 241a, b
Gold, Anne, Aachen, 152a, b, c, 289a, b, 270a, b, 383
Herre, Volkmar, Leipzig, 249a, b, 278a, b, 286a, b, 291a, b, 331a, b, 342a, b
Ineichen, Auktionshaus, 141a, b, 231
Kiefer, Studio, Düsseldorf, 11, 14, 16, 36a, b, 40a, b, 45, 46, 62a, b, c, 105a, b, 119a, b, 120a, b, 126a, b, 149a, b, 165a, b, 166a, b, 167a, b, c, 171a, b, 173, 181, 190a, b, c, d, 207a, b, 229, 253a, b, 281a, b
Koch, Studio 9, Lucerne, 178a, b, 187a, b, 193a, b
Kronow, Rolf, Copenhagen, 153a, b, c
Landesbildstelle Rheinland, Düsseldorf, 213a, b
Lösche, Rudolstadt, 214a, b
Meis, Konstanz, 349
Mercer, Dr. R. V., Cookham, Berkshire, 218, 236, 275, 298a, b
Mercer, Tony, St. Albans, 220, 296, 356, 357a, b, 358, 359
Mercier, Claude, Geneva, 114a, b, c, d, 271a
Miklosch, Aachen, 215
Müller, Hamburg, 350a, b, c, d
Pelster, Helmut, Frankfurt, 247a, b, 263a, b
Randall, A. G., Birmingham, 386a, b, 389
Rompel & Wölffler, Hamburg, 348a
Stender, Sint Michelsgestel, 53a, b, c, 169a, b, 189a, b, c, 192a, b, c, 199a, b, c, d, 237a, b, c, 238a, b, c
Stone, London, 96a, b, c, d, 98a, b, 127a, b, 138a, b, c, 230, 233, 301, 386a, b
Tweedy, London, 139a, b
Vogel, Düsseldorf, 200a, b
von Voithenberg, Munich, 205a, b, 240a, b, c, d, 244a, b, c, 245a, b, 246a, b, c, d, 256a, b, 258a, b, c, 259a, b, c, 260a, b, 266a, b, 279a, b, 284a, b, c, d, 288a, b, 308a, b, 334a, b, 339a, b, 343a, b, 344a, b, 352a, b, 354a, b, 355a, b
Wempe, Hamburg, 348b
Wood, Edinburgh, 133a, b, 151, 157a, b, 159a, b, 177a, b, 224a, b, 226a, b, 261a, b, 292a, b, 306a, b
Wurthmann, Bremen, 250a, b, 272a, b, 283a, b, 307a, b